CONTROL AND
DYNAMIC SYSTEMS

*Advances in Theory
and Applications*

Volume 44

CONTRIBUTORS TO THIS VOLUME

HUA CHEN
G. S. CHRISTENSEN
PETER E. CROUCH
MOHAMED A. EL-SHARKAWI
R. S. GORUR
LESLIE F. JARRIEL
KWANG Y. LEE
CHEN-CHING LIU
M. Y. MOHAMED
YOUNG MOON PARK
ARUN G. PHADKE
S. A. SOLIMAN
JAMES S. THORP
DANIEL J. TYLAVSKY
SHIH-MING WANG

CONTROL AND DYNAMIC SYSTEMS

ADVANCES IN THEORY AND APPLICATIONS

Edited by
C. T. LEONDES

Department of Electrical Engineering
University of Washington
Seattle, Washington
and
School of Engineering and Applied Science
University of California, Los Angeles
Los Angeles, California

VOLUME 44: ANALYSIS AND CONTROL SYSTEM TECHNIQUES FOR ELECTRIC POWER SYSTEMS
Part 4 of 4

ACADEMIC PRESS, INC.
Harcourt Brace Jovanovich, Publishers
San Diego   New York   Boston
London   Sydney   Tokyo   Toronto

Academic Press Rapid Manuscript Reproduction

Academic Press, Inc.
San Diego, California 92101

United Kingdom Edition published by
ACADEMIC PRESS LIMITED
24-28 Oval Road, London NW1 7DX

Library of Congress Catalog Card Number: 64-8027

ISBN   0-12-012744-X   (alk. paper)

PRINTED IN THE UNITED STATES OF AMERICA
91 92 93 94   9 8 7 6 5 4 3 2 1

CONTENTS

CONTRIBUTORS vii
PREFACE ix

Computer Relaying in Power Systems 1
    James S. Thorp and Arun G. Phadke

Advanced Control Techniques for High Performance Electric Drives 59
    Mohamed A. El-Sharkawi

High Voltage Outdoor Insulation Technology 131
    R. S. Gorur

Power System Generation Expansion Planning Using the Maximum
Principle and Analytical Production Cost Model 193
    Kwang Y. Lee and Young Moon Park

Development of Expert Systems and Their Learning Capability for
Power System Applications 239
    Chen-Ching Liu and Shih-Ming Wang

Advances in Fast Power Flow Algorithms 295
    Daniel J. Tylavsky, Peter E. Crouch, Leslie F. Jarriel, and Hua Chen

v

Power Systems State Estimation Based on Least Absolute
Value (LAV)   345

   G. S. Christensen, S. A. Soliman, and M. Y. Mohamed

INDEX   489

CONTRIBUTORS

Numbers in parentheses indicate the pages on which the authors' contributions begin.

Hua Chen (295), *Center for Systems Science and Engineering and Department of Electrical Engineering, Arizona State University, Tempe, Arizona 85287*

G. S. Christensen (345), *Department of Electrical Engineering, University of Alberta, Edmonton, Alberta, Canada T6G 2G7*

Peter E. Crouch (295), *Center for Systems Science and Engineering and Department of Electrical Engineering, Arizona State University, Tempe, Arizona 85287*

Mohamed A. El-Sharkawi (59), *Department of Electrical Engineering, University of Washington, Seattle, Washington 98195*

R. S. Gorur (131), *Department of Electrical Engineering, Arizona State University, Tempe, Arizona 85287*

Leslie F. Jarriel (295), *Center for Systems Science and Engineering and Department of Electrical Engineering, Arizona State University, Tempe, Arizona 85287*

Kwang Y. Lee (193), *Department of Electrical and Computer Engineering, Pennsylvania State University, University Park, Pennsylvania 16802*

Chen-Ching Liu (239), *Department of Electrical Engineering, University of Washington, Seattle, Washington 98195*

M. Y. Mohamed (345), *Electrical Power and Machines Department, Ain Shams University, Cairo, Egypt*

Young Moon Park (193), *Department of Electrical Engineering, Seoul National University, Seoul 151, Korea*

Arun G. Phadke (1), *Department of Electrical Engineering, Virginia Polytechnic Institute and State University, Blacksburg, Virginia 24061*

S. A. Soliman (345), *Electrical Power and Machines Department, Ain Shams University, Cairo, Egypt*

James S. Thorp (1), *School of Electrical Engineering, Cornell University, Ithaca, New York 14853*

Daniel J. Tylavsky (295), *Center for Systems Science and Engineering and Department of Electrical Engineering, Arizona State University, Tempe, Arizona 85287*

Shih-Ming Wang (239), *Department of Electrical Engineering, University of Washington, Seattle, Washington 98195*

PREFACE

Research and development in electric power systems analysis and control techniques has been an area of significant activity for decades. However, because of increasingly powerful advances in techniques and technology, the activity in electric power systems analysis and control techniques has increased significantly over the past decade and continues to do so at an expanding rate because of the great economic significance of this field. Major centers of research and development in electrical power systems continue to grow and expand because of the great complexity, challenges, and significance of this field. These centers have become focal points for the brilliant research efforts of many academicians and industrial professionals and for the exchange of ideas between these individuals. As a result, this is a particularly appropriate time to treat advances in the many issues and modern techniques involved in electric power systems in this international series. Thus, this is the fourth volume of a four volume sequence in this series devoted to the significant theme of "Analysis and Control System Techniques for Electric Power Systems." The broad topics involved include transmission line and transformer modeling. Since the issues in these two fields are rather well in hand, although advances continue to be made, this four volume sequence will focus on advances in areas including power flow analysis, economic operation of power systems, generator modeling, power system stability, voltage and power control techniques, and system protection, among others.

The first contribution to this volume is "Computer Relaying in Power Systems," by James S. Thorp and Arun G. Phadke. The field of computer relaying started with conceptual and feasibility studies in the 1960s. As the digital computer was being used in planning for short circuit studies, load flow, and stability problems, it was hoped that relaying would be the next promising and exciting field for computerization. In spite of the high cost and the lack of speed of computers of that period, visionary rsearchers were attracted to the challenge of developing algorithms for digital protection. The rapid evolution of computers over the past few decades along with contributions in relaying algorithms have brought computer relaying into reality. This contribution reviews the historical developments in the field of computer relaying, describes the structure of a typical protection computer, and discusses the critical hardware components and the influence they have on the relaying tasks.

The next contribution is "Advanced Control Techniques for High Performance Electric Drivers," by Mohamed A. El-Sharkawi. This is an issue of significant importance not only for the provision of servo control in the power control of electric power systems, but also in the steady proliferation of the requirements for high performance electric drives. Any high performance drive system must consist of three basic components: an electric motor, a broad performance high speed solid-state switching converter, and an advanced controller. In this contribution an overview of high performance drive systems is presented. Several types of electric motors that are considered among the best options for high torque, high performance applications are discussed. Appropriate control strategies for tracking control are developed and evaluated via laboratory testing and/or computer simulations.

The next contribution is "High Voltage Outdoor Insulation Technology," by R.S. Gorur. A review of high voltage insulation technology used for outdoor power transmission and distribution is presented. Emphasis is placed on the state-of-the-art polymer materials which are finding increasing use in today's highly competitive and advanced power industry. The contribution begins with a review of the problems with porcelain and glass which have been used for many decades for outdoor insulation applications, such as insulators, cable terminations, and surge arresters. It then addresses how polymers have alleviated some of the problems, while creating a few of their own. The typical constructional features of polymer insulating devices are presented. Next, the service experience of utilities, both worldwide and within the United States, with polymeric insulators is presented. Their performance in outdoor test stations, which is valuable for advancing the technology, is discussed. Laboratory tests which are required to obtain a better understanding of the service performance in a relatively short time are reviewed. The current research efforts and significant findings have been reviewed. Topics which need further research are proposed.

The next contribution is "Power System Generation Expansion Planning Using the Maximum Principle and Analytical Production Cost Model," by Kwang Y. Lee and Young Moon Park. Historically, electric utility demand in most countries has increased rapidly, with a doubling of approximately 10 years in the case of developing countries. In order to meet this growth in demand, planners of expansion policies are concerned with obtaining expansion plans which dictate what new generation facilities to add and when to add them. However, the practical planning problem is extremely difficult and complex and requires many hours of the planner's time even though the alternatives examined are extremely limited. In this connection, increased motivation for more sophisticated techniques of evaluating utility expansion policies have been developed during the past decade. Among them, the long-range generation expansion planning aims to select the most economical and reliable generation expansion plans in order to meet future power demand over a long period of time subject to a multitude of technical, economical, and social constraints. These techniques generally follow these difficult and complex problems:

1. What kind of optimization technique is to be utilized?

2. How can the uncertainties such as future load growth, fuel cost and availability, hydrological quantities, availability of facilities, and economic fluctuations be predicted?

3. How can the social constraints such as environmental, legal, and political factors and social impact by supply shortage be quantified?

This contribution is an in-depth analysis and presentation of the issues and techniques in this broadly complex area of such great significance for modern electric power systems.

The next contribution is "Development of Expert Systems and Their Learning Capability for Power System Applications," by Chen-Ching Liu and Shih-Ming Wang. This contribution provides a tutorial on expert system applications to electric power systems and reports on recent results in the development of learning capability. Part I starts with a tutorial on expert systems. Issues such as knowledge acquisition, inference engine, development tools, and maintenance are discussed. Following the tutorial, a brief survey of the state of the art is given based on an extended power system model. An on-line operational expert system, CRAFT (Customer Restoration and Fault Testing), is used to illustrate the expert system components and development facilities, including knowledge base, inference engine, maintenance, and explanation facilities. Besides the components and facilities of expert systems described in Part I, an important intelligent function, learning, is discussed in this contribution. Part II contains an explanation of an inductive learning (decision-tree) technique. The learning algorithm is extended to allow the decision tree to be modified when a misclassification occurs. The extended inductive learning method is applied to power systems. One application deals with determination of the control amount to eliminate voltage violations. The other application uses the learning algorithm to acquire high-level knowledge about security of a power system with respect to contingencies. The area of the possible application of expert systems in electric power systems is a relatively recent area of research activity on the international scene, and so this contribution is a rather important element of these volumes.

The next contribution is "Advances in Fast Power Flow Algorithms," by Daniel J. Tylavsky, Peter E. Crouch, Leslie F. Jarriel, and Hua Chen. The object of this contribution is to survey the recent literature on fast power flow algorithms in order to determine the direction of research which is most likely to lead to new, possibly faster and/or more robust power flow algorithms. The analysis here identifies the central idea running through these algorithms. The orientation provided by this review is used to propose some new algorithms. One of the many candidate algorithms proposed is tested and is shown to be robust and to have low execution times when compared with other algorithms. It is expected that there exist more algorithms or variations of current algorithms which have similar properties.

The final contribution to this volume is "Power Systems State Estimation Based on Least Absolute Value (LAV)," by G.S. Christensen, S.A. Soliman, and M.Y. Mohamed. State estimation is the process of assigning a value to unknown electric power system state variables and filtering out erroneous measurements before they enter in the calculating process. As a result, the system conditions presented to the control room operators are guaranteed to be corrected even if some bad measurements exist. A commonly used and familiar criterion in state estimation is that of minimizing the sum of the squares of the difference between the estimated and true (measured) value of a function. Another valuable technique of state estimation is based on minimizing the absolute value of the difference between the measured and calculated quantities, and it is called least absolute value (LAV). Among other objectives, this contribution shows how least absolute value estimators are a valuable alternative to least squares estimators, especially in regard to the robustness in the presence of imperfect measurements in electric power systems.

This fourth volume is a particularly appropriate one with which to conclude this four volume sequence on analysis and control techniques in electric power systems. The authors are all to be commended for their superb contributions, which will provide a significant reference source for workers on the international scene for years to come.

COMPUTER RELAYING IN POWER SYSTEMS

JAMES S. THORP
Cornell University
Ithaca, New York, 14853

ARUN G. PHADKE
Virginia Tech
Blacksburg, Virginia, 24061

I. INTRODUCTION

The field of computer relaying started with conceptual and feasibility studies in the 1960's [1]. As the digital computer was being used in planning for short circuit studies, load flow, and stability problems, it was hoped that relaying would be the next promising and exciting field for computerization. In spite of the high cost and the lack of speed of computers of that period, visionary researchers were attracted to the challenge of developing algorithms for digital protection. The rapid evolution of computers over the past few decades along with contributions in relaying algorithms has brought computer relaying into reality. In this section we will briefly review the historical developments in the field of computer relaying, describe the structure of a typical protection computer, and discuss the critical hardware components and the influence they have on the relaying tasks.

Given the price of computers in the 1960's one of the early papers in computer relaying [1] considered handling all the relaying in the

CONTROL AND DYNAMIC SYSTEMS, VOL. 44

1

substation by a single computer. Putting aside the difficulties presented by the possible failure of that single computer, the paper introduced many of the important problems and investigated a number of algorithms. Other papers [2–3] began the development of algorithms for high voltage transmission line protection. Transmission line protection was a first choice for study because of its complexity and potential payoff. Recognizing that the computational burden was considerably lighter than in transmission line protection, work was also begun on algorithms for apparatus protection using the differential relaying principle [4–6]. The calculation of the harmonic restraint function does add complexity to transformer protection, and there are problems associated with current transformer saturation that may be even harder to solve in a digital environment than in an analog setting. Nevertheless, by the early 1970's it was clear that computer based relays could, in principle, provide performance at least as good as conventional relays. Examination of the hundreds of references [7] in digital relaying over the past 20 years indicates that roughly two–thirds of the publications have been on algorithms. It seems unlikely that new algorithms will be found or that much will be learned by additional algorithm comparisons. Advances in the field are more likely to come from the use of improved computer hardware to implement well–understood relaying algorithms.

The advances in computer hardware during this period have been remarkable. The speed of computation had more than doubled while the size, power consumption, and cost of computers had gone down by an order of magnitude. The appearance of 16 bit and 32 bit microprocessors has made high speed computer relaying technically achievable, while at the same time cost of computer based relays began to become comparable to that of conventional relays. This trend has continued to the point that it is clear that the most economical and technically superior way to build relay systems of the future (except possibly for some functionally simple and inexpensive relays such as overcurrent relays) is with digital computers.

More recently, attention shifted from algorithms for stand–alone relays, to the opportunities offered by integration of the microprocessor based devices into a substation–wide, or even a system–wide, network. The ability of digital relays to communicate with each other and with other computers in the power system may be one of their most important benefits [8–9]. Among the benefits flowing from computer relays are:

A. COST

While the computational power of the microprocessors used to build digital relays has increased substantially, the cost has steadily declined. Due to general inflation and a relatively low volume of production and sales combined with some design improvements, the cost of conventional (analog) relays has steadily increased. The cost of the most sophisticated digital computer relays (including software costs) would be about the same as that of conventional relaying systems with the same performance. There are some conventional relays such as overcurrent relays which are so inexpensive that they are unlikely to be replaced in the near future. However, for major protection systems, issues such as lower wiring costs for an integrated digital system make the digital system the low cost alternative.

B. SELF CHECKING AND RELIABILITY

In the majority of cases of major power system disturbances one of the contributing factors in the chain of events that leads to the breakdown is relay misoperation [10]. A computer relay can be programmed to monitor its hardware and software subsystems continuously, in an attempt to detect failures before they lead to relay malfunctions. The digital relay can be programmed to take itself out of service if a failure is detected – and send a service request alarm to a central location. Although computer based relays are less reliable because of their complexity, the self–monitoring feature can be used to produce an inherently more reliable system.

C. SYSTEM INTEGRATION

Computer relays will fit in naturally in substations of the future. They can be integrated with new computer based control and monitoring systems and accept digital signals obtained from transducers and fiber optic channels that are under development. In the area of control the digital relay can be regarded as a measuring device which can provide new inputs for a variety of control functions. For a discussion of this subject see the chapter "Improved Control and Protection of Power Systems Through Synchronized Phasor Measurements" in this volume. The ability of a digital device to change the relay characteristics (settings) on command or for changing system conditions (adaptive relaying) is also discussed in that chapter.

There are a few issues that have limited complete acceptance of digital relays. One is the rapid change in computer hardware. Existing relays have performed well for as long as thirty years in some cases. There is concern in the industry about the maintainability of digital hardware which seems to change significantly every few years. Solutions may lie in the use of products from a single family in which modules can be replaced during the service life without having to change the relay itself. Another issue is the lack of a system–wide communications network. A fiber optic network connecting all substations and control centers would benefit many areas including relaying. Until such communication networks are in place, some of the expected benefits of computer relaying will remain unrealized.

Software is another area of concern. Because relaying calculations are particularly intense in the immediate post–fault interval, large portions of the programs for relaying applications are written in assembly language. More transportable higher level languages – such as FORTRAN, PASCAL, C etc. – can not be used because of their inefficiency. The non– transferability of software makes software costs a significant part of computer relaying development.   The solution probably will be obtained by the development of even faster processors which will be fast enough to allow for the inefficiency of higher level languages.   Finally, there is the issue of the harsh environment in which the relays must function.   Extremes of temperature and humidity, as well as severe EMI must be anticipated in the typical utility substation. It is believed that all of these problems can be solved (at some expense).   There are ongoing attempts to develop environmental specifications suitable for computer based relays [11].

II. COMPUTER RELAY ARCHITECTURE

The diagram in Figure 1 shows the main subsystems of a computer relay.   The shaded region in the figure represents the out–door   switch

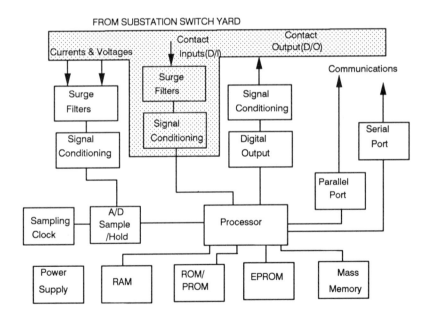

Figure 1   Relaying Computer Subsystems The shaded
area is the out-door switch yard.

yard. Digital inputs derived from contacts within the yard are shown shaded for illustration. Digital inputs from other subsystems within the control house might also exist but are not shown. If the other subsystems were digital then no special processing would be necessary. In some designs it has also been proposed that the currents and voltages be processed in the switch yard and that samples of current and voltage be brought by fiber optic link into the control house.

Each of the various types of memory shown in Figure 1 serves a specific need for the digital relay. In many cases the Random Access Memory (RAM) holds the relaying software as well as the input samples as they are brought in and processed. It also serves as a buffer for data for later storage elsewhere. In addition, RAM is used as a scratch pad to be used during relay algorithm execution. The Read Only Memory (ROM) or Programmable Read Only Memory (PROM) is for permanent program storage. In some cases the programs may execute directly from the ROM if its read time is short enough. The Erasable PROM (EPROM) is needed for storing parameters (such as the relay settings) which may be changed from time to time, but once set must remain fixed even if the power supply to the computer is interrupted. If the EPROM were large enough it could be used

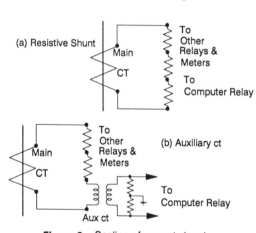

Figure 2   Scaling of current signals

for storing fault related data tables, time–tagged event logs, and audit trails of interrogations and setting changes made in the relay.

The relay inputs are primarily current and voltage signals obtained from current and voltage transformer secondary windings. Two techniques for obtaining appropriate current signals are shown in Figure 2. Since normal current transformer secondary currents may be as high as hundreds of amperes, shunts of resistance of a few milliohms can be used to produce the desired voltage for the ADC's. A second possibility is to use an auxiliary current transformer which also provides electrical isolation between the main ct secondary and the computer input system. Figure 3 shows connections to the voltage transformer. A fused circuit

is provided for the computer relay as well as for every other instrument or relay. The normal voltage of 67 volts rms for a phase to neutral connection can be reduced to the desired level by a resistive divider as shown.

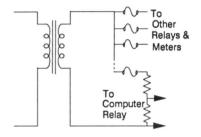

Faults and switching operations in the switch yard and certain types of switching operations within

Figure 3   Voltage Transformer

the control house can create high voltage and energy surges that can be coupled into the wiring which connects current, voltage, and digital inputs to the protection system. Sparking contacts in inductive protection and control circuits within the control house have been found to be a source of significant disturbances [12]. Careful grounding and shielding of leads and equipment, as well as low–pass filtering is used to suppress such surges. Surge suppression filters are necessary for power supply leads, as well as for the input and output wiring [11].

After isolation the current and voltage signals are scaled to the level required for analog to digital conversion. Usually the input to an Analog to Digital Converter (ADC) is limited to a full scale value of ± 10 volts. In addition, filtering to reduce the high frequency content of the signal takes place before the A/D conversion. The cut–off frequency of these filters is dependent on the sampling frequency through the Nyquist sampling theorem. To prevent aliasing, the signal to be sampled must be band limited to a frequency band less than one–half the sampling frequency. The actual design of the anti–aliasing filter involves considerable compromise. A sharp cut–off seems implied by the Nyquist theorem; phase shift at 60 Hz will translate into a delay in making a relaying decision; and the step response of the filter will be seen in the post – fault waveforms. The sharp cut–off suggests high order Butterworth or Chebyshev filters while the phase shift and step response argue for lower order softer cut–off filters. (The Butterworth filter has significant overshoot in its step response.) Figure 4 shows a second order RC filter used for sampling at 720 Hz along with an active realization of a second order Butterworth filter. The passive filter has an additional advantage in being more robust to component errors such as those that might be encountered in aging.

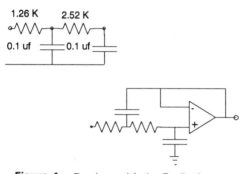

Figure 4   Passive and Active Realizations
of Anti-Aliasing Filters

At a fixed rate, defined by the sampling clock, the analog input signals are sampled to produce a digital quantity which is made available to the processor. Since a number of inputs are required, several conversions are performed at each sampling instant. In order to achieve simultaneous sampling it is necessary that the conversion process be very fast, or that all the signals be sampled at the same instant and held for processing by a relatively slow conversion – transmission cycle for each sample. This is typical of a multiplexed analog input system. A third possibility is to use individual ADC's for each input channel. Trends are toward high speed multiplexing without sample and hold circuits.

The desirability of simultaneous sampling becomes clear if differential protection is considered. In a differential relaying application all input current samples could be added directly to form samples of the differential current only if samples had been taken simultaneously. In other relaying applications such as impedance relaying, simultaneous measurement of two or more phasor quantities (voltage and current) is required. It should be noted that at 16.67 milliseconds, a period at 60 Hz, 10 microseconds (typical conversion time for a 12 bit conversion) corresponds to about 0.2 degree. In other words, true simultaneity is not required and total scan times of up to 50 microseconds are acceptable. These facts, plus the relative ease of achieving local synchronization, has led to the general practice of simultaneous sampling of all input signals by each relaying computer (and in fact, each computer in the substation). Another advantage of synchronized sampling is the possibility of sharing measurements between protection modules in the event of a transducer failure. The benefits of system wide synchronization are considered in the Chapter "Improved Control and Protection of Power Systems Through Synchronized Phasor Measurements".

To complete a description of the computer relaying subsystems in Figure 1, the digital output subsystem must be described. Digital outputs used to activate external devices such as alarms, breaker trip coils etc. must be generated. A parallel output port of the processor provides one word for these outputs typically with each bit used as a source for one contact. The computer output bit is a TTL (Transistor to Transistor Logic) level signal, and should be optically isolated before driving a high speed multi-contact relay or thyristors. Finally, the power supply is usually a single dc input multiple dc output converter powered by the station battery. The input is generally 125 volts dc, and the output is 5 volts dc and ± 15 volts dc. The 5 volt supply is needed to power the logic circuits, while the 15 volt supply is needed for the analog circuits. The station battery is of course continuously charged from the station ac service.

A. ANALOG TO DIGITAL CONVERTERS

The Analog to Digital Converter (ADC) converts an analog voltage level to its digital representation with a accuracy depending on its word length expressed in bits. Using two's complement notation, an ADC with 12 bit word length has as a maximum output the binary number 0111 1111 1111 (7FF in hexadecimal notation), while 1000 0000 0000 (800 in hexadecimal notation) represents the smallest (negative) number. In decimal notation, hexadecimal 7FF is equal to $(2^{11}-1) = 2047$, and hexadecimal 800 is equal to $-2^{11} = -2048$. Considering that the analog input signal may range between ±10 volts, it is clear that each bit of the 12 bit ADC word represents 10/2048 volts or 4.883 millivolts. This is the quantization error of the ADC. Normalized to the largest possible input voltage, the per unit quantization error is

$$\text{per unit quantization error} = 2^{-N}$$

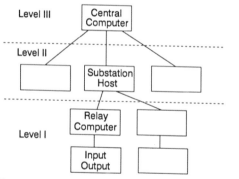

Figure 5 System wide hierarchical computer system

where N is the number of bits in the converter.

The quantization error component is only one part of the total error in the analog to digital conversion. There are also errors depending upon the gain, offset and non-linearity of the device and the technique (successive approximation converters are common) of conversion. If we consider the ensemble of all conversions made by the ADC we can regard the total conversion error as a random process. The contribution of this error along with others on the relaying process will be considered in the sequel.

B. SUBSTATION COMPUTER HIERARCHY

A proposed hierarchy of various relaying and other computers in a substation is shown in Figure 5. It has been suggested that computer relays are at the lowest level of a vast system wide protection and control computer hierarchy [13–14]. The functions performed at level I include: input/output to the switch yard, measurements, relaying, control, diagnostics, man–machine interface, and communication with level II. At level II the functions are: data acquisition and storage, sequence of events analyses and coordination, assignment for back–up in case of failures,man–machine interface, and communication with the level I and level III computers. The level III computers initiate control actions, collect and collate system wide sequence of event analyses, communication with level II, prepare reports and possibly coordinate adaptive relaying functions. Adaptive relaying principles and their relationship to computer relaying are also discussed in the chapter "Improved Control and Protection of Power Systems Through Synchronized Phasor Measurements".

III. RELAYING PRACTICES

Protective relaying itself is too large a subject to be included in this chapter. There are excellent references in the area [15–17] which provide a complete discussions of traditional relaying and application practices. For our discussion of digital relaying there are a few areas of relaying practice that must be mentioned in order to understand how the digital relay fits into an overall protection system. Faults (in most cases short circuits) occurs due to weather related events such storms (lightning, tree limbs blown down) or equipment failures and are essentially unpredictable. If the faulted power system component (line, bus, transformer, etc.) is not isolated from the system quickly, it may be damaged or lead to power system instability or break–up of the system. The role of the protection system is to remove the faulted element from the rest of the power system as quickly as possible. The relay is a device which responds to its inputs (voltages and currents from transducers or contact status) to provide output signals to trip circuit breakers for faults for which the relay is designed to operate. The relay must be designed such that it produces a trip output for all fault conditions for which it is responsible and fails to produce a trip output for any other conditions. It can be seen that a relay could malfunction in two basic modes; it might fail to trip for a fault for which it was responsible, or it could trip for a condition for which it should not trip. The concept of **reliability** in relaying reflects these two sides of relay performance. To a relay engineer, reliability has two components: **dependability**, and **security**. Dependability implies that the relay will always operate for conditions for which it is designed to operate. A relay is said to be secure if the relay will not operate for any other power system disturbance. Traditionally, failure to trip had more serious consequences than false trips. As a result most relaying systems have been designed to be dependable at the expense of security.

Responsibility for protection of portions of the power system is defined in terms of zones of protection. A zone of protection (usually defined by circuit breakers) is a region defined by an imaginary boundary line on the power system one line diagram. One or more relays are responsible for faults occurring within the zone. Given such a fault, the relays activate trip coils of circuit breakers, hopefully de–energizing the fault. Zones of protection overlap to ensure that there are no gaps in the protection system. Zones are protected by several protection systems in order to make sure that failure of the protection system itself will not leave the power system unprotected. The use of independent primary protection systems along with backup protection systems (which are designed to function only when the primary protection fails) reinforces the dependability of the overall protection system. It is clear, however, that security suffers when so many relays may cause a line to trip. The possibility of altering the relative security and dependability of a protection system is another attractive possibility of digital protection systems.

IV. TRANSMISSION LINE ALGORITHMS

Most of the early activity in computer relaying was in the development of algorithms for transmission line protection. Almost from the outset there were two basic approach. One class of algorithms computes the fundamental frequency components of both voltages and currents from the samples. The ratios of appropriate voltages and currents then provide the impedance to the fault. These algorithms can be thought of as emulating an analog impedance relay. All are based on the ability to estimate the fundamental frequency components of a signal from a few samples. The distinction between algorithms of this type is in their ability to estimate the fundamental frequency components when signals other than the fundamental frequency are present. The second type of algorithm is based on a circuit model of the faulted transmission line, for example a series $R - L$ circuit. This approach has the apparent advantage of not using a single frequency model such as an impedance and thus requiring less filtering of the original current and voltage.

Certain notation and concepts common to all algorithms will be established before presenting specific algorithms. The following notation will be used in discussing all of the algorithms:

$x(t)$ = The instantaneous value of an ac waveform, a voltage or a current

x_k = The kth sample value of $x(t)$

ω_0 = The fundamental power system frequency in radians per second

Δt = The fixed interval between samples, i.e.
$$x_k = x(k\Delta t)$$

θ = The fundamental frequency angle between samples, i.e.
$$\theta = \omega_o \Delta t$$

To illustrate some of the common features of algorithms based on a waveform model suppose $x(t)$ is assumed to be of the form

$$x(t) = X_c \cos \omega_o t + X_s \sin \omega_o t \tag{1}$$

where X_c and X_s are real numbers. Further, assume samples are taken at 0 and Δt

$$x_o = x(0) \tag{2}$$
$$x_1 = x(\Delta t)$$

The samples are related to the amplitudes X_c and X_s through

$$\begin{bmatrix} x_o \\ x_1 \end{bmatrix} = \begin{bmatrix} 1 & 0 \\ \cos\theta & \sin\theta \end{bmatrix} \begin{bmatrix} X_c \\ X_s \end{bmatrix} \tag{3}$$

where θ is the fundamental frequency angle between samples. It is clear that two samples are sufficient to determine X_c and X_s if the signal is described by Eq. (1). The solution to Eq. (3) is

$$X_c = x_0 \tag{4}$$
$$X_s = (x_1 - x_0 \cos\theta)/\sin\theta$$

If samples at $k\Delta t$ and $(k+1)\Delta t$ are used in Eq. (4) with

$$x_k = x(k\Delta t) \tag{5}$$
$$x_{k+1} = x((k+1)\Delta t)$$

Then

$$X_c^{(k)} = x_k \tag{6}$$
$$X_s^{(k)} = (x_{k+1} - x_k \cos\theta)/\sin\theta$$

where the superscript k indicates calculations beginning at the kth sample. If $x(t)$ were described by Eq. (1) then

$$X_c^{(k)} = X_c\cos k\theta + X_s\sin k\theta \tag{7}$$

$$X_s^{(k)} = X_s\cos k\theta - X_c\sin k\theta \tag{8}$$

In polar form

$$|X^{(k)}| = \sqrt{(X_c^{(k)})^2 + (X_s^{(k)})^2} \tag{9}$$

$$\varphi^{(k)} = \tan^{-1}\left[\frac{X_s^{(k)}}{X_c^{(k)}}\right] = \tan^{-1}\left[\frac{X_s}{X_c}\right] - k\theta \tag{10}$$

It can be seen from Eq. (10) that the computed phasor has the correct amplitude but rotates, i.e. the angle $\varphi^{(k)}$ decreases by the angle θ at each sample point. If the ratio of voltage and current phasors is to be used for an impedance calculation, the rotation will cancel in the division. In some applications it may be necessary to correct for the rotation.

The use of only two samples in Eq. (4) makes the equations simple but is impractical when any kind of error signals are involved. The division by $\sin\theta$ in Eq. (6) would amplify any error in the value of x_{k+1} for example. Small values of θ could result in large errors in X_s. Using more samples provides some averaging and improved rejection of errors. If more samples are considered in Eq. (3), the equation similar to Eq. (3) becomes

$$\begin{bmatrix} x_1 \\ x_2 \\ x_3 \\ \vdots \\ x_N \end{bmatrix} = \begin{bmatrix} \cos\theta & \sin\theta \\ \cos 2\theta & \sin 2\theta \\ \cos 3\theta & \sin 3\theta \\ \vdots & \vdots \\ \cos N\theta & \sin N\theta \end{bmatrix} \begin{bmatrix} X_c \\ X_s \end{bmatrix} \tag{11}$$

or

$$\mathbf{x} = \mathbf{S}\ \mathbf{X} \tag{12}$$

where $\mathbf{x}$ is an vector of N sample values, $\mathbf{X}$ is the a vector of the two unknowns, and $\mathbf{S}$ is an N by 2 matrix. Equation (12) is a set of "overdefined equations" and has solution [18]

$$\mathbf{X} = (\mathbf{S}^T\mathbf{S})^{-1}\mathbf{S}^T\mathbf{x} \tag{13}$$

Thus samples from x_0 to x_N (or from x_k to x_{k+N-1} as in Eq. (6)) are processed to obtain estimates of X_c and X_s. Equation (13) will be examined in more detail after considering some general features of all such algorithms. The algorithm described by Eq. (6) has a *data window* of two samples, while the algorithm in Eq. (11) has a data window of N samples. In each case, as a new sample becomes available, the oldest of the (N) sample values is discarded and the new sample value is included in the calculation. The calculation in Eqs. (6) or (13) must be completed by the microprocessor before the next sample is produced. In addition the calculations for the relay logic must also fit in the time interval between samples.

A moving data window of six samples is shown in Figure 6 for a current waveform sampled at 12 samples per cycle. The current increases after the fault instant. Window 1 contains six samples of pre–fault data, while, for example, Windows 4 contains three pre– and three post–fault samples, and Window 7 has only post–fault data. The calculations Eq. (13) will produce the correct values for X_c and X_s for Windows 1 and 7 while the data in the windows in between cannot be fitted by a pure sinusoid. If the values of X_c and X_s computed by Eq. (13) were inserted in Eq. (1), the resulting time function would not fit the six samples.

Several issues are raised by Figure 6. The first is the time between samples, Δt, which determines the amount of time the microprocessor has to complete its calculations. Very high sampling rates will provide

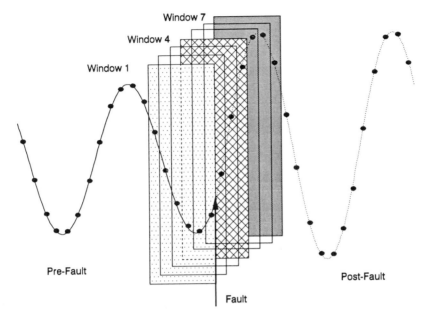

Figure 6   The Moving Data Window

less time for the calculations implied by Eq. (13) and will require more powerful processors.   The rate of 12 samples per cycle in Figure 12 gives Δt = 1.3889 msec for a 60 Hz system.   Existing algorithms use sampling rates from 4 to 64 samples a cycle.   The second issue is that of the length of the data window.   It seems reasonable to wait until the results are reliable (when the window contains only post–fault data) before making relaying decisions.   It is clear that faster decisions can be made by short window algorithms which pass over the pre–fault post–fault interval more quickly.   Unfortunately, the length of the data window determines the ability of an algorithm to reject non–fundamental frequency signals.   In other words, there is an inherent inverse relationship between relaying speed and accuracy.   While the algorithms represented by Eqs. (6) and (13) yields the correct phasor if the signal $x(t)$ is given by Eq. (1), we must recognize that the signal to be sampled is more accurately given by:

$$x(t) = X_c \cos\omega_o t + X_s \sin\omega_o t + \epsilon(t) \tag{14}$$

It is the nature of the signal $\epsilon(t)$ in Eq. (14) that must be understood in order to evaluate line relay performance.

A. SOURCES OF ERROR

The post–fault current and voltage waveforms are not the pure sinusoid in Eq. (1) for a number of reasons. The most predictable non–fundamental frequency term is the decaying exponential which can be present in the current waveform. For a series R – L model of the faulted line (assuming zero pre–fault current) the instantaneous current for a fault at time t_0 is given by

$$i(t) = I \cos(\omega_0 t - \varphi) - [I \cos(\omega_0 t_0 - \varphi)] e^{-(t - t_0)R/L} \tag{15}$$

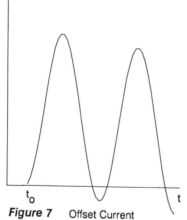

Figure 7   Offset Current

The second term in Eq. (15) decays exponentially with the time constant of the line. The time constant is in the range of 30–50 ms for a typical EHV line. The initial offset current can be almost twice the amplitude of the steady state fault current, as shown in Figure 7. The situation can be even more complicated near a large generator. For algorithms which see the offset as an error the decay can be removed with an external filter or even in software (if the time constant of the line is known). The dependence of the time constant on fault resistance makes the removal less effective for high resistance faults.

Other non–fundamental frequency terms are more difficult to remove because they are unpredictable. The current and voltage transducers contribute some of these error signals. For example, current transformers can produce nonlinear effects due to flux left in the core. High frequency signals associated with the reflection of waveforms between the bus and fault may be present. The fault arc itself may produce harmonic signals. In addition, the A/D converter contributes errors due to the least significant bit in the conversion and due to timing errors, i.e., the samples are not exactly Δt sec apart. The anti–aliasing filter attenuates the high frequency content of these signals but will contribute a transient response of its own. Drift over time of the component values in such filters (particularly active realizations of such filters) are also sources of error.

Finally, the power system itself is a source of non–fundamental frequency signals. The voltage waveform shown in Figure 8 is typical. The post – fault voltage is made up of at least two non – fundamental frequency signals which, in general, are not harmonics of the fundamental. The non–fundamental frequencies are natural frequencies of the system

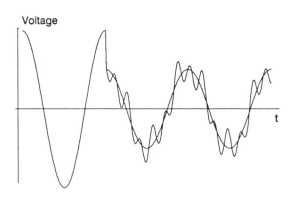

Figure 8   Voltage Waveforms

which are excited by the application of the fault. Since the network is fixed if the fault location is held constant, it follows that the natural frequencies change as the fault location or the network structure behind the fault is changed. The phase of the non–fundamental frequency components is a function of the fault incidence angle. Experiments on a model power system combining the effect of changing the fault type

and location and the structure of the network feeding the fault have
been reported [19].   The conclusion is that an important part of the
non–fundamental frequency signal $\epsilon(t)$ in Eq. (14) is due to the network
itself.   Since these signals depend on the fault location and on the
nature of the system feeding the fault they are random in the sense
that, considering the ensemble of times that the relay is expected to
operate, the frequency and phase of the signal cannot be predicted.

If the signal $\epsilon(t)$ in Eq. (14) is taken as a random process, the
anti–aliasing filter and the algorithm taken together filter the random
process.   One way of evaluating an algorithm then would be to compute
the frequency response of the algorithm.   To obtain the frequency
response we should compute the phasor Eq. (4) when the input signal is
of the form $e^{j\omega t}$.

Example 1 With $x(t) = e^{j\omega t}$, $x_0 = 1$, $x_1 = e^{j\Delta t\omega}$

$$\hat{X}_c = 1$$

$$\hat{X}_s = (e^{j\Delta t\omega} - \cos\theta)/\sin\theta = \frac{\cos\omega\Delta t - \cos\theta}{\sin\theta} + j\frac{\sin\omega\Delta t}{\sin\theta}$$

With $x(t) = \text{Re}\{e^{j\omega_0 t}\} = \cos\omega_0 t$, $\hat{X}_c = 1$, and $\hat{X}_s = 0$. In general,
if $\omega \neq \omega_0$, i.e. if $x(t) = \text{Re}\{e^{j(\omega t+\varphi)}\} = \cos(\omega t + \varphi)$ then

$$\hat{X}_c = \cos\varphi$$

$$\hat{X}_s = \cos\varphi\left[\frac{\cos\omega\Delta t - \cos\theta}{\sin\theta}\right] - \sin\varphi\left[\frac{\sin\omega\Delta t}{\sin\theta}\right]$$

The magnitude of the phasor $\hat{X}_c + j\hat{X}_s$ depends on the angles φ and
θ. The magnitude $\sqrt{(\hat{Y}_c^2 + \hat{Y}_s^2)}$ is shown in Figure 9 for $\varphi = 0$, $\pi/4$
and $\pi/2$ for a sampling rate of twelve samples per cycle ($\theta = 30^\circ$).

The frequency responses shown in Figure 9 are not very impressive. Although the response is always unity at 60 Hz, there is amplification of high frequencies. The problem, of course, is that only two samples are being used to form the estimate. While the two –sample algorithm can reach decisions quickly (it moves quickly through the pre – fault post – fault region) it has little ability to reject non–fundamental frequency signals. We will see that longer window algorithms have a greater ability to reject non – fundamental frequencies at the expense of a longer decision time (the longer window takes longer to clear the instant of fault inception)

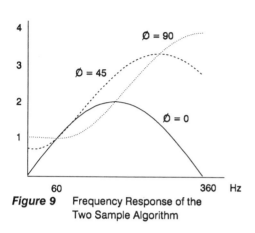

Figure 9   Frequency Response of the
Two Sample Algorithm

If Eq. (11) is generalized to include other known signals (in addition to the fundamental frequency components), the solution from Eq. (13) can be extended to a number of relaying algorithms. The voltage or current is written as

$$x(t) = \sum_{n=1}^{N} X_n \, s_n(t) + \epsilon(t) \tag{16}$$

or

$$x_k = \sum_{n=1}^{N} X_n \, s_n(k\Delta t) + \epsilon(k\Delta t) \tag{17}$$

where the signals $s(t)$ are assumed known but the coefficients X_n are unknown. One set of choices for the signals would be:

$$s_1(t) = \cos\omega_o t \quad \Big\} \text{ the fundamental}$$

$$s_2(t) = \sin\omega_o t$$

$$s_3(t) = \cos 2\omega_o t \quad \Big\} \text{ the second harmonic}$$

$$s_4(t) = \sin 2\omega_o t$$

$$\vdots \quad \Big\} \text{ other harmonics}$$

$$s_N(t) = e^{-(R/L)t} \quad \} \text{ the exponential offset}$$

The estimation of the coefficients X_n from the measurements x_k is similar to Eq. (11) except for the error terms. If we write

$$
\begin{bmatrix} x_1 \\ x_2 \\ \vdots \\ x_k \end{bmatrix}
=
\begin{bmatrix} s_1(\Delta t) & s_2(\Delta t) & \cdots & s_N(\Delta t) \\ s_1(2\Delta t) & s_2(2\Delta t) & \cdots & s_N(2\Delta t) \\ \vdots & \vdots & & \vdots \\ s_1(k\Delta t) & s_2(k\Delta t) & \cdots & s_N(k\Delta t) \end{bmatrix}
\begin{bmatrix} X_1 \\ X_2 \\ \vdots \\ X_N \end{bmatrix}
+
\begin{bmatrix} \epsilon_1 \\ \epsilon_2 \\ \vdots \\ \epsilon_k \end{bmatrix}
$$

or

$$x = S\,X + \epsilon \tag{18}$$

Equation (18) represents k equations in N ($k \geq N$ is required to estimate X). If we assume that ϵ is a vector of random numbers with zero mean, i.e. $E\{\epsilon\} = 0$, and a covariance matrix

$$E\{\,\epsilon\,\epsilon^T\} = W \tag{19}$$

then the solution for the estimate $\hat{X}$ is

$$\hat{X} = (\,S^T\,W^{-1}S\,)^{-1}S^T W^{-1} x \tag{20}$$

It is common [20–21] to assume that W is a multiple of a unit matrix or more simply to assume the form of Eq. (12) for the solution, i.e.,

$$\hat{X} = (S^T S)^{-1} S^T \, x \tag{21}$$

The matrix $(S^T S)^{-1} S^T$ can be computed off–line and stored. In fact, only the first two rows of the matrix corresponding to the fundamental frequency ($\cos\omega_0 t$ and $\sin\omega_0 t$) are needed to estimate the fundamental frequency phasor (for impedance relaying). If the exponential decay in the fault current is included, the matrix $(S^T S)^{-1} S^T$ is full, so that a full set of 2k numbers is needed. If only harmonics are included (the offset is removed with an analog prefilter or separate algorithm) then a familiar form is obtained.

B. FOURIER ALGORITHMS

Two different versions are of interest. If only the fundamental frequency components are included as in Eq. (11) then $(S^T S)$ is a 2×2 matrix and Eq. (13) or (21) becomes:

$$\begin{bmatrix} \hat{X}_c \\ \\ \hat{X}_s \end{bmatrix} = \begin{bmatrix} \sum\limits_{k=1}^{K} \cos^2(k\theta) & \sum\limits_{k=1}^{K} \cos(k\theta)\sin(k\theta) \\ \sum\limits_{k=1}^{K} \cos(k\theta)\sin(k\theta) & \sum\limits_{k=1}^{K} \sin^2(k\theta) \end{bmatrix}^{-1} \begin{bmatrix} \sum\limits_{k=1}^{K} x_k \cos(k\theta) \\ \sum\limits_{k=1}^{K} x_k \sin(k\theta) \end{bmatrix} \tag{22}$$

It should be emphasized that the estimates in Eq. (22) are optimal estimates for X_c and X_s if the signal $x(t)$ is made up of pure fundamental plus additive noise which is independent at sample times and has a constant variance. If $\theta = 2\pi/K$, (K samples span a full period) Eq. (22) corresponds to the full–cycle Fourier algorithm. The matrix $(S^T S)$ is diagonal with diagonal entries $2/K$ and Eq. (22) becomes

$$\hat{Y}_c = \frac{2}{K} \sum_{k=1}^{K} y_k \cos(k\theta) \tag{23}$$

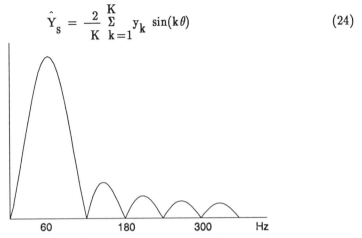

$$\hat{Y}_s = \frac{2}{K} \sum_{k=1}^{K} y_k \, \sin(k\theta) \tag{24}$$

Figure 10 Frequency Response of the Full Cycle Fourier

The frequency response of the full–cycle Fourier algorithm is shown in Figure 10. The fact that the algorithm rejects DC plus all other harmonics is due to the orthogonality of the trigonometric functions. It can be seen that the matrix (S^TS) is still diagonal if $\theta = \pi/K$ and Eq. (22) and Eq. (23) are appropriate at both a full–cycle and a half–cycle. The frequency response of the half–cycle Fourier algorithm is shown in Figure 11. It can be seen that the half–cycle algorithm is less selective than the full–cycle version particularly at DC and even harmonics. The algorithm was intended to be used with an external filter to eliminate the exponential offset in the current.

The second case of interest is produced if the fundamental and harmonics are included in the signal set $\{s_n(t)\}$ and an even number of samples spanning a full period is used. In this case Eq. (21) becomes a rectangular

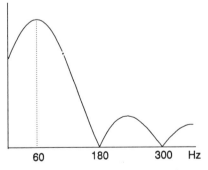

Figure 11 Frequency Response of the 1/2 cycle Fourier

form of the Discrete Fourier Transform (DFT). With K samples per cycle (K/2–1) harmonics can be computed. The fundamental frequency components are given by Eqs. (23) and (24) while for the pth harmonic

$$\hat{X}_{cp} = \frac{2}{K}\sum_{k=1}^{K} x_k \cos(pk\theta) \tag{25}$$

$$\hat{X}_{sp} = \frac{2}{K}\sum_{k=1}^{K} x_k \sin(pk\theta) \tag{26}$$

The estimates of the harmonics given by Eqs. (25) and (26) are used in transformer protection but have no role in transmission line protection.

C. RECURSIVE FORMS

The calculations in Eqs. (23) and (24) represent more calculations than are actually necessary in practice. The resulting phasor also rotates as in Eq. (10). A solution to the rotation also provides a more efficient computation. Let super M denote the complex estimate formed with K samples ending at M and drop the normalizing factor of 2/K ,

$$X^{(M)} = \sum_{k=M-K+1}^{M} x_k \, e^{-j(k+K-M)\theta}$$

and rotate by an angle of $(K-M)\theta$ to keep the result stationary

$$\tilde{X}^{(M)} = X^{(L)} e^{\,j(K-M)\theta} = \sum_{k=M-K+1}^{M} x_k e^{-jk\theta} \tag{27}$$

Note that

$$\tilde{X}^{(M-1)} = \sum_{k=M-K}^{M-1} x_k e^{-jk\theta} \tag{28}$$

so that the difference between Eqs. (27) and (28) is the last term of Eq. (28) and the first term of Eq. (27), i.e.

$$\tilde{X}^{(M)} = \tilde{X}^{(M-1)} + [x_M - x_{M-K} \ e^{\ jK\theta}] \ e^{-jM\theta} \qquad (29)$$

Equation (29) is valid for any θ or any K. If $K\theta$ is not a multiple of a half–cycle then the real and imaginary parts of Eq. (29) must be multiplied by the matrix $(S^T S)^{-1}$ to form the estimates. With $K\theta = 2\pi$ (a full –cycle window), the recursive form of the algorithm becomes

$$\hat{X}_c^{(M)} = \hat{X}_c^{(M-1)} + [x_M - x_{M-K}] \cos(M\theta) \qquad (30)$$

$$\hat{X}_s^{(M)} = \hat{X}_s^{(M-1)} + [x_M - x_{M-K}] \sin(M\theta) \qquad (31)$$

while with $K\theta = \pi$

$$\hat{X}_c^{(M)} = \hat{X}_c^{(M-1)} + [x_M + x_{M-K}] \cos(M\theta) \qquad (32)$$

$$\hat{X}_s^{(M)} = \hat{X}_s^{(M-1)} + [x_M + x_{M-K}] \sin(M\theta) \qquad (33)$$

The bracketed quantity in Eqs. (30) and (31) is zero if the signal is purely periodic while the same term is zero in Eqs. (32) and (33) for a signal made up of only odd harmonics. The reduction in computation compared with Eqs. (23) and (24) is striking. Additional simplification is achieved by choice of sampling rate. Sampling rates of four, eight, and 12 times a cycle would only require the sines and cosines of multiples of $90°$, $45°$, or $30°$ respectively. The multiplication by the irrational numbers $1/\sqrt{2}$ or $\sqrt{3}/2$ in the last two cases can be achieved in analog hardware or approximated by a sequence of shifts and adds.

C. PHASORS

A final point about sign convention and normalization of phasor quantities should be made. The phasor associated with the sinusoid

$$x(t) = X_m \cos(\omega t + \varphi) = X_m \cos\varphi \ \cos\omega t - X_m \sin\varphi \ \sin\omega t \quad (34)$$

is $\quad X = \dfrac{X_m}{\sqrt{2}}\, e^{j\varphi} = \dfrac{X_m}{\sqrt{2}}\, \cos\varphi + j\, \dfrac{X_m}{\sqrt{2}}\, \sin\varphi$ $\qquad$ (35)

which implies that the proper phasor computed from $\hat{X}_c$ and $\hat{X}_s$ is

$$X = \frac{1}{\sqrt{2}}\, (\, \hat{X}_c - j\, \hat{X}_s) \qquad (36)$$

D. DIFFERENTIAL–EQUATION ALGORITHMS

The algorithms described so far estimate the fundamental frequency components of currents and voltages and can be used to compute impedances in line relaying or harmonics of currents in transformer protection. The differential equation algorithms, on the other hand, are based on a model of the circuit seen by the relay. Assuming a single–phase model of the faulted line and writing the differential equation relating the voltage and current seen by the relay, we obtain

$$v(t) = R\, i(t) + L\, \frac{di(t)}{dt} \qquad (37)$$

A more tractable form of Eq. (37) can be obtained by integrating over 2 consecutive intervals:

$$\int_{t_0}^{t_1} v(t)\ dt = R \int_{t_0}^{t_1} i(t)\ dt + L\, [\, i(t_1) - i(t_0)] \qquad (38)$$

$$\int_{t_1}^{t_2} v(t)\ dt = R \int_{t_1}^{t_2} i(t)\ dt + L\, [\, i(t_2) - i(t_1)] \qquad (39)$$

The trapezoidal rule for integration with equally spaced samples at an interval Δt gives

$$\int_{t_0}^{t_1} v(t) \, dt = \frac{\Delta t}{2} [v(t_1) + v(t_0)] = \frac{\Delta t}{2} [v_1 + v_0] \qquad (40)$$

Using Eq. (40) in Eqs. (38) and (39) with samples at k, k+1, and k+2

$$\begin{bmatrix} \frac{\Delta t}{2}(i_{k+1}+i_k) & (i_{k+1}-i_k) \\ \frac{\Delta t}{2}(i_{k+2}+i_{k+1}) & (i_{k+2}-i_{k+1}) \end{bmatrix} \begin{bmatrix} R \\ L \end{bmatrix} = \begin{bmatrix} \frac{\Delta t}{2}(v_{k+1}+v_k) \\ \frac{\Delta t}{2}(v_{k+2}+v_{k+1}) \end{bmatrix} (41)$$

Thus three consecutive samples of current and voltage are sufficient to compute estimates of R and L as

$$\hat{R} = \left[\frac{(v_{k+1}+v_k)(i_{k+2}-i_{k+1}) - (v_{k+2}+v_{k+1})(i_{k+1}-i_k)}{(i_{k+1}+i_k)(i_{k+2}-i_{k+1}) - (i_{k+2}+i_{k+1})(i_{k+1}-i_k)} \right] \qquad (42)$$

and

$$\hat{L} = \frac{\Delta t}{2} \left[\frac{(i_{k+1}+i_k)(v_{k+2}+v_{k+1}) - (i_{k+2}+i_{k+1})(v_{k+1}+v_k)}{(i_{k+1}+i_k)(i_{k+2}-i_{k+1}) - (i_{k+2}+i_{k+1})(i_{k+1}-i_k)} \right] \qquad (43)$$

There is a degree of approximation involved in Eqs. (42) and (43) involving the trapezoidal integration that is not present in Fourier calculations. Aside from that, Eqs. (42) and (43) can be compared with taking three samples of voltage and current, finding the phasors for voltage and current using (22) (the 2x2 inverse would be required), and evaluating a complex division of the form.

$$\frac{a + j b}{c + j d} = \frac{(a c + b d)}{c^2 + d^2} + j \frac{(b c - a d)}{c^2 + d^2}$$

to find the impedance.

The algorithm using only three samples is a short window algorithm and has a relatively poor frequency response. A variety of techniques have been suggested to produce a longer window algorithm. One possibility is to make the intervals $[t_0, t_1]$ and $[t_1, t_2]$ in Eqs. (38) and (39) longer and select the intervals so that certain harmonics are rejected [22]. The integrals can be evaluated with a number of samples per integral rather than just the end points as in Eqs. (38) and (39). Another approach would be to include more than two intervals so that Eq. (41) became an overdefined set. The solution of the overdefined equations would involve a large number of multiplications in forming the equivalent of the matrix $(S^T S)$. As an alternate, a technique of using a sequence of estimates, each of which is obtained from the three–sample algorithm, has been developed [23]. The scheme involves a counter and a zone–one characteristic. If the result of the three–point calculation lies in the characteristic, a counter is increased by one and if the computed values lie outside the characteristic, the counter is reduced by one. By setting a large value for the required counter value for a trip signal, the window length can be effectively increased. A possible reason to use such counters will be seen in the next section.

F. ERROR ANALYSIS FOR DIFFERENTIAL–EQUATION ALGORITHMS

The apparent advantage of the differential–equation algorithms is that the exponential offset in the current does not have to be removed in order to estimate the values of R and L. Indeed it seems that the offset is an aid in the estimation. The validity of these impressions can be checked by considering the sources of error in the calculations. Currents and voltages which do not satisfy the differential equation (Eq. (37)) will cause errors in the estimated R and L. Included in the sources of such signals are the effects produced by the shunt capacitances which would be included in the π section model of the transmission lines along with transducer errors, A/D errors, and errors in the anti–aliasing filters. The shunt capacitors are more important in models of longer, higher voltage lines. If we let $v(t)$ and $i(t)$ satisfy the differential equation, and denote the measured current and voltage by $i_m(t)$ and $v_m(t)$ respectively, then

$$i_m(t) = i(t) + \epsilon_i(t) \tag{44}$$

$$v_m(t) = v(t) + \epsilon_v(t) \tag{45}$$

It can be seen that the current $i_m(t)$ satisfies the differential equation

$$R\; i_m(t) + L\; \frac{di_m(t)}{dt} = v(t) + R\; \epsilon_i(t) + L\; \frac{d\epsilon_i(t)}{dt} \tag{46}$$

and the current $i_m(t)$ and voltage $v_m(t)$ are related by:

$$R\; i_m(t) + L\; \frac{di_m(t)}{dt} = v_m(t) + R\; \epsilon_i(t) + L\; \frac{d\epsilon_i(t)}{dt} - \epsilon_v(t) \tag{47}$$

In other words, the measured voltage and current satisfy a differential equation such as Eq. (37) where the "currents" are without error and the "voltage" has an equivalent error made up of the voltage error plus a processed current error term. The current term is equivalent to the output of a filter designed to remove the offset and includes the derivative term which will amplify high frequency noise terms.

With the error terms in the expressions for $\hat{R}$ and $\hat{L}$ confined to the voltage terms in the numerator it is necessary to examine the denominator of Eqs. (42) and (43). The denominator can be simplified to

$$(i_{k+1}+i_k)(i_{k+2}-i_{k+1})-(i_{k+2}+i_{k+1})(i_{k+1}-i_k) = -2\;(i_{k+1}^2-i_k i_{k+2}) \tag{48}$$

For a current of the form (for simplicity there are no measurement errors)

$$i(t) = I\; \cos(\omega_o t) - I\; \cos(\omega_o t_o)\; e^{-\frac{R}{L}(t-t_o)} \tag{49}$$

with sixteen samples per cycle and a line time constant of 40 ms, the current (Eq. (49)) and reciprocal of the denominator (1/Eq. (48)) are shown in Figure 12 as a function of the time of the kth sample. The amplification of the error terms in Eq. (47) is striking. The counting algorithm will be delayed in issuing the trip signal by the poor estimates. The fact that the denominator is a constant when the offset is absent suggests that the differential–equation algorithm is not as immune to the offset as initially imagined. There have been attempts to include the shunt capacitances in the line model to reduce the non–modeled terms [24]. The resulting differential equation is second order and the amount of computation required is quite large.

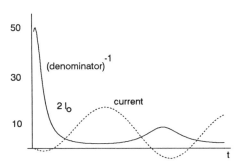

Figure 12   Offset Current and the reciprocal of the denominator of (42) and (43)

G. KALMAN FILTER ALGORITHMS

The Kalman filter was applied to line relaying because of its ability to handle measurements whose quality changes in time [25–26]. Simulation studies on a simple system led to the conclusion that the covariance of the noise in the voltage and current was not a constant but rather decayed in time. Although this could be accommodated by letting the Matrix **W** in Eq. (19) be diagonal with decaying diagonal entries the Kalman filter is also a possibility. The equations necessary for a Kalman filter formulation are:

$$x_{k+1} = \varphi_k \, x_k + \Gamma_k \, w_k \qquad (50)$$

$$z_k = H_k \, x_k + \epsilon_k \qquad (51)$$

where the quantity to be estimated is the state x_k which evolves according to the state equation (Eq. (50)) while the measurements z_k are used to estimate x. The terms ϵ_k and w_k are discrete time random processes representing measurement errors, and state noise, i.e. random inputs in the evolution of the state, respectively. Typically w_k and ϵ_k are assumed to be white (uncorrelated from sample to sample), to be independent of each other, to have zero means, and to have covariance matrices

$$E \{ w_k \, w_j^T \} = \delta_{kj} \, Q_k \tag{52}$$

$$E \{ \epsilon_k \, \epsilon_j^T \} = \delta_{kj} \, R_k \tag{53}$$

where $\delta_{kj} = 0$; $k \neq j$, and $\delta_{kj} = 1$ $k = j$. If we imagine sampling a pure sinusoid of the form

$$x(t) = X_c \cos(\omega t) + X_s \sin(\omega t)$$

at equal intervals corresponding to $\omega \Delta t = \theta$ then we could form a two state model of the phasor estimation problem for voltages by taking

$$x = \begin{bmatrix} X_c \\ X_s \end{bmatrix}, \quad \varphi = \begin{bmatrix} 1 & 0 \\ 0 & 1 \end{bmatrix}, \text{ and } \quad H_k = [\cos(k\theta) \; \sin(k\theta)] \tag{54}$$

while a three state model (to account for the offset) for the current is formed as;

$$x = \begin{bmatrix} X_c \\ X_s \\ X_o \end{bmatrix}, \quad \varphi = \begin{bmatrix} 1 & 0 & 0 \\ 0 & 1 & 0 \\ 0 & 0 & e^{-\alpha \Delta t} \end{bmatrix} \tag{55}$$

and

$$H_k = [\cos(k\theta) \; \sin(k\theta) \; 1] \tag{56}$$

The covariance of the measurement noise was taken as

$$R_k = K \, e^{-k\Delta t / T} \tag{56}$$

where different K's were used for voltage and current models and T was one–half the expected time constant of the protected line. The optimal estimate obtained by the filter is

$$\hat{x}_{k+1} = \varphi_k \, \hat{x}_k + K_{k+1} \, [\, z_{k+1} - H_{k+1} \, \varphi_k \, \hat{x}_k \,] \qquad (57)$$

where **K** is given by

$$K_{k+1} = P(k+1|k) \, H_{k+1}^T \, [\, H_{k+1} \, P(k+1|k) \, H_{k+1}^T + R_{k+1} \,]^{-1} \qquad (58)$$

$$P(k+1|k) = \varphi_k \, P(k|k) \, \varphi_k^T + \Gamma_k \, Q_k \, \Gamma_k^T \qquad (59)$$

and

$$P(k+1|k+1) = [\, I - K_{k+1}H_{k+1}] \, P(k+1|k) \qquad (60)$$

The Kalman gain matrix K_k, the covariance of the one–step prediction $P(k+1|k)$ and the covariance matrix in the estimate at time $k+1$, $P(k+1|k+1)$ are given by Eqs. (58)–(60). The recursion in Eqs. (57)–(60) is initialized with an *a–priori* estimate of **x** and an *a–priori* covariance matrix $P(0|0)$. The prior estimate actually distinguishes the Kalman filter solution from the solution in Eq. (20) more than the changing error covariance in Eq. (56). A version of the Kalman filter was proposed which included an eleven state model with two states for each of the fundamental through the fifth harmonic plus one state for the offset [27]. As a constant error covariance was used, the only justification for the Kalman filter (as opposed to a solution such as Eq. (21)) was the use of a prior estimate. The prior estimate seems hard to justify in the relaying environment. Interestingly a technique for eliminating dependence on the prior estimate also provides a connection between the Kalman and Fourier algorithms.

Define the inverse of the covariance matrix as

$$F_k = P(k|k)^{-1} \qquad (61)$$

and assume that there are no random inputs (i.e. the phasor is constant while we attempt to estimate it) hence $Q_k = 0$, then using the matrix inversion lemma, viz

$$(A^{-1} + B^T C^{-1} B)^{-1} = A - A B^T (B A B^T + C)^{-1} BA \qquad (62)$$

where A, B, and C are matrices of appropriate dimension and where all the indicated inverses exist,it can be shown that

$$P H^T (H P H^T + R)^{-1} = (P + H^T R^{-1} H)^{-1} H^T R^{-1}. \qquad (63)$$

With $P = P(k+1|k)$ Eq. (63) is seen as an alternate for Eq. (58) which can be substituted into Eq. (60)

$$P(k+1|k+1) = P - P H^T [H P H^T + R] H P \qquad (64)$$

which from Eq. (62) is

$$P(k+1|k+1) = [P^{-1}(k+1|k) + H_{k+1} R_{k+1}^{-1} H_{k+1}]^{-1} \qquad (65)$$

or

$$P^{-1}(k+1|k+1) = P^{-1}(k+1|k) + H_{k+1} R_{k+1}^{-1} H_{k+1} \qquad (66)$$

or from Eq. (59) with $Q_k = 0$

$$F_{k+1} = \varphi_k^{-T} F_k \varphi_k^{-1} + H_{k+1}^T R_{k+1}^{-1} H_{k+1} \qquad (67)$$

Equation (67) can be initialized with F_0 taken as zero (corresponding to no information or an infinite a-priori covariance). The matrix F_k is referred to as the information matrix and can also be used to compute the gain from

$$F_{k+1} K_{k+1} = H_{k+1}^T R_{k+1}^{-1} \qquad (68)$$

It is not possible to use Eq. (68) unless F_{k+1} is invertible although the recursion in Eq. (67) can be continued until F_{k+1} is invertible.   For the specific form of the two–state system in Eq. (54): $\varphi_k = I$ and $H_k = [\ \cos(k\psi)\quad \sin(k\psi)\ ]$ with $R_k = I$, the recursion for the information matrix becomes

$$F_{k+1} = F_k + \begin{bmatrix} \cos^2(k+1)\theta & \cos(k+1)\theta \ s\ i\ n(k+1)\theta \\ \cos(k+1)\theta \ s\ in(k+1)\theta & s\ in^2(k+1)\theta \end{bmatrix} \tag{69}$$

or

$$F_N = \sum_{k=1}^{N} \begin{bmatrix} \cos^2 k\theta & \cos k\theta \ \sin k\theta \\ \cos k\theta \ \sin k\theta & \sin^2 k\theta \end{bmatrix} \tag{70}$$

and Eq. (68) simplifies to

$$K_{k+1} = F_{k+1}^{-1} \ H_{k+1}^{T} \tag{71}$$

and using Eq. (71) in Eq. (57)

$$\hat{x}_{k+1} = \hat{x}_k + F_{k+1}^{-1} \ H_{k+1}^{T}[\ z_{k+1} - H_{k+1} \ \hat{x}_k] \tag{72}$$

or

$$\hat{x}_{k+1} = F_{k+1}^{-1} \left[\ [\ F_{k+1} - H_{k+1}^{T}H_{k+1}]\hat{x}_k + H_{k+1}^{T} \ z_{k+1}\right] \tag{73}$$

With $\varphi = I$ and $R = I$ Eq. (67) in Eq. (73) yields

$$\hat{x}_{k+1} = F_{k+1}^{-1} [\ F_k \ \hat{x}_k + H_{k+1}^{T} \ z_{k+1}] \tag{74}$$

With $F_0 = 0$, Eq. (74) yields

$$\hat{x}_N = F_N^{-1} [\ H_1^{T} \ z_1 + H_2^{T} \ z_2 + \cdots + H_N^{T} \ z_N] \tag{75}$$

Observing that $F_N = S^T S$ (compare Eqs. (11)–(13) with Eqs. (70) and (75)) and that the bracketed quantity in Eq. (75) is

$$\begin{bmatrix} \Sigma \cos(k\theta)z_k \\ \Sigma \sin(k\theta)z_k \end{bmatrix}$$

it can be seen that the two–state Kalman filter with no initial information and a measurement error with a constant covariance is identical to the equivalent least–squares solution. In fact, in this situation, if $N\theta$ is a multiple of a half–cycle, then the Kalman filter is the same as the appropriate Fourier estimates. The same conclusions apply to Kalman filter solution with more terms [27] since S in Eqs. (20) or (21) can also model more than two terms. It must be noted that there is a subtle difference between the Kalman solution and the least–square and Fourier. The Fourier and least–squares algorithm have a fixed length window which moves across the data while the Kalman filter begins at some instant and then examines all data with an ever growing window. In the two–state model the Kalman estimates will agree with some Fourier calculation when the time since the Kalman filter began calculations is a multiple of a half cycle. While the lengthening window may have some appeal, it comes at the expense of increased computation and the new problem of deciding when to initiate the Kalman calculations (serious problem will be encountered if the Kalman filter calculations do not begin just after fault inception).

The possible motivations for the use of a Kalman filter in line protection are: a desire to use an initial estimate, an interest in a data window which grows in length, or a concern that the measurement error is of changing quality. Of these, the non–constant measurement error covariance would seem the most persuasive. Other studies [28] and field data [29], however, do not support an error model with decay that is noticeable in the short time during which a relay decision must be made (during the first cycle after the fault).

H. REMOVAL OF THE DC OFFSET

Both the half–cycle Fourier algorithm and the two–state Kalman algorithm require that the DC offset in the current be removed prior to processing. If a fault occurred at $t = 0$ in the circuit shown in Figure 13 with a source voltage given by

$$v(t) = \sqrt{2}\ V\ \cos\ (\omega_o t + \phi) \tag{76}$$

the fault current $i(t)$ (assuming no pre–fault current in the circuit) is given by

$$i(t) = \frac{\sqrt{2}\ V}{\sqrt{(R^2 + X^2)}}\ \cos(\omega_o t + \phi - \psi) - \frac{\sqrt{2}\ V\ e^{-(R/L)t}}{\sqrt{(R^2 + X^2)}}\ \cos\ (\phi - \psi) \tag{77}$$

where $X = \omega_o L$. If the current $i(t)$ (or its analog in the secondary winding of a current transformer) is passed through a impedance of $(r + jx)$ where $x/r = X/R$ then the voltage across the impedance $(r + jx)$ is given by

$$e(t) = \frac{\sqrt{2}\ V}{\sqrt{(R^2 + X^2)}}\ \sqrt{(r^2 + x^2)}\ \cos\ (\omega_o t + \psi) \tag{78}$$

This voltage (which is used to represent the fault current) is without the dc offset and is in phase with the line voltage $v(t)$ as shown in Figure 13. The impedance $(r + jx)$ as a load on a current transformer is known as a

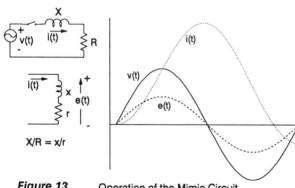

Figure 13   Operation of the Mimic Circuit

mimic circuit. For a digital relay the mimic effect can be combined with the anti–aliasing filter [30] or achieved digitally [31].

V. TRANSMISSION LINE RELAYING

Although there continues to be a great deal of interest in algorithms and the comparison of algorithms, the integration of particular algorithms into a total protection package is the ultimate test. Comparisons of relaying algorithms is difficult at best and requires a great deal of intimate knowledge of all of the algorithms involved. Further, the most important reason for selecting a particular algorithm may lie completely beyond the domain of the comparison criteria. Issues which must be considered beyond the algorithm's ability to estimate parameters include: how the determination of fault type is connected to the parameter estimation, monitoring of the quality of the estimates, and speed–reach considerations. In this section we will concentrate on issues related to the algorithms. It should be recognized, however, that a complete relaying program could also include: phase and ground distance protection, some directional comparison scheme, three–zone stepped distance protection, provision for single phase–tripping, high speed reclosing, auto–reclosing with sync check, breaker failure protection, sequence of events recording, and memory voltage for three–phase bus fault.

A. FAULT CLASSIFICATION

The algorithms discussed in section IV are single–phase algorithms involving a single current and/or voltage. A line relay using one of these algorithms could protect only one terminal of a three–phase transmission line. Since there are ten distinct possible faults (three single phase–to–ground faults, three phase–to–phase faults, three phase–to–phase–to–ground, and one three–phase fault) the relay must select between, some practical technique for fault classification must be found. Note that if the fault type were known that samples of the appropriate voltage and current can be used to compute phasors and the impedance to the fault or used in the single–phase differential equation algorithm. For example, phase a current with zero sequence compensation and phase a voltage for an a–ground fault or phase b current and the voltage between phases a and b for an a–b fault.

One solution is to make the distance calculation for all the required possibilities (three phase–ground faults and three phase–phase faults). This is difficult today but may become possible in the future with faster processors. An obvious advantage would result if some fault classification could be performed before beginning. By comparing the voltage or current waveforms (samples) with those a cycle ago it is possible to determine which phases are involved in the fault. The fact that such schemes work most of the time is, unfortunately, not sufficient. Voltage based classifiers are bothered when the source is so strong that the voltage does not change or so weak that all the voltages change. Current based classifiers are confused by large load currents.

A second approach uses transformations of the phase variables. The Clarke components [32] have been used with the differential equation algorithm [23] while symmetrical components have been used with the Fourier algorithm [33]. The symmetrical component approach [33] is quite successful in that it avoids fault classification entirely and produces an expression for the distance to the fault which is independent of fault type. The symmetrical component transformation (with phase a as reference) of phasor quantities is obtained by multiplying the phase quantities by the matrix T

$$
T = \frac{1}{3} \begin{bmatrix} 1 & 1 & 1 \\ 1 & \alpha & \alpha^2 \\ 1 & \alpha^2 & \alpha \end{bmatrix} \tag{79}
$$

where $\alpha = e^{j2\pi/3}$. The phasors in question are the fundamental frequency phasors obtained from the Fourier calculations. The phase quantities are transformed by T to the zero, positive, and negative sequence quantities (0,1,2), for example, for currents

$$
\begin{bmatrix} I_o \\ I_1 \\ I_2 \end{bmatrix} = T \begin{bmatrix} I_a \\ I_b \\ I_c \end{bmatrix} \tag{81}
$$

For a fault at a fraction k of the line length away from the relay the following dimensionless quantities are computed.

$$
k_0 = \frac{E_0}{\Delta E_0}, \quad k_1 = \frac{E_1}{\Delta E_1}, \quad k_2 = \frac{E_2}{\Delta E_2}, \text{ and } k_l = \frac{Z_1 I_1}{\Delta E_1} \tag{81}
$$

where the denominators in Eq. (81) are voltage drops given by

$$
\Delta E_0 = \Delta I_0 \, Z_0, \quad \Delta E_1 = \Delta I_1 \, Z_1, \text{ and } \Delta E_2 = \Delta I_2 \, Z_2 \tag{82}
$$

where Z_0, Z_1, and Z_2 are the sequence impedances of the entire line, and the Δ quantities represent differences from the pre–fault values of I_0, I_1, and I_2.

$$
\Delta I_0 = I_0 - I_0 \,, \quad \Delta I_1 = I_1 - I_1, \text{ and } \Delta I_2 = I_2 - I_2 \tag{83}
$$

Typically only I_1 will be significant. The ratios given by Eq. (81) can be used to determine fault location for all fault types. An expression for the distance to the fault can be obtained [33] in the form

$$
k = \frac{k_1 + k_2 k_2' + k_0 k_0'}{1 + k_0' + k_2' + k_l} \tag{84}
$$

where k_0' and k_2' are

$$k_0' = \left| \frac{\Delta E_0}{\Delta E_1} \right| \tag{85}$$

and

$$k_2' = \begin{cases} 1 & \text{if } |\Delta E_2| \gtrsim |\Delta E_1| \\ 0 & \text{otherwise} \end{cases} \tag{86}$$

Eq. (84) combined with an efficient technique for computing the symmetrical components, is the nucleus of an attractive relaying program.

B. SYMMETRICAL COMPONENT CALCULATION

If a sampling frequency of 12 times per cycle is used ($\theta = 30°$), the symmetrical component calculation can be combined with the Fourier calculation. The factor α in Eq. (79) is equivalent to four shifts by the angle θ while α^2 is eight shifts forward or four backward. If we let

$$\Delta x_M = (x_M - x_{M-12}) \quad \text{for the full–cycle algorithm}$$

and

$$\Delta x_M = (x_M + x_{M-6}) \quad \text{for the half–cycle}$$

then the full– or half–cycle versions of the SCDFT (Symmetrical Component Discrete Fourier Transform) can be written in a single expression [30]:

$$\hat{X}_{0c}^{(M+1)} = \hat{X}_{0c}^{(M)} + (\Delta x_{a,M} + \Delta x_{b,M} + \Delta x_{c,M}) \cos(M\theta)$$

$$\hat{X}_{0s}^{(M+1)} = \hat{X}_{0s}^{(M)} + (\Delta x_{a,M} + \Delta x_{b,M} + \Delta x_{c,M}) \sin(M\theta)$$

$$
\hat{X}_{1c}^{(M+1)} = \hat{X}_{1c}^{(M)} + \Delta x_{a,M}\cos(M\theta) + \Delta x_{b,M}\cos(M+4)\theta \\
+ \Delta x_{c,M}\cos(M-4)\theta
$$

$$
\hat{X}_{1s}^{(M+1)} = \hat{X}_{1s}^{(M)} + \Delta x_{a,M}\sin(M\theta) + \Delta x_{b,M}\sin(M+4)\theta \qquad (87) \\
+ \Delta x_{c,M}\sin(M-4)\theta
$$

$$
\hat{X}_{2c}^{(M+1)} = \hat{X}_{2c}^{(M)} + \Delta x_{a,M}\cos(M\theta) + \Delta x_{b,M}\cos(M-4)\theta \\
+ \Delta x_{c,M}\sin(M+4)\theta
$$

$$
\hat{X}_{2s}^{(M+1)} = \hat{X}_{2s}^{(M)} + \Delta x_{a,M}\sin(M\theta) + \Delta x_{b,M}\sin(M-4)\theta \\
+ \Delta x_{c,M}\sin(M+4)\theta
$$

where the subscripts a, b, and c are the phase quantities. The expressions in Eq. (87) are not correctly scaled. The factor of $1/3$ from Eq. (79) and the $2/K$ from Eqs. (23) and (24) along with the $\sqrt{2}$ from Eq. (35) should be considered if the actual phasors are needed.

C. TRANSIENT MONITOR

The difficulty with the unreliability of the estimates when the data window spans the pre–fault post–fault interval can be overcome by disabling the relay when the estimate does not fit the data. If we compute the samples that would result from the estimate denoted by the vector $\tilde{x}$ then

$$
\tilde{x} = S\,\hat{X} = S\,(S^T S\,)^{-1} S^T\,x \qquad (88)
$$

The samples and the reconstructed samples $\tilde{x}$ are shown in Figure 14. If x is close to $\tilde{x}$ then the close fit would imply that the estimate was trustworthy. Large differences would imply that the estimate should not be used for relay decisions. Let Δx denote the difference

$$\Delta x = \tilde{x} - x = [\ S\ (S^T S)^{-1} S^T - I\ ]\ x \qquad (89)$$

$$\Delta x = M\ x \qquad (90)$$

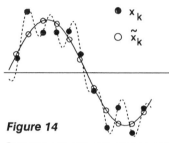

Figure 14

Samples and reconstructed samples

If a *transient monitor* function is defined as the sum of the absolute values of the Δx_k. i.e.

$$t = \sum_{k=1}^{K} |\ \Delta x_k| \qquad (91)$$

then t can be computed recursively using the fact that the Δx_k also satisfy a recursive relationship. In practice the transient monitor function is computed for each phase current and the relay is inhibited from tripping if any t exceeds some threshold.

D. SPEED REACH LIMITATIONS

The performance of the estimates given by Eq. (20) is given by the covariance of the estimate

$$E\{(\hat{X} - X)(\hat{X} - X)^T\} = (\ S^T W^{-1} S\ )^{-1} \qquad (92)$$

If $W = \sigma_\epsilon^2\ I$ then Eq. (92) gives

$$E\{(\hat{X} - X)(\hat{X} - X)^T\} = \sigma_\epsilon^2 \begin{bmatrix} \sum_{k=1}^{K} \cos^2(k\theta) & \sum_{k=1}^{K} \cos(k\theta)\sin(k\theta) \\ \sum_{k=1}^{K} \cos(k\theta)\sin(k\theta) & \sum_{k=1}^{K} \sin^2(k\theta) \end{bmatrix}^{-1} \qquad (93)$$

where $\mathbf{X}^T = [X_c \; X_s]$ and

$$\sum_{k=1}^{K} \cos^2 k\theta = \frac{(K-1)}{2} + \frac{\cos^2 K\theta}{2} + \frac{1}{2} \sin 2K\theta \; \frac{\cos\theta}{\sin\theta}$$

$$\sum_{k=1}^{K} \sin^2 k\theta = \frac{(K+1)}{2} - \frac{\cos^2 K\theta}{2} - \frac{1}{2} \sin 2K\theta \; \frac{\cos\theta}{\sin\theta} \qquad (94)$$

$$\sum_{k=1}^{K} \cos(k\theta)\sin(k\theta) = \frac{1}{4} \sin 2K\theta + \frac{\sin^2 K\theta}{2} \; \frac{\cos\theta}{\sin\theta}$$

For $K\theta$ a multiple of π the matrix in Eq. (93) is diagonal with equal diagonal entries of $2\sigma_e^2/K$. If we add the diagonal entries to obtain a real number for the quality of the estimate of the phasor $\hat{X}_c + j\hat{X}_s$ we obtain

$$\sigma_{\hat{X}}^2 = \frac{4\sigma_\epsilon^2}{K} \qquad (95)$$

The term σ_e^2 represents the noise per sample at the output of the anti–aliasing filter. If we model the non–fundamental frequency signals that are being sampled to produce the error ϵ as a wide–sense stationary random process with a power density spectrum which is constant i.e., for the voltage

$$S(\omega) = S_v \qquad \text{for all } \omega$$

and assume an ideal low–pass filter with a Nyquist cut–off frequency of

$$\omega_c = \frac{K}{2} \omega_0 = \frac{\pi \, \omega_0}{\theta} \qquad (96)$$

then the noise at the output of the anti–aliasing filter has a spectrum

$$S_s(\omega) = S_v \quad ; \quad |\omega| < \pi\, \omega_o/\theta$$

$$S_s(\omega) = 0 \quad ; \quad |\omega| > \pi\, \omega_o/\theta. \tag{97}$$

From the relationship between the correlation function and the spectrum we obtain

$$\sigma_\epsilon^2 = R_v(0) = \frac{1}{2\pi}\int_{-\pi\omega_o/\theta}^{\pi\omega_o/\theta} S_v \, d\omega = \frac{S_v \omega_o}{\theta} \tag{98}$$

or

$$\sigma_{\hat{V}}^2 = \frac{4\, S_v \omega_o}{K\theta} = \frac{4\, S_v}{T} \tag{99}$$

where T is the length of the window in seconds. Assuming the errors are small the variance of the impedance error is given by

$$\sigma_{\hat{Z}}^2 = \sigma_{\hat{V}}^2 + \sigma_{\hat{I}}^2 = \frac{4\,(\, S_v + S_I)}{T} \tag{100}$$

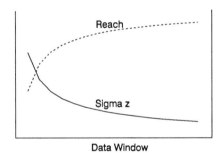

Figure 15   Speed - Reach Characteristic

Equation (100) is only valid for a data window that is a multiple of a half–cycle of the fundamental frequency. The inverse time characteristic is preserved but the expression is more complicated than Eq. (100) when T is not a multiple of a half cycle. The standard deviation (σ_z) of the impedance estimate is shown in Figure 15.

Note that the sampling rate does not appear in Eq. (100). The explanation lies with the assumption about the noise and anti–aliasing filter. The assumption that the noise before filtering is white ($S(\omega)$ =

S_v for all ω) coupled with the ideal anti–aliasing filter produces a situation where a higher sampling rate yields more samples each of which has a larger variance. Since the noise is ultimately band–limited, for example, by the transducers, there is a sampling frequency at which there would be a gain in faster sampling. Sampling frequencies in the kHz range are probably required before this effect is noticeable [19]. Non–ideal anti–aliasing filters have a phase shift which translates into a delay. Lower sampling frequencies would mean a longer delay introduced by the anti–aliasing filter. Qualitatively, Eq. (100) is appropriate for all of the algorithms described in this chapter. All algorithms are faced with noise–like signals, all produce better estimates with longer windows, all algorithms use anti–aliasing filters, and higher sampling rates produce more noise per sample. In general, we conclude that the variance of the estimated fault location is inverse to the length of the data window.

Given that the estimated fault location is not exact, there is a probability of false trip or failure to trip. Equation (100) indicates that the spread in the uncertainty decreases as the length of the data window increases. The reach or setting of the relay is also shown in Figure 15. It can be argued that the maximum setting of the relay can only be $(1 - 2.5\ \sigma)$ where σ is from Eq. (100) [19]. The speed–reach relationship shown in Figure 15 is inherent in electromechanical relays and is recognized in relaying practice [34] For a fixed window digital relay (such as a half–cycle Fourier) some technique must be employed to adjust the effective window length to recognize the inverse time effect. An accumulator has been suggested for this purpose [35].

VI. POWER TRANSFORMER ALGORITHMS

Most of the power transformer protection algorithms are based on the principle of percentage differential protection. The principle is appropriate to transformer, bus, and generator protection. Consider the single–phase transformer shown in Figure 16. If the turns ratios of the ct's can be appropriately chosen so that

$$N_1 n_1 = N_2 n_2$$

and the tap changer is at the nominal position $(T = 1)$, then $i_1 = i_2$ when there is no internal fault. Unfortunately, the ct's can not usually be selected to match exactly and the tap changer is commonly in use. The percentage differential

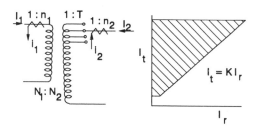

Figure 16   Single Phase Transformer
Percent Differential Characteristic

characteristic shown in Figure 16 offers a solution to theses problems. The trip current I_t and restraint current I_r are defined as:

$$I_t = i_1 + i_2 \tag{101}$$

and

$$I_r = (i_1 - i_2)/2 \tag{102}$$

The differential relay is set to trip when the current I_t and I_r are in the shaded area in Figure 16. The slope K (typically in the range of 10 to 40 percent) allows the ct mismatch and tap changer errors discussed.

A. CURRENT BASED RESTRAINTS

During energization of the transformer abnormal currents flow in the winding being energized These "magnetizing inrush" currents appears to be an internal fault to the percentage differential characteristic. Overexcitation has a similar effect. The conventional solution to these two effects is the use of harmonic restraint. Inrush current has a high second harmonic content and overexcitation a high fifth harmonic content, while fault current is primarily fundamental. For multi –

winding transformers there are a number of restraint currents. If we consider a three winding transformer with the current polarities consistent with Figure 16, the trip current would be given by

$$I_t = I_1 + I_2 + I_3 \tag{103}$$

while there would be two restraining currents given by

$$I_{r1} = I_1 - I_2 - I_3 \tag{104}$$

and

$$I_{r2} = -I_1 + I_2 - I_3 \tag{105}$$

The second restraint current is necessary to protect the transformer operating with the primary breaker open. There would be three trip currents (one per phase) and six restraint currents (two per phase) for a three–phase three–winding transformer.

While the percentage differential characteristic of Figure 16 could be applied to each phase on a per sample basis, the calculation of harmonics for harmonic restraint requires a one–cycle data window [36] and provides an effective algorithm [37]. Similar to Eqs. (30)–(31) the complex–recursive form of the full–cycle Fourier calculation of the nth harmonic for samples ending at sample M is given by

$$\hat{X}_n(M) = \hat{X}_n(M-1) + \frac{2}{K} [x_M - x_{M-K}] e^{jnM\theta} \tag{106}$$

where $K\theta = 2\pi$.

At a sampling rate of twelve samples per cycle the harmonic calculation is no more complicated than the fundamental. For a three–phase three–winding transformer the fundamental, second, and fifth harmonic must be computed for each of nine phase currents. The harmonic restraints are formed by combining the harmonics from all three phases [38]. The algorithm would form two harmonic restraints: one from the sum of the second harmonics magnitudes and one from the sum of fifth harmonics magnitudes. Since the harmonics develop more rapidly than the fundamental during a transient the relay is restrained for approximately a full cycle by the harmonic restraints.

B. VOLTAGE BASED RESTRAINTS

In an integrated substation protection system it is reasonable to assume that bus voltage measurements would be available for transformer protection. That is, in the computer hierarchy of Figure 5, relays that need voltage measurements (line relays) could share those measurements with a transformer protection unit. In a stand–alone system the requirement of additional voltage measurements would increase cost but such voltage measurements would be inexpensively in an integrated system.

1. Digital Tripping Suppressor

An early solution to the inrush problem, the so called "tripping suppressor" [38], used voltage measurements to restrain a percentage differential. A voltage relay was used to suppress tripping if the voltage was high. The analog form the "tripping suppressor" was slower than the harmonic restraint devices which replaced it. Since the voltage phasors can be computed in as little as a quarter of a cycle, it has been suggested that a digital "tripping suppressor" may be faster than the digital harmonic restraint algorithm [39].

The algorithm [39] uses a one–half cycle window for the calculation of the primary voltage for each phase along with the fundamental frequency components of trip and restraint currents. Since harmonic development can not be used to make the relay secure, it was necessary to include a transient monitor (Section V C) for the voltage signals. The total computation is comparable to the harmonic restraint algorithm [39] and potentially provides faster clearing of faults.

2. Flux Restraint

Other uses of measured voltages produce the transformer equivalent of the differential equation algorithms for line protection, that is algorithms which are based on physical models as opposed to waveform models. Using a linear model of the transformer [40] it was proposed that the measured currents be used to compute the terminal voltages. By comparing the computed voltages with the measured voltages it could be determined whether saturation had begun. Another possibility is to determine the internal flux of the transformer using the measured voltages and currents [41].

The voltage at the terminals of a transformer winding, $v(t)$, the current through that winding, $i(t)$, and the flux linkage, $\Lambda(t)$ of the transformer are related by

$$v(t) - L \frac{di(t)}{dt} = \frac{d\Lambda(t)}{dt} \tag{107}$$

where L is the leakage inductance of the winding and the resistance of the winding has been neglected, Using trapezoidal integration as in Eq. (40) and assuming sampling at Δt second interval

$$\Lambda(t_2) = \Lambda(t_1) + \frac{\Delta t}{2} [v(t_2) + v(t_1)] - L [i(t_2) - i(t_1)] \tag{108}$$

With subscript k denoting the kth sample

$$\Lambda_{k+1} = \Lambda_k + \frac{\Delta t}{2} [v_{k+1} + v_k] - L [i_{k+1} - i_k] \tag{109}$$

The voltage in Eq. (109) is the winding voltage while the current in equation is the trip current for each phase of the three–winding transformer. Given the initial flux linkage, Eq. (109) can be used to track the flux–current plot of the transformer at each sample time. An open circuit magnetizing curve of a transformer is shown in Figure 17 along with the flux–current curve that would be associated with an internal fault. The difficulty, of course, it that it is not possible to know the initial flux and hence it is not possible to discriminate between fault and no–fault situations. The solution lies in the slope of the flux curve. From Eq. (109) we can form an expression for the slope $(d\Lambda/di)$ as

$$\left(\frac{d\Lambda}{di} \right)_k = \frac{\Lambda_k - \Lambda_{k-1}}{i_k - i_{k-1}} = \frac{\Delta t}{2} \left[\frac{e_k + e_{k-1}}{i_k - i_{k-1}} \right] - L \tag{110}$$

From Figure 17 it can be seen that the the slope $(d\Lambda/di)$ is larger in the unsaturated region of the open circuit magnetizing curve while the slope is smaller in the fault or no –fault (saturated) regions. For an internal fault the $(d\Lambda/di)$ samples are all small while the $(d\Lambda/di)$

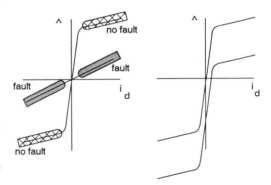

Figure 17   Flux - Current Relationships

samples alternate between large and small values as the magnetizing curve is traced during inrush. A counter similar to the counter for the differential equation algorithm for transmission line protection can be defined. Let c_r be defined as follows:

$$c_r = c_r + 1 \quad \text{if the current differential indicates trip} \\ \text{and if } (d\Lambda/di)_k < \zeta$$

$$c_r = \begin{cases} c_r - 1 & \text{if } c_r > 0 \text{ and } (d\Lambda/di)_k > \zeta \\ \\ c_r & \text{if } c_r = 0 \end{cases} \tag{111}$$

where the threshold value ζ separates the high and low slopes of the magnetizing curve.   The index c_r grows almost monotonically for internal faults but shows a saw–tooth behavior for non–fault cases [41]. The decision to trip is made when c_r exceeds some threshold c_{rmax}.   If c_{rmax} is too small the relay may trip on the top of a saw–tooth associated with a non–fault.   A larger c_{rmax} means a slower but more secure relay in the familiar trade–off between speed and security.

VII. BUS PROTECTION

Apart from some early interest [4,42], bus protection with digital computers has been slow in developing.   Part of the reason for the slow development is that the analog high impedance relay works well and has no obvious digital counterpart.   The need to include bus protection in an integrated protection and control system for the entire substation [7,13,35] has rekindled interest in digital bus protection.   Digital bus protection in an integrated system is dependent on the currents in all the circuit breakers and switches connected to the bus being available.

The primary concern in the design of a bus protection system is with saturation of the current transformer.   A saturated secondary current due to a remnant flux in the core is shown in Figure 18.   If the saturation did not take place for the approximate one–half cycle shown in Figure 18, a variety of digital solutions might be used.   The possibility of saturation in much less than a half cycle is a more serious problem.   It is clear that either a sample–by–sample percentage differential

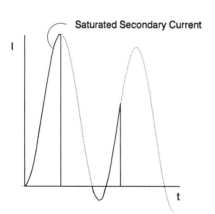

Figure 18   Primary and secondary current

relay, or one based on phasors computed from more samples can be used. A phasor based percentage differential scheme can only be used if there is no *ct* saturation. A quarter cycle phasor calculation coupled with a quarter cycle transient monitor would provide a suitable computer based bus differential relay. Modifications of the percentage differential characteristic itself which make the characteristic nonlinear have also been proposed [13]. It is also clear that higher sampling rates are indicated if there is an attempt to make a decision before the *ct* reaches saturation. A sample based differential relay could be used if the relay were disabled as soon as one of the *cts* saturates. This could be accomplished by a transient monitor or through the use of an analog circuits which would provide a trigger indicating the start of *ct* saturation.

A different approach to digital bus protection has been proposed using percentage differential principles and phase comparison techniques [43]. A saturation detector based on the change of slope of the current samples is used in a scheme using a percentage differential as the primary protection and phase comparison as a backup. Since the computational burden is light, a sampling rate of 24 times per cycle is possible. The saturation detector works by disabling the relay and holding the trip/block signal at its previous state until the *ct* comes out of saturation. The phase comparison scheme also uses the saturation detector to establish a polarity for phase comparison.

VIII. CONCLUSION

Twenty years of algorithm development, field experience and advances in computer hardware have brought the field of computer relaying to its current mature state. A number of early problems have been solved by the sheer speed of a new generation of processors while others have yielded to innovations in algorithms and application. The attempt to develop digital protection has itself clarified issues. We have gained understanding of the fundamental limitations of the protection process in terms of speed–reach limitations and in terms of the trade–offs between security and dependability. Some relaying areas such as protection of series compensated lines and improved fault locators remain the subject of investigation. More generally however, the digital relay is taking on the role as the lowest level of a vast computer hierarchy devoted to the protection and control of the system. Applications of digital relays in system wide monitoring and control offer exciting possibilities as described in other chapters in this volume. Adaptive protection techniques that exploit the digital relay's ability to communicate are also on the horizon.

IX. REFERENCES

1.   G.D. Rockefeller, "Fault Protection with a Digital Computer," *IEEE Transactions on Power Apparatus and Systems*, Vol. 88, No. 4, April 1969, pp 438–461.

2.   B.J. Mann and I.F. Morrison, "Digital Calculation of Impedance for Transmission Mine Protection," *IEEE Transactions on Power Apparatus and Systems*, Vol. 90, No. 1, January/February 1971, pp 270–279.

3.   R. Poncelet, "The Use of Digital Computers for Network Protection," CIGRE Paper No. 32–08, August 1972.

4.   B.J. Cory and J.F. Moont, "Application of Digital Computers to Busbar Protection," IEE Conference on the Application of Computers to Power System Protection and Metering, Bournemouth, England, May 1970, pp 201–209.

5.   J.A. Sykes and I.F. Morrison, "A Proposed Method of Harmonic Restraint Differential Protection of Transformers by Digital Computers," *IEEE Transactions on Power Apparatus and Systems*, Vol. 91, No. 3, May/June 1972, pp 1266–72.

6.   M.S. Sachdev and D.W. Wind, "Generator Differential Protection Using a Hybrid Computer," *IEEE Transactions on Power Apparatus and Systems*, Vol. 92, No. 6, November/December 1973, pp 2063–2072.

7.   "Microprocessor Relays and Protection Systems," M.S. Sachdev (Co–ordinator), IEEE Tutorial Course Text, Publication No.88EH0269–1–PWR, February 1988.

8. A.G. Phadke, S.H. Horowitz, J.S. Thorp, "Integrated Computer System for Protection and Control of High Voltage Substations," CIGRE Colloquium, Tokyo, Japan, November 1983.

9. J.S. Deliyannides and E.A. Udren, "From Concepts to Reality – The Implementation of an Integrated Protection and Control System," Developments in Power System Protection, IEE Conference Publication NO. 249, London, April 1985, pp 24–28.

10. North American Reliability Council, "System Disturbance," 1983, 1984, etc. Research Park, Terhune Road, Princeton, New Jersey.

11. "Surge Withstand Capability (SWC) Tests for Protective Relays and Relay Systems," ANSI/IEEE C37.90.1–1989

12. W.C. Kotheimer and L.L. Mankoff, "Electromagnetic Interference and Solid State Protective Relays," *IEEE Transactions on Power Apparatus and Systems*, Vol. PAS–96, No. 4, July/August 1977, pp 1311–1317.

13. E.A. Udren, "An Integrated, Microcomputer Based system for Relaying and Control of Substations – Design Features and Testing Program," 12th Annual Western Protective Relaying Conference, Spokane, Washington, October 1985.

14. A.G. Phadke (Convener), "CIGRE Working Group 34.02 Final Report," Paris, 1985.

15. C.R. Mason, *"The Art and Science of Protective Relaying"*, John Wiley & Son, 1956

16. *"Applied Protective Relaying"*, Westinghouse Electric Corporation, 1982

17. *"Protective Relaying for Power Systems"*, Stanley H. Horowitz, Editor, IEEE Press, New York, 1980

18. A.G. Phadke and J.S. Thorp, *"Computer Relaying for Power Systems"*, Research Studies Press Ltd., John Wiley & Sons Inc., Second Printing, April 1990.

19. J.S. Thorp, A.G. Phadke, S.H. Horowitz, and J.E. Beehler, "Limits to Impedance Relaying" *IEEE Transactions on Power Apparatus and Systems*, Vol. PAS–98, No. 1, January/February 1979, pp 246–260.

20. M.S. Sachdev and M.A. Baribeau, "A Digital Computer Relay for Impedance Protection of Transmission Lines," Trans. of Engineering and Operating Division Canadian Electrical Association, Vol. 18, Part 3, No. 79–SP–158, 1979, pp 1–5.

21. R.G. Luckett, P.J. Munday, and B.E. Murray, "A Substation–Based Computer for Control and Protection," Developments in Power System Protection, IEE Conference Publication No. 125, London, March 1975, pp 252–260.

22. A.M. Ranjbar and B.J. Cory, "An Improved Method for the Digital Protection of High Voltage Transmission Lines," *IEEE Transactions on Power Apparatus and Systems*, Vol. PAS–94, No. 2, March/April 1975, pp 544–550.

23. W.D. Breingan, M.M. Chen and T.F. Gallen, "The Laboratory Investigation of a Digital System for the Protection of Transmission Lines," *IEEE Transactions on Power Apparatus and Systems*, Vol. PAS–98, No. 2, March/April 1979, pp 350–368.

24. W.J. Smolinski, "An Algorithm for Digital Impedance Calculation Using a Single Pi Section," *IEEE Transactions on Power Apparatus and Systems*, Vol. PAS–98, No. 5, September/October 1979, pp 1546–1551, and PAS–Vol. 99, No. 6, November/December 1980, pp 2251–2252.

25. A.A. Girgis and R.G. Brown, "Application of Kalman Filtering in Computer Relaying," *IEEE Transactions on Power Apparatus and Systems*, Vol. PAS–100, No. 7, July 1981, pp 3387–3397.

26. A.A. Girgis and R.G. Brown, "Modeling of Fault–Induced Noise Signals for Computer Relaying Applications," *IEEE Transactions on Power Apparatus and Systems*, Vol. PAS–102, No. 9, September 1983, pp 2834–2841.

27. M.S. Sachdev, H.C. Wood and N.G. Johnson, "Kalman Filtering Applied to Power System Measurements for Relaying," *IEEE Transactions on Power Apparatus and Systems*, Vol. PAS–104, No. 12, December 1985, pp 3565–3573.

28. G.W. Swift, "The Spectra of Fault Induced Transients," *IEEE Transactions on Power Apparatus and Systems*, Vol. PAS–98, No 3, May/June 1979, pp 940–947.

29. A.G. Phadke, T. Hlibka, M.G. Adamiak, M. Ibrahim, and J.S. Thorp, "A Microcomputer Based Ultra–High Speed Distance Relay: Field Tests," *IEEE Transactions on Power Apparatus and Systems*, Vol. PAS–100, No. 4, April 1981, pp 2026–2036.

30. A.G. Phadke, T. Hlibka, M. Ibrahim and M.G. Adamiak, "A Microprocessor Based Symmetrical Component Distance Relay," Proceedings of PICA, May 1979, Cleveland.

31. V. Centeno, "Mimic Circuit Simulation in Real Time," M.S. Thesis, Virginia Tech, 1988.

32. Edith Clarke, *Circuit Analysis of A-C Power Systems*, Vol I, John Wiley & Sons, New York, 1943.

33. A.G. Phadke, T. Hlibka and M. Ibrahim, "Fundamental Basis for Distance Relaying with Symmetrical Components," *IEEE Transactions on Power Apparatus and Systems*, Vol. PAS-96, No. 2, March/April 1977, pp 635–646.

34. J.G. Andrichak and S.B. Wilkinson, "Considerations of Speed, Dependability and Security in High Speed Pilot Relaying Schemes," Minnesota Power Systems Conference, October 1976.

35. E.A. Udren and M. Sackin, "Relaying Features of an Integrated Microprocessor–Based Substation Control and Protection System," IEE Conference Publication No. 185, London, U.K., April 1980, pp 97–101.

36. J.S. Thorp and A.G. Phadke, "A Microprocessor–Based Three–Phase Transformer Differential Relay, " *IEEE Transactions on Power Apparatus and Systems*, Vol. 102, No. 2, February 1982, pp 426–432.

37. M. Habib and M.A. Marin, "A Comparative Analysis of Digital Relaying Algorithms for the Differential Protection of Three Phase Transformers," PICA May 1987, Montreal Canada.

38. E.L. Harder and W.E. Marter, "Principles and Practices of Relaying in the United States," AIEE Transactions, Vol. 67, Part II, 1948, pp 1005–1022.

39. J.S. Thorp and A.G. Phadke, "A Microprocessor Based Voltage–Restrained Three–Phase Transformer Differential Relay," Proceedings of the South Eastern Symposium on Systems Theory, April 1982, pp 312–316.

40. J.A. Sykes, "A New Technique for High Speed Transformer Fault Protection Suitable for Digital Computer Implementation," IEEE PES Summer Meeting, 1972.

41. A.G. Phadke and J.S. Thorp, " A New Computer Based, Flux Restrained, Current Differential Relay for Power Transformer Protection," *IEEE Transactions on Power Apparatus and Systems*, Vol. 102, No. 11, November 1983, pp 3624–3629.

42. "Computer Relaying, " M.S. Sachdev (Co–ordinator), IEEE Tutorial Course Text, Publication No.79 EH0148–1–PWR, Chapt 6. 1979.

43. L. Yang, P.A. Dolloff, and A.G. Phadke, "A Microprocessor Based Bus Relay Using a Current Transformer Saturation Detector," North Americal Power Conference, Auburn, 1990

Advanced Control Techniques for High Performance Electric Drives[*]

Mohamed A. El-Sharkawi

Department of Electrical Engineering
University of Washington
Seattle, WA 98195

1. DEFINITION OF HIGH PERFORMANCE DRIVES

In modern high performance electric drives applications, such as robotic, guided manipulation and supervised actuation, controlling the rotor position is no longer the only goal. It is essential that the rotor of the electric drive motor follows a preselected track at all time. For most robots, for example, just moving the end effector of the robot arm from point "A" to point "B" is insufficient. It is, however, essential that the end effector, while travelling, follows a previously determined trajectory. To achieve this, every motor in the robot arm must follow a specific track so that the aggregated motion of all motors keeps the

---

[*] Portions of this chapter are reprints, with permission, from IEEE Transactions on Energy Conversion [1,2,3,9,33], 1988-1990.

CONTROL AND DYNAMIC SYSTEMS, VOL. 44

end effector alongside its trajectory at all time. This is quite different from the positional control applications where only the final value of the rotor position and/or speed are controlled with no or minimal control on the travelling time or overshoots.

Any high performance drive system must consist of three basic components: an electric motor; a broad performance high speed solid-state switching converter and an advanced controller.

Unlike most of the conventional positional controllers, the controllers for high performance tracking do not employ constant parameters. The structure and/or the parameters of the controllers must be adaptively tuned to achieve two basic objectives: 1) to provide the best possible tracking performance without overstressing the hardware; and 2) to enhance the system robustness. In the applications where the parameters of the load or the drive system are changing, the robustness of the controller is a basic requirement [1-3]. Fixed parameters controllers, such as the PID can not be considered robust.

In this chapter, an overview of high performance drive systems is presented. Several types of electric motors that are considered among the best options for high torque, high performance applications are discussed. Appropriate control strategies for tracking control are developed and evaluated via laboratory testing and/or computer simulations.

2. MODELING OF HIGH TORQUE MOTORS

Due to the rapid improvements in magnetic materials and core technologies, newer motors with high power to volume ratios have emerged. These motors can be used in high torque, high performance applications such as robotic or guided manipulations. In addition, the rapid development of solid state power electronics technology results in broad and enhanced performances of electric motors.   For example, a synchronous motor can be made to operate in a mode that resembles that of a dc motor.   Or, an induction motor that can operate in a synchronous mode. These flexibilities eliminate the early design restriction where the load characteristic and the type of electric source restrict the selection of the electric motor in the drive system.

In this chapter basic dynamic models of three types of electric motors commonly used in industrial drive applications are presented: 1) dc Motors; 2) Brushless Motors and 3) Induction Motors. Each of these motors is treated from the drive, rather than machinery, prospective.

2.1 MODEL OF DC MOTOR DRIVE

Direct current motors are historically considered the only option for high performance drives. This is due to their intrinsic properties such as: 1) the ease by which they can be controlled; 2) their ability to deliver high starting torques and 3) their near linear performance, especially with fixed field current. DC motors are widely used in such applications as actuations, tractions and manipulations.

The basic model of a separately excited dc motor drive can be expressed in the state space continuous time domain

$$\frac{dx}{dt} = A_s X + B_s U \tag{1}$$

$$X = [i_a(t) \quad \omega_m(t) \quad \delta_m(t)]^T$$

$$U = [V_{in}(t) \quad T_L(t)]^T$$

$$A_s = \begin{bmatrix} \dfrac{-r_a}{l_a} & -\dfrac{k_m}{l_a} & 0 \\ \dfrac{k_m}{J_m} & \dfrac{D}{J_m} & 0 \\ 0 & 1 & 0 \end{bmatrix} \; ; \quad \text{and } B_s = \begin{bmatrix} \dfrac{1}{l_a} & 0 \\ 0 & \dfrac{1}{J_m} \\ 0 & 0 \end{bmatrix}$$

where X is a vector composed of the state variables of the motor. U is the load and control vector. A_s and B_s are the system matrix and the input/output vector, respectively. The system parameters are defined as follows:

J_m : rotor inertia ($Kg.m^2$)

K_m : torque or back emf constant (Nm/A)

r_a : armature resistance (Ω)

l_a : armature inductance (H)

$i_a(t)$: armature current (A)

$\omega_m(t)$: rotor speed (rad/s)

$\delta_m(t)$: rotor position (rad)

$V_{in}(t)$: input voltage or controlled variable (V)

$T_L(t)$: load torque (Nm)

D : damping constant

Most of the adaptive control strategies require the system model to be in a discrete time domain which can be written in the following form:

$$X(t+T) = F\,X(t) + G\,U(t) \tag{2}$$

where,

$$F = \exp(A_s T)$$

$$G = \int_0^T e^{[A_s(T-\tau)]} B_s \, d\tau$$

Where T is the sampling period

2.2 MODEL OF BRUSHLESS DC MOTOR DRIVE

Although dc motors are the most common machines for electric drives, they have drawbacks that restrict their use in some applications. For example, dc motors are relatively high maintenance machines, large in sizes and expensive compared to other options such as induction motors. They are not suitable for high speed applications due to the limitation of their commutator arrangements. Also, because of the electrical discharging between the commutator segments and the brushes, dc machines can not be used in clean or explosive environments, unless encapsulated.

Although dc motor has many popular features for drives applications, the presence of the mechanical commutator limits its use. A newer type of motor has been developed to perform somehow like a dc machine but with electronically commutated windings. It is called brushless dc machine. It is basically a synchronous machine with a permanent magnet in the rotor circuit. The armature windings, which are mounted on the stator, are electronically switched according to the position of the rotor. Thereby, it mimics the commutator switching action of the conventional dc machines.

Brushless dc motors with permanent magnetic material such as Samarium-Cobalt are currently considered among the best new options for electric drives applications, especially high performance [4-7]. These machines are characterized by their high power/volume ratios, high starting torques and the ease by which they can be controlled. Since these machines are brushless, they are ideal to use in clean and explosive environments such as food processing, aeronautics or chemical industries.

Because of the increasing market for these motors; the continual improvement in the manufacturing process of the dc brushless motors; and the advancements in the solid-state power electronic devices and circuits, the prices of these motors are becoming competitive with other conventional options. These motors are rapidly replacing the existing hydraulic as well as the conventional electric drive systems in a number of applications that require high torque and precision control.

A dc brushless drive system consists of four basic components [6]: 1) Brushless motor; 2) Position encoder to provide information on the rotor position; 3) Switching converter; and 4) Controller. In the three-phase systems, the switching converter provides three-phase waveforms with frequency proportional to the speed of the rotor. In recent designs, instead of using a positional encoder, a high frequency signal is injected in the armature winding to indirectly measure the rotor position.

The model of the brushless machine is given in equation (3). Details of this model can be found in reference [8]. The model is based on the following assumptions: 1) all electric quantities are referred to direct and quadrature reference frames mounted on the rotor; 2) the airgap is uniform; 3) the inverter is a balanced three-phase voltage source; 4) The inverter is a 180°, six-step switching; 5) the fundamental frequency has the dominant effect on system dynamics.

$$\frac{d}{dt}\begin{bmatrix} i_q \\ i_d \\ \omega \\ \delta \end{bmatrix} = \begin{bmatrix} \dfrac{-r}{L} & -\omega & \dfrac{-\lambda}{L} & 0 \\[2mm] \omega & \dfrac{-r}{L} & 0 & 0 \\[2mm] \dfrac{3P\lambda}{4J} & 0 & \dfrac{-D}{J} & \omega \\[2mm] 0 & 0 & 1 & 0 \end{bmatrix}\begin{bmatrix} i_q \\ i_d \\ \omega \\ \delta \end{bmatrix}$$

$$+ \begin{bmatrix} \cos\phi_v \\ -\sin\phi_v \\ 0 \\ 0 \end{bmatrix}\frac{2\,V_i}{\pi L} - \begin{bmatrix} 0 \\ 0 \\ \dfrac{1}{J} \\ 0 \end{bmatrix}T_m(t) \qquad\qquad (3)$$

Where i_q and i_d are the direct and quadrature components of the stator current. ω and δ are the rotor speed and the rotor angle respectively. r and L are the stator resistance and self inductance. λ is the amplitude of the flux linkages established by the permanent magnet as viewed from the stator windings. D and J are the damping factor and inertia constant of the machine. P is the number of poles. V_i is the voltage of the dc link. The angle ϕ_v is defined by

$$\phi_v = \Theta_v(0) + \Theta_r(0);$$

$\Theta_v(0)$ is the initial phase angle of the stator voltage. $\Theta_r(0)$ is the initial rotor position (quadrature axis) with respect to the stator axis. $\Theta_v(0)$ can be adjusted to advance or delay the switching of the inverter relative to the rotor position.

All the above states can be directly or indirectly measured. δ and ω can be obtained by a position encoder [9]. i_q and i_d can be calculated from the stator current using Park's transformation [8]

$$i_q = i\cos\varphi$$

$$i_d = -i\sin\varphi$$

where φ is the phase angle between the stator voltage and rotor angle.

λ can be computed by a simple open circuit test, where the rotor circuit is driven by a prime mover, and the open circuit voltage across the stator winding is measured [8]. By this test, λ equals to

$$\lambda = \frac{V_o}{\omega_o}$$

where V_o is the open circuit peak phase voltage of the stator windings. ω_o is the electrical speed of the rotor.

Equation (3) can also be written in the state space form

$$\frac{dX(t)}{dt} = A(t) X(t) + B V_i(t) - G T_m(t) \tag{4}$$

Where A is the state matrix, B is the control vector and G is the load torque vector. X is the state variable vector of the system. Matrix A is time dependent because it contain one of the state variables (ω). Since the control strategy is applied in a discrete form, Matrix A can be assumed constant during a single interval of time, i.e. Matrix $A(t)$ at the time interval (t) is a function of $\omega(t-1)$ of the previous time step. The model of equation (4) may also be written in discrete form similar to that given in equation (2)

2.3 MODEL OF INDUCTION MOTOR DRIVE SYSTEM

Another viable machine, that has recently been considered for high performance drives, is the induction motor [10-14]. Induction motors are low maintenance machines, rugged and relatively smaller in size compared to dc machines. The challenging problems of controlling the induction motors have recently been lessened by a combination of advanced power electronic devices, innovative power electronic circuits and microcomputers [11-14].

Although the potentials for induction motors in the area of high performance drives are very promising, designing such a system is still a very difficult task to accomplish due to two key reasons: 1) the induction motor model is highly nonlinear and complex; and 2) the parameters of the motor drift during operation due to temperature changes, magnetic saturation, etc. [15,16].

To lessen the problems associated with the model complexity, a method known as "*vector control*", or "*field oriented control*", has been developed for induction motor drives [15-22]. The word "*control*" is probably imprecise since no traditional control action is utilized. Therefore, it will be called in this chapter "*vector transformation.*" The electric drive system, by this method, is composed of an induction motor, a converter and a vector transformation block. With vector transformation, the stator current is resolved into a torque producing component and a flux producing component, each can be independently specified. This decoupling makes an induction motor drive system behaves somehow like a dc drive.

Because of its potentials and simplicity, the vector transformation method is growing in popularity, and several successful results are reported [15-21]. The key limiting factor, however, is the sensitivity of this method to parametric variations such as rotor resistance and core saturation. Variation of these, and other, parameters may result in inaccurate transformation; hence inadequate performance [15,16,18,22]. To reduce the effects of the parameter variations, a combination of elaborate parameter adaptation algorithms and/or direct measurement of the airgap flux may be needed [16,22].

The dynamic model of the induction motor can be developed based on the following equations [8]:

$$V_{qs} = R_s i_{qs} + \frac{d\phi_{qs}}{dt} + \omega_s \phi_{ds}$$

$$V_{ds} = R_s i_{ds} + \frac{d\phi_{ds}}{dt} - \omega_s \phi_{qs}$$

$$0 \;.= R_r i_{qr} + \frac{d\phi_{qr}}{dt} + \Delta\omega \, \phi_{dr}$$

$$0 \; = R_r i_{dr} + \frac{d\phi_{dr}}{dt} - \Delta\omega \, \phi_{qr}$$

$$T_e \; = \frac{3\,P}{4} L_m \, (i_{qs} i_{dr} - i_{ds} i_{qr})$$

Where R_s and R_r are the stator and rotor resistances respectively. L_m is the mutual inductance. ϕ is the flux. ω_s is the synchronous speed, and $\Delta\omega$ is the difference between the synchronous and shaft speeds (slip speed). The subscript (s) is for the stator and (r) is for the rotor. d and q are the direct and quadrature axes of the reference frame, which are rotating at synchronous speed.

The above equations show no obvious decoupling of the current components. However, if the machine is fed by impressed currents and the direct axis is aligned with rotor flux, the machine model can be transformed into the following:

$$\frac{d\phi_r}{dt} = \frac{1}{\tau_r \, (L_m i_{ds} - \phi_r)}$$

$$T_e = K_t i_{qs} \phi_r$$

$$\Delta\omega = \frac{L_m i_{qs}}{\tau_r \phi_r}$$

where

$$\tau_r = \frac{L_r}{R_r}; \quad \text{rotor circuit time constant}$$

$$K_t = \frac{3\,P\,L_m}{4\,L_r}$$

From the above equations, if the flux and the electromechanical torque are specified, the direct and quadrature components of the current, as well as the slip speed, can be calculated. These quantities can be transformed into stator currents as shown in Figure 1.

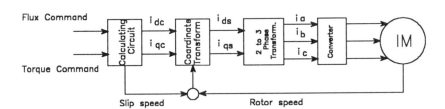

Figure 1: Induction Motor Drive System with Vector Transformation

The rotor dynamics of an induction motor can be represented by the following equation:

$$\begin{bmatrix} \dfrac{d\,\delta}{dt} \\ \dfrac{d\,\omega}{dt} \end{bmatrix} = \begin{pmatrix} 0 & 1 \\ 0 & \dfrac{-D}{J} \end{pmatrix} \begin{bmatrix} \delta \\ \omega \end{bmatrix} + \begin{bmatrix} 0 \\ \dfrac{1}{J} \end{bmatrix} [T_e - T_m] \qquad (5)$$

Where D and J are the total damping and total inertia of the system, respectively. T_e is the developed electric torque. T_m is the total mechanical torque. T_e is the input command to the vector transformation block.

In state space form, equation (5) can be expressed as follows:

$$\frac{d\,X(t)}{dt} = A\,X(t) + B\,T_e - B\,T_m \tag{6}$$

3. SELECTION OF REFERENCE TRACKS

Any proposed controller should not be restricted to any shape(s) of track(s). Any restriction, however, must only be imposed by the physical limitations of the hardware (motor, electronic converter and power source).

In most of the electric drives applications, several performance properties should be maintained to avoid "*premature fatality*" of the hardware, especially for large size systems. Among these properties, which might be correlated, are the following:

1) The system should have the property of "*soft transition*"; e.g. soft starting, soft speed change and soft braking. Abrupt large changes in speed may eventually results in ruinous effects on the mechanical integrity of the motor or load, and unnecessary electrical stresses on the motor or converter. A soft transition does not necessarily mean a slow transition;

2) The system should have a sufficient damping for speed oscillations at all time, including at the equilibrium state (holding state);

3) The large abrupt changes in the supply voltage should be avoided;

4) The magnitude of the inrush current should be kept under control at all time. The overshoots of the inrush current should be limited to some tolerated values;

5) Natural electromechanical oscillations should be avoided. They usually occur at low speeds when the electrical modes of the system correspond to the natural frequencies of the load and supporting structure [23].

The basis for selecting the tracks should always be to achieve the needed performance for the desired application without compromising any of the above properties.

To provide a smooth transition of position, speed, current and other variables, it is proposed that the rotor position track be selected as a Sigmoid Function (SF) with the desired time constants and terminal conditions. Any arbitrary complex track can be constructed from composed segments of SF.

To ensure that all the tracks are traceable, the speed, current or torque tracks must all be computed after the position track is selected. If this is not the case, the position and other tracks will not conform with the system dynamics.

A single segment of a Sigmoid Function for the position track is shown in Figure 2, and can be expressed by the following equation:

$$\delta_T(t) = \delta_0 + \frac{\Delta\delta_f}{[1 + e^{\left(\frac{-\Delta t}{\tau}\right)}]} \quad ; \quad \Delta t = t - t_{th} \qquad (7)$$

Where $\delta_T(t)$ is the desired segment of the position track at an arbitrary time (t). δ_0 is the initial rotor angle. $\Delta\delta_f$ is the desired change in the rotor angle at the final time of the track segment. t_{th} is the threshold time which centers the transition region of $\delta_T(t)$. τ is the desired time constant of the track segment which determines the abruptness of the transition.

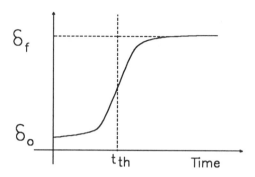

Figure 2: Sigmoid Function

The speed track $\omega_T(t)$ can be calculated by differentiating equation (7). The track for the electrical torque can also be calculated by using the equation of the rotor dynamics

$$T_{eT}(t) = T_m(t) + J \frac{d\,\omega_T}{dt} + D\,\omega_T$$

Where $T_{eT}(t)$ is the track for the electric torque. T_m is the mechanical torque. $\dfrac{d\,\omega_T}{dt}$ is the acceleration of the track which is equal to the second derivative of equation (7). The reference vector $R(t)$ containing all tracks can be written as follows:

$$R(t) = [\ \delta_T(t) \quad \omega_T(t) \quad T_{et}(t)\ ]^T$$

It should be noted that all tracks and their derivatives are expressed in closed form equations. This results in a fast computation of the control signal (U) which facilitates the on-line applications. This in specially true if the mechanical load torque is known apriori. If not, the reference vector (R) should not contain a track for the torque.

4. CONTROL STRATEGIES FOR HIGH PERFORMANCE DRIVES

As mentioned earlier, the utilization of electric motors in complex applications such as robotic or manipulations requires not only a position control at the final time, but also tracking or trajectory control. This may require that each motor in the manipulator follows its predetermined speed or position track during starting, speed change and breaking.

The construction of a high performance drive system requires, in addition to a fast converter, an advanced controller which must be "*adaptive*" and "*robust*". The tracking performance of the system must not suffer due to system variations, such as parameter variations of the motor, load changes, disturbances, measurement noise or model errors [24-26].

In many drives applications, the mechanical load varies considerably during operation. Robots and machine tools with changing load inertia are two typical examples. When a fixed controller setting is used in a drive system with a widely changing load, unsatisfactory performance is often produced. This is quite obvious in position-controlled drives where slightly misaligned controllers can cause considerable overshoot and oscillations [26]. In addition, controllers with fixed parameters are not always robust unless unrealistically high gains are used [26].

Robustness is of particular importance in many applications that involve actuation. Robustness can be greatly enhanced if the controller is adaptive in nature, where the controller parameters are continually adjusted to counter any change in the system operating conditions. Under this class of adaptive control are "*Self-Tuning*" [2,26-32], "*Model Reference*" [24,25,31,32], and "*Variable structure*" [1,33] controllers.

5. VARIABLE STRUCTURE TRACKING CONTROL

Variable Structure System Control (VSC) has been investigated by researchers for over ten years [34,35], and much of the initial theories have been translated into English [34,36,37]. The idea behind the VSC control can be demonstrated by the phase portrait of the second order system shown in Figure 3 [1,33-37]. Let the system states represent errors to be compensated. Hence, the point of origin is the desired final destination of the states. A switching surface, or switching hyperplane, is constructed from the system states in a weighted linear or nonlinear form. With the VSC control, the structure of the control loop is kept fixed until the states hit (crosses) the switching hyperplane. Then the structure is adaptively adjusted to slide the states along this hyperplane until they reach their final destination. This change of structure is achieved by switching actions where the sign and magnitude of the control loop gain is adaptively changed.

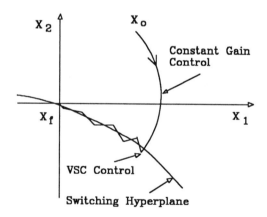

Figure 3: Variable Structure Control

The switching actions are only applied when the operating point of the systems is in the vicinity of the switching hyperplane. However, the motion of the system from the initial time until it hits this hyperplane is accomplished by a fixed gain feedback loops [34-41]. This makes the system susceptible to parametric variations such as inertia and armature resistance. Hence, the robustness of the controller is not ensured during this period, but may be improved if a high feedback gain is used. Unfortunately, this high gain could be impractical to implement [42].

To enhance the robustness of the controller, some methods are proposed in the literatures [42], and described in Figures 4 and 5. In Figure 4, several hyperplanes with different slopes are used. The system initially slides on the nearest hyperplane, then switches to the next, and so on. The system does not stay in the constant feedback gain region for a lengthy period; therefore, some improvement to the system robustness is achieved. In Figure 5, four hyperplanes are used, each with a specific property. The first hyperplane represents a constant acceleration. The second is a constant velocity. The third is a constant deceleration. The forth is a linear combination of the speed and position. With these hyperplanes, the VSC control signal is designed by using closed form equations and lookup tables.

The robustness of the controller by these methods could be enhanced. However, the operation of the system is very much limited to the mentioned specific shapes of the hyperplanes. In addition, a VSC position control, by these methods, can be very difficult to design when the armature inductance is relatively large. If such inductance is neglected, the effectiveness of the controller would be very much diminished since the supply voltage (or current) in the VSC environment is rapidly changing.

With this form of VSC, the travelling time from the initial point to the final destination can neither be specified nor controlled. Also, the states of the system may overshoot several times before reaching the final destination. This may cause unnecessary stress on the system hardware.

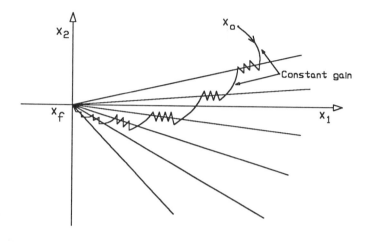

Figure 4: Multi Switching Hyperplane

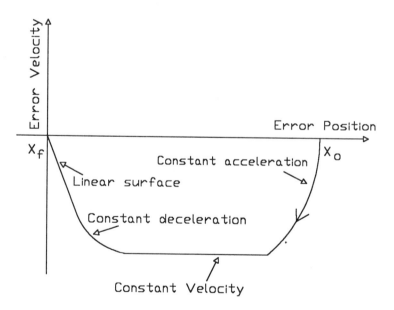

Figure 5: Switching Hyperplane with Special Property

To avert these drawbacks, a modified version of the VSC can be used [1,33]. It is called Variable Structure Tracking Control (VSTC). To briefly describe the basic structure of the VSTC, let the system under control be expressed by the following state variables equations:

$$\frac{d\,X(t)}{dt} = A\,X(t) + B\,U(t) - G\,T_m$$

$$Y(t) = H\,X(t)$$

Where X is the system state vector, U is the control vector, T_m is the load torque and Y is output vector. A, B and H are system matrix, control vector and output matrix respectively. G is the load torque vector.

Assume that R(t) is a desired track vector containing the desired time history of the variables under control. The control signal U(t) should be selected to ensure that the following error function is reduced to a minimum at all time:

$$E(t) = Y(t) - R(t)$$

The first step in applying the VSTC technique is to select a hyperplane $\sigma(t)$ in the following form [1,33]:

$$\sigma(t) = C\,E(t) \qquad\qquad (8)$$

Where C is a weighting vector. The elements of C can be selected to give a more emphasis on tracking one particular output over another. It should also be selected to ensure the stability of the controller [37].

A "*switching hyperplane*" is defined as $\sigma(t) = 0$. It represents any operating point of the system at which the weighted magnitudes of the errors in E(t) are summed up to zero. It is also known as a "*sliding hyperplane*".

The objective of the VSTC is to maintain the system operating point in the vicinity of the switching hyperplane while "*sliding*" the states to their final destination. A sufficient condition is given as follows [1,33,37]:

$$\sigma(t) \frac{d\,\sigma(t)}{dt} \leq 0 \tag{9}$$

hence,

$$\sigma(t)\ C\ \frac{[d[H\ X(t) - R(t)]}{dt} =$$

$$\sigma(t)\ C\ [H\ \{A\ X(t) - G\ T_m + B\ U(t)\} - \frac{dR(t)}{dt}] \leq 0$$

To satisfy the condition of equation (9), U(t) must be selected as follows:

$$U(t) \leq (CHB)^{-1}\ C\ [\frac{dR(t)}{dt} - HA\ X(t) + HG\ T_m]\ ; \qquad iff \quad \sigma(t) \geq 0 \tag{10}$$

and

$$U(t) > (CHB)^{-1}\ C\ [\frac{dR(t)}{dt} - HA\ X(t) + HG\ T_m]\ ; \qquad iff \quad \sigma(t) < 0 \tag{11}$$

It appears from the above equations that the control signal is dependent on the load torque. This implies that the torque should be measured or identified, which could be a very complex, costly and/or time consuming process. However, because of the inequalities in equations (10) and (11), the exact value of the load torque is not really needed. Both equations can be met if the load torque in (10) is replaced by its minimum expected value, and that of equation (11) is replaced by its maximum expected value.

6. SELF-TUNING CONTROL

Self-Tuning control has two main forms: Explicit and Implicit. Both schemes require valid digital models to represent the plant under control. In the explicit scheme, the parameters of the plant are directly identified. Then, the controller parameters are computed based on the selected control law, and on the digital model of the plant. On the other hand, with the implicit scheme, the controller parameters are directly identified. Hence, the implicit method requires less computational time as compared to the explicit method. The saving in the computational time is significant in fast response applications, especially for higher order systems.

In this section, the theory of implicit self-tuning control is applied to the problem of tracking drives. The theory is based on the principle of Generalized Minimum Variance Adaptive Control [2,3,27-32].

The general concept of Self-Tuning control algorithm is shown in Figure 6. The plant under control has an input $U(t)$ and output $Y(t)$ at a sampling instant "t". The sequential input and output data, collected over each sampling instant, are utilized to identify a discrete model of the plant. The model in conjunction with the specified reference track (Y_{ref}), and the input/output history of the motor are used to generate the control signal at the following time step.

A Self-Tuning controller should be designed taking into account the practical limitation of the system under control. Self-Tuning, if not adequately designed, could by itself become unstable, or could generate an unrealistically large control signals. To avert these problems, the Self-Tuning tracking controller should be designed with the following features:

1)   A closed form method should be used to determine the
     structure of the model (order and delay of the difference
     equation) describing the system under control. This results in
     an accurate representation of the drive system, which is
     essential for effective control design.

2)   The control signal should be included in the cost index.
     .Appropriate weighting factors should be introduced to restrain
     the magnitude and oscillations of the control signal. These
     properties are essential to avert excessive stresses on the
     converter, motor and the mechanical load.

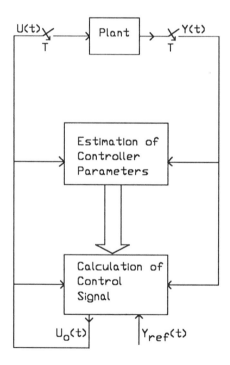

Figure 6: Self-Tuning Control Algorithm

3)   The system should achieve good tracking performance even when the plant under control exhibits non-minimum phase property.

4)   The system must be robust - variations in system parameters or operating conditions must be negated by the controller.

6.1 MATHEMATICAL FORMULATION

Under a stochastic measurement environment with uncertainties, such as random disturbances, quantization and truncation errors, an Autoregressive Moving Average Exogeneous (ARMAX) model is often used to represent the system under control [31,32,43,44]. A motor operating in such conditions (load changes or any other disturbance) can also be represented by the ARMAX model:

$$A(q^{-1})\, Y(t) = q^{-k}\, B(q^{-1})\, U(t) + C(q^{-1})\, \epsilon(t) \qquad (12)$$

where   Y: measured output
       U: control input
       ϵ: uncorrelated white noise sequence
       t   sampling instant
       q^{-1} backward shift operator
       k   process time delay ($k \geq 1$, integer)

A, B and C of equation (12) are polynomials of q^{-1} defined as follows:

$$A(q^{-1}) = 1 + a_1\, q^{-1} + a_2\, q^{-1} + ... + a_{n_a}\, q^{-n_a}$$

$$B(q^{-1}) = b_0 + b_1\, q^{-1} + b_2\, q^{-2} + ... + b_{n_b}\, q^{-n_b} ; \quad b_0 \neq 0 \qquad (13)$$

$$C(q^{-1}) = 1 + c_1\, q^{-1} + c_2\, q^{-2} + ... + c_{n_c}\, q^{-n_c}$$

The constants n_a, n_b and n_c are the orders of polynomials A, B and C respectively. Once these constants, in addition to (k), are known, model parameters a_1 to a_{n_a}, b_0 to b_{n_b} and c_1 to c_{n_c} can be identified by using one of the established methods of parameter identification [43,44]. An accurate model, however, depends on the proper choice of n_a, n_b, n_c and k, which are often obtained by trial and error methods [44]. To achieve accurate representation of the system under control, a systematic method should be used to determine the plant orders and the time delay of the model. This could be done by transforming the time domain model of the motor into a transfer function in the "z" domain. Then, comparing it to the transfer function of the ARMAX model of equation (12).

The z-transformation of equation (12) can be written in the following form:

$$Y(z) = z^{-k}\frac{B(z^{-1})}{A(z^{-1})} U(z) + \frac{C(z^{-1})}{A(z^{-1})} \epsilon(z) \tag{14}$$

The first term in the right hand side of equation (14) corresponds to the transfer function of the system under control. The second term represents external disturbances, noise, etc. Any changes in the system, such as load torque variations, noisy measurements or other disturbances are adaptively included in the coefficients of the polynomial A, B and C by the recursive identification method.

After a proper model for the motor is selected, the control signal can be designed. Under the Generalized Minimum Variance Control, the following cost index (J) is minimized in order to calculate an optimal control signal:

$$J = E\{\gamma(t+k)\} = E\{[Y(t+k) - Y_{ref}(t+k)]^2 + \mu\,[Q\,U(t)]^2 | t\}$$

$$\tag{15}$$

where $E\{.\}$ is the expectation at time $(t+k)$, which is conditioned for data acquired up to time t [31]. $Y_{ref}(t)$ is the specified trajectory (track). By minimizing this cost index, the difference between the system response $Y(t+k)$ and the track $Y_{ref}(t+k)$ is minimized. The value of μ can be selected to adjust the trade off between the tracking accuracy and the magnitude of the control signal. To limit the abrupt changes in the control signal, Q can be selected as

$$Q = 1 - q^{-1}$$

J is minimized if the derivative of $\gamma(t)$ with respect of the control signal, at time t, is set equal to zero.

$$\frac{d\gamma(t)}{dU(t)} = 2 [Y(t+k|t) - Y_{ref}(t+k)] \frac{dY(t+k|t)}{dU(t)} + 2\mu (1 - q^{-1}) U_0(t)$$
$$= 0$$

$U_0(t)$ is the optimal control signal that minimizes the cost index of equation (15). Based on equation (12), the derivative $\dfrac{dY(t+k|t)}{dU(t)}$ is equal to b_0; hence,

$$Y(t+k|t) - Y_{ref}(t+k) + \lambda (1 - q^{-1}) U_0(t) = 0 \qquad (16)$$

Where $\lambda = \mu/b_0$

Let equation (12) be rewritten as

$$Y(t+k) = \frac{C(q^{-1})}{A(q^{-1})} \epsilon(t+k) + \frac{B(q^{-1})}{A(q^{-1})} U(t) \qquad (17)$$

By expanding $1/A(q^{-1})$ as a power series in q^{-1}, it can be seen that the first term of equation (17) (the disturbance term) has two components: one related to future disturbances; and the other related to past disturbance [32]. The past can be reconstructed using input-output measurements, but the future disturbances can not be predicted. In order to separate the two disturbance components, the following identity (known as a Diphontine equation) is introduced [31]:

$$\frac{C(q^{-1})}{A(q^{-1})} = F(q^{-1}) + q^{-k}\frac{G(q^{-1})}{A(q^{-1})} \tag{18}$$

where F is a polynomial of order (k-1) and G is a polynomial of order (n_a-1). Substituting $C(q^{-1})$ of equation (18) in equation (17) gives the following:

$$Y(t+1) = F(q^{-1}) \, \epsilon(t+1) + \frac{G(q^{-1})}{A(q^{-1})} \epsilon(t) + \frac{B(q^{-1})}{A(q^{-1})} U(t) \tag{19}$$

Substituting the value of $\epsilon(t)$ of equation (12) into (19) yields

$$Y(t+1) = F(q^{-1}) \, [\, \epsilon(t+1) + \frac{B(q^{-1})}{C(q^{-1})} U(t) \,] + \frac{G(q^{-1})}{C(q^{-1})} Y(t) \tag{20}$$

The first term in the right hand side of equation (20) is future disturbance which can not be predicted since the noise sequence is uncorrelated. Therefore, it is not considered in the minimization process of J. Hence, if Y(t+1|t) is a "one-step-ahead predictor" of Y, given the input/output history up to time t, $F(q^{-1})$ is equal to unity. Hence,

$$Y(t+1|t) = \frac{B(q^{-1})}{C(q^{-1})} U(t) + \frac{G(q^{-1})}{C(q^{-1})} Y(t) \tag{21}$$

Substituting equation (21) into (16) gives

$G(q^{-1}) Y(t) +$

$[B(q^{-1}) + \lambda (1 - q^{-1}) C(q^{-1})] U_o(t) - C(q^{-1}) Y_{ref}(t+1) = 0$ (22)

The closed loop poles of the Controller described by equation (22) are the roots of the following equation [7,8]:

$B(q^{-1}) + \lambda (1 - q^{-1}) A(q^{-1}) = 0$

which are not only dependent on the roots of the polynomial B; but also on λ. Thereby, a stable controller of non-minimum phase plants [28-32] can be achieved by on-line adjustment of λ.

Expanding equation (22) as a difference equation gives the optimal control signal $U_o(t)$ as,

$$U_o(t) = \frac{1}{b_0 + \lambda} [Y_{ref}(t+1) + c_1 Y_{ref}(t) - g_0 Y(t) - g_1 Y(t-1) -$$

$$(b_1 + \lambda c_1 - \lambda) U_o(t-1) + \lambda c_1 U_o(t-2)]$$ (23)

6.2 RECURSIVE PARAMETER ESTIMATION

The function of the recursive estimation process is to identify the parameters of polynomials B, C and G by using single input single output measurement of the drive system.

The system model can be rearranged for parameter identification by multiplying both sides of equation (21) by $C(q^{-1})$.

$C(q^{-1}) Y(t+1|t) = G(q^{-1}) Y(t) + B(q^{-1}) U(t)$

hence,

$$Y(t+1|t) = g_0 Y(t) + g_1 Y(t-1) + b_0 U(t) + b_1 U(t-1)$$
$$- c_1 Y(t|t-1) \tag{24}$$

Define the measurement vector $X(t)$, and the unknown parameter vector $\theta(t)$ as follows:

$$X(t) = [Y(t) \ Y(t-1) \ U(t) \ U(t-1) \ Y(t|t-1)]$$

$$\theta(t) = [g_0 \ g_1 \ b_0 \ b_1 \ -c_1]^T \tag{25}$$

Then equation (24) can be written as

$$Y(t+1|t) = X(t) \ \theta(t)$$

Assuming the actual output of the system is contaminated with noise ξ,

$$Y(t+1) = X(t) \ \theta(t) + \xi(t+1|t)$$

$$\theta(t+1) = \theta(t) + e(t)$$

Where e represents a parameter error vector between two time steps.

The parameters in θ can be identified by a Kalman Filtering method [44] as follows:

$$\theta(t) = \theta(t-1) + K(t) \ [Y(t) - X(t-1) \ \theta(t-1)] \tag{26}$$

$K(t)$ is the Kalman gain given by,

$$K(t) = \frac{P(t-1) \ X(t-1)^T}{1 + X(t-1) \ P(t-1) \ X(t-1)^T} \tag{27}$$

and

$$P(t) = \frac{1}{\alpha} [P(t\text{-}1) - \frac{P(t\text{-}1)\, X(t\text{-}1)^T\, X(t\text{-}1)\, P(t\text{-}1)}{\alpha + X(t\text{-}1)\, P(t\text{-}1)\, X(t\text{-}1)^T}]\; ; \quad 0 < \alpha \leq 1$$

$$(28)$$

The purpose of α, which is known as the forgetting factor [31], is to give more weight to more recent data. For rapid system variations, α can be selected slightly less than one (about 0.95).

7. LABORATORY IMPLEMENTATION FOR DRIVES CONTROL

In the not-so-distant past, the development of advanced control theories for drives applications were very much limited to computer simulations without adequate laboratory implementation or verification. This is because building a versatile hardware control circuit that operates in various control modes was a technically difficult and costly task, especially for electric drives applications.

During the past few years, however, enormous improvement were made in the areas of power electronics, digital electronics and microprocessors. Solid state devices for power switching are now much more reliable than their predecessors. More advanced performances can be obtained from improved devices as well as from new types of solid-state switches. The prices of these devices are becoming less and less expensive; although, much of them are still beyond reach for any "*openhanded*" laboratory.

Among the important development in the solid-state power electronics technology is the integrated modules. Solid state switches can now be found in various configurations such as H-bridge or six-pack modules. They are normally equipped with some snubbing capabilities. Also, complete driving circuits can now be found in very sophisticated and elegant designs. Most of them have built-in Pulse Width Modulation (PWM) options for speed control, and also are equipped with overcurrent protection and speed feedback. Building such a module, in the not-so-distant past, was a several months job at a multitude of the cost.

With the above mentioned developments, a versatile laboratory setup for electric drive control can be build out of off-the-shelf modular hardware and general purpose software packages. Next section describe one of these setups.

7.1 SYSTEM DESCRIPTION

The key functional blocks of the laboratory setup are shown in Figure 7 [9]. The main modules of the setup are: an electric motor (M); a power switching module (or converter (C)); a driving module (D); a position encoder (E); and a PC based computer (PC). In this section, a setup for dc motor drive is presented. For ac , or brushless motors, the switching module and driving modules must be selected (or designed) accordingly.

Since the system is electromechanical, a 10 MHz PC is quite adequate. The computer is interfaced with the external hardware via an input/output (I/O) card.

The converter circuit is shown in Figure 8. It is a MOSFET chopper circuit that can be found in a single module. It consists of four transistors in an H-bridge configuration to operate the motor in all four quadrants. Four external fast recovery diodes (D) are added to provide the needed reverse path for the current during commutation.

The terminal voltage of the motor is controlled by using Pulse Width Modulation technique (PWM). By this method, the motor terminal voltage is proportional to the duty ratio of the PWM cycle.

This PWM signal is generated by a multipurpose driving module. One of the inputs to the driving module is an analog signal proportional to the control command provided by the PC. This analog signal is converted into a PWM signal with proportional duty ratio.

The output of the driving module is amplified by using opto-coupler/buffer module. The output of each opto-coupler triggers one power switch in the converter module. The opto-couplers, also, provide the needed isolation between the high voltage switching circuit and the low voltage control circuit.

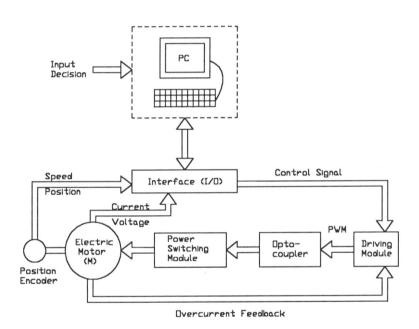

Figure 7: Main Functional Blocks of Lab Setup

From [9], courtesy of IEEE, (C) IEEE 1990

The driving module also provides overcurrent protection. When the armature current exceeds a predetermined value, the driving module disables its PWM output. Thus the power transistors are turned off.

For feedback purposes, several variables can be measured such as currents, voltages, speed and rotor angle. The armature current is measured by using a dc current probe. The voltage is measured by a voltage divider. The rotor speed and position are measured by an optical speed encoder mounted onto the shaft of the motor.

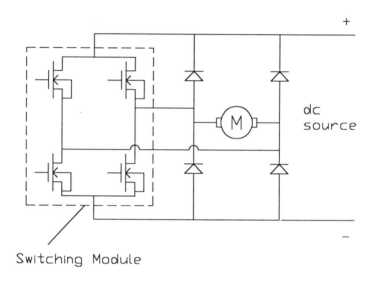

Figure 8: Converter Circuit

7.2 SOFTWARE BLOCKS

No special purpose software should be needed for this setup. There are many commercially available control oriented simulation software packages that can fulfill the needs at reasonable prices. Most of these commercial software have modules to interface with external hardware. The basic requirement of the software is to perform two tasks: real time control and data acquisition.

A number of these control oriented software, such as that reported in [9], mimic analog computers where the basic functions are represented by element blocks; such as integrator, differentiator, attenuator, sum, delay, inverter, input, and output. Most controllers can be easily built by these blocks. The input and output blocks allow for the interfacing with the external hardware.

7.3 OPERATION OF SETUP

Before using the setup in on-line control, the control strategy and the system under study can be simulated at various conditions and design parameters. The topology of the control circuit can be built by using the necessary blocks in the software library. Figure 9 shows an example for speed (or position) control of a separately excited dc motor. The dashed area represents the software portion of the system which consists of the following blocks: IN and OUT to address the I/O card; SUM for comparing the measured speed (or position) to its corresponding command; and a controller to carry out the desired control strategy.

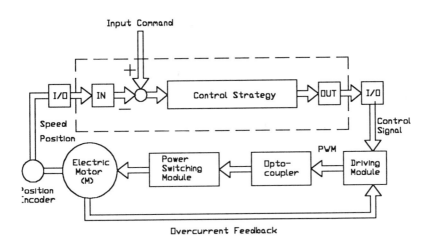

Figure 9: Setup with Controller

From [9], courtesy of IEEE, (C) IEEE 1990

The motor speed (or position) is sensed by the encoder (E) and inputted to the PC via the I/O card. The error between the measurement and the desired track is processed by the controller block. The output of the controller determines the necessary voltage that should be applied to the motor. This control signal is transmitted to the driving module via the I/O card. The driving module provides the corresponding PWM signals to trigger the transistors of the converter (power switching). The converter, in turn, provides the necessary terminal voltage for the motor.

While executing the control function, the PC can also do data acquisition. The user can see the results of the experiment displayed on the screen as the experiment progresses. The user can also tune the parameters and/or change the structure of the controller to modify the system performance.

8. APPLICATIONS OF VARIABLE STRUCTURE TRACKING TO BRUSHLESS MOTORs

Figure 10 shows a schematic of the basic components of the laboratory setup for brushless motor drive. The motor is a 6-pole, 3-hp Samarium-Cobalt brushless machine with the following parameters::

Rated voltage = 270 V (dc link)
No load speed = 10,000 r/min
Stator resistance (at $65^{\circ}F$) = 0.43 Ω
Stator inductance (L) = 8.26 mH
Rotor inertia = 0.00021 kg m^2

The position encoder mounted on the rotor shaft produces 2000 pulses per revolution. The motor is driven by a three-phase, six-step dc/ac switching converter. The switching devices are MOSFETs. The switching converter is driven by an ac/dc converter. The triggering logics of the switching devices are generated based on the rotor position to achieve the brushless dc motor operation. The armature voltage of the stator windings is adjusted by means of Pulse Width Modulation (PWM) method.

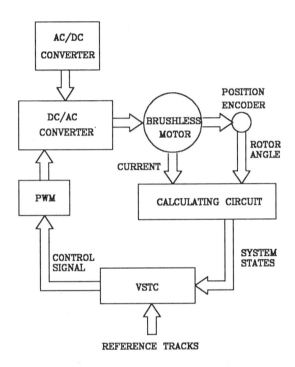

Figure 10: Schematic of Laboratory Setup

From [33], courtesy of IEEE, (C) IEEE 1990

The measurements of the system are: 1) rotor position; 2) rotor speed; 3) armature current; 4) armature voltage; and 5) phase angle of the armature current. All these measurements are directly sensed via a dynamic filtering circuit to provide information on only the fundamental frequency. The *pass* frequency of the filters is adjusted to match the speed of the rotor.

These measurements are injected into a calculating circuit to compute the states of the system. It also updates the state space matrix A(t) at every time interval. The output of the calculating circuit is given to the VSTC block. The VSTC block also receives information on the reference tracks. The output of the VSTC, which is the control signal, is inputted to the PWM chip that adjusts the duty cycles of the switching MOSFETs, and in turn adjusts the magnitude of the armature voltage.

Details of the VSTC block is shown in Figure 11. There are three inputs to the VSTC block: 1) the reference track vector R(t); 2) the derivatives of the reference tracks $\frac{dR(t)}{dt}$; and 3) the state vector of the system X(t).

Two reference tracks are used: a track for rotor position and a track for rotor speed. The rotor position track can be arbitrarily selected but the rotor speed track must be computed as a derivative of the position track. If not, the position and speed tracks will not conform with the system dynamics. The derivative of the reference tracks can be obtained in a closed form if the rotor position track is represented by a set of equations.

The dashed box in Figure 11 shows the implementation of the right hand side of equations (10) or (11). The hyperplane is calculated by comparing the measurements Y(t) to the corresponding tracks at every time step as given in equation (8). The sign of the hyperplane determines the switching action of blocks S_1 and S_2. Based on the sign of the hyperplane, and on the sign of t e output signal of the dashed box, the switching action in block S_1 satisfies the conditions of the control equations (10) or (11). k_1 and k_2 are constants replacing the inequalities in these control equations.

If the mechanical torque is directly monitored, its value can be injected into the VSTC block. However, as seen from equations (10) or (11), and based on the discussions that followed these equations, only the minimum and maximum values of the mechanical torque are needed. In this case, the torque limit is selected by the switching action of block S_2.

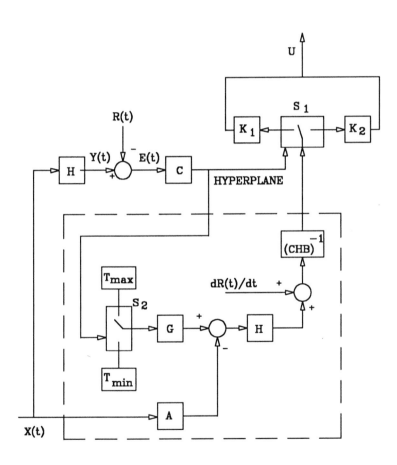

Figure 11: Details of the VSTC Block

8.1 TEST RESULTS

The VSTC technique is not restricted to *any shape* of tracks. Any restriction, however, is imposed by the system's hardware limitations. As mentioned earlier, to reduce any unnecessary stress, the tracks should conform with the dynamics of the system. The tracks should also result in *soft transition*; e.g. soft starting, soft speed change and soft braking. Abrupt and repeated large changes in speed may eventually result in ruinous effects on the mechanical integrity of the motor or load, and unnecessary electrical stresses on the motor or converter. A soft transition does not necessarily mean a slow transition.

The track selected in this test is composed of three segments of sigmoid functions. The first is to advance the rotor position by 100 radian; the second is to further advance the rotor position by another 100 radian; and the third is to return the rotor to its original position by means of counter current braking. There are also three holding positions: the first is at 100 radian; the second is at the 200 radian and the third is at the original position. This track includes starting, speed change, braking and holding. These are the modes of operation of any electric drive system.

The motor is loaded by a constant weight of 10 Kg. A pulley arrangement is designed to produce varying load torque in a single direction of rotation. The minimum is about 2.4 Nm, and the maximum is about 6.2 Nm.

Figure 12 shows the track for the rotor position and also the measured rotor angle. As seen in this figure, the brushless motor with the VSTC achieves a high degree of tracking accuracy. This is especially encouraging since the load torque is not continually monitored.

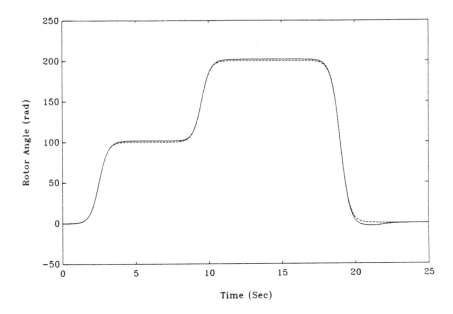

Figure 12: Position Tracking

From [33], courtesy of IEEE, (C) IEEE 1990

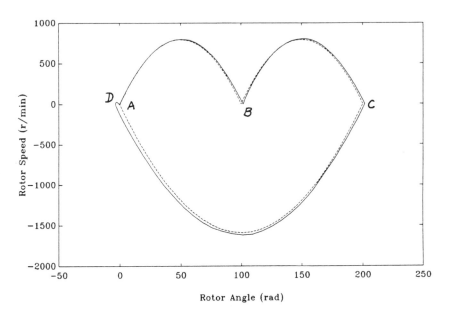

Figure 13: Phase Portrait of Speed Tracking Vs Position Tracking

From [33], courtesy of IEEE, (C) IEEE 1990

MOHAMED A. EL-SHARKAWI

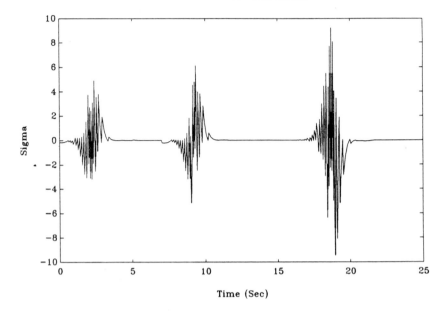

Figure 14: System Hyperplane

From [33], courtesy of IEEE, (C) IEEE 1990

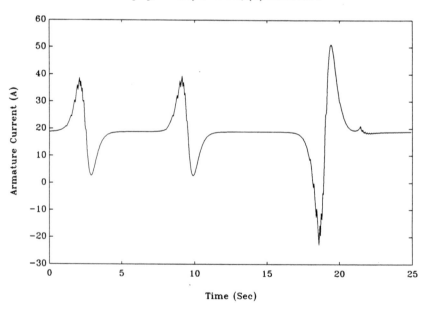

Figure 15: Current in the dc Link

From [33], courtesy of IEEE, (C) IEEE 1990

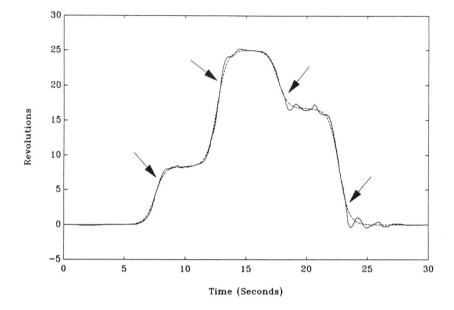

Figure 16: Position Tracking with Load Disturbance

From [33], courtesy of IEEE, (C) IEEE 1990

Figure 13 shows the phase portrait of the speed versus rotor position. Point A represents the starting point, point B represents the first holding position at 100 radian, point C represents the second holding position at 200 radian, and point D represents the final destination. Again the track (dashed line) and the measurements (solid line) are in very good agreement.

The hyperplane profile is shown in Figure 14. The switching hyperplane is a horizontal line at zero value. It is shown in this figure that the VSTC is continually forcing the system to move along the switching hyperplane. Any deviation from the switching hyperplane is corrected in the following time step(s).

Figure 15 shows the current in the dc link. The switching action of the VSTC can be seen in this plot. This switching action is minimized to reduce unnecessary stresses on the machine by adjusting the constants

C, K_1 and K_2 of Figure 11. A trial-and-error method is used until acceptable switching is achieved.

Another interesting test is shown in Figure 16. In this case the load is intentionally disturbed several times while the motor is tracing its trajectories. The disturbance is done by hand lifting the load then suddenly dropping it. This is equivalent to a disturbance in excess of 100%. Both of the mechanical load and the system inertia are disturbed. This disturbance is repeated several times, the locations of the disturbance are marked in the figure by arrows. The track in this case is specified by the number of revolutions versus time. The figure shows that the system maintain excellent tracking capability even in the face of unexpected large disturbances.

9. APPLICATIONS OF VARIABLE STRUCTURE TRACKING TO DC MOTORS

The block diagram of the dc drive system with VSTC is shown in Figure 17. The outputs of the VSTC controller and the overcurrent protection circuit are augmented and converted into armature voltage. A ceiling limit on the magnitude of the armature voltage is set to 110% of the rated value. The block outside the dashed area is the VSTC software control block.

The VSTC block is similar to that shown in figure 11. The output control signal is the result of switching actions taking place at each time interval of about 20 mSec. The input measurements are the rotor angle, rotor speed and armature current. Each state is compared to its corresponding track. The error (E) of each state is multiplied by its weighting constant (C) to form the hyperplane σ as given in equation (8). Based on the sign of σ, the control signal is determined by the switching action in block (S). k_1 and k_2 are constants replacing the inequalities in equations (10) and (11).

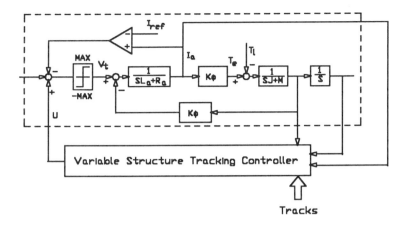

Figure 17: Block Diagram of dc Drive System

From [1], courtesy of IEEE, (C) IEEE 1989

9.1 TEST RESULTS

In the following figures, the solid lines represent actual measurements, and the dashed lines represent reference tracks.

Figures 18 to 20 show several motor responses when the following three-segment track for the rotor position is selected: 1) forward rotation ; 2) holding for 2 seconds and 3) reverse rotation to original position. The position track in forward and reverse rotations are selected as sigmoid functions. The data of the sigmoid functions are as follows:

$\Delta\delta_f = 100^O$ radian.

$t_{th} = 5$ second.

$\tau = 600$ mSec.

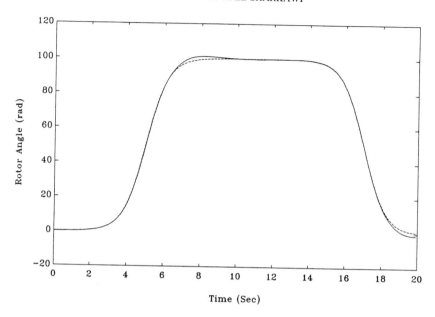

Figure 18: Position Tracking

From [1], courtesy of IEEE, (C) IEEE 1989

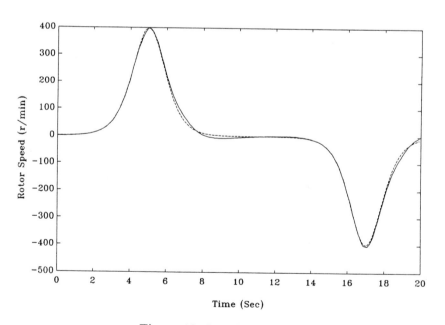

Figure 19: Speed Tracking

From [1], courtesy of IEEE, (C) IEEE 1989

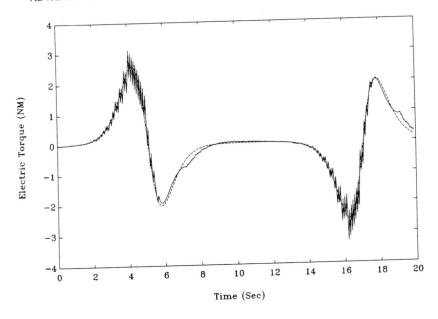

Figure 20: Torque Tracking

From [1], courtesy of IEEE, (C) IEEE 1989

Figures 18 to 20 show, respectively, the tracks and the measurements of the rotor position, speed and electric torque. The figures show that the motor with the VSTC has a very good tracking performance. In Figure 20, the electric torque (hence armature current) always pursues the track; no large overshoots or excessive inrush currents are seen.

10. APPLICATIONS OF VARIABLE STRUCTURE TRACKING TO INDUCTION MOTORS

Figure 21 Shows the main functional blocks of the entire drive system. The induction motor is driven by a converter. The measured output of the motor are shaft speed and rotor position. These measurements are compared to their corresponding tracks. The differences (errors) are processed by the Variable Structure Tracking Controller (VSTC). The output of the VSTC is the electromechanical torque command, which in addition to a flux command are the two inputs of the vector transformation block. These input commands are utilized by the "Calculating Circuit" to compute the slip speed ($\Delta\omega$), and the direct and quadrature components of the stator current along a synchronously rotating field vector (i_{dec} and i_{qec}). By the "Coordination Transformation" block, these current components, in addition to the slip speed and the rotor speed, are used to obtain the direct and quadrature components of the stator current in the fixed frame (i_{doc} and i_{qoc}). They, in turn, are transformed into three phase current commands via the "Phase Transformation" block (i_{ac}, i_{bc} and i_{cc}). These current commands are then inputted to the converter. The flux command is usually kept constant unless one more degree of freedom is needed [15,45].

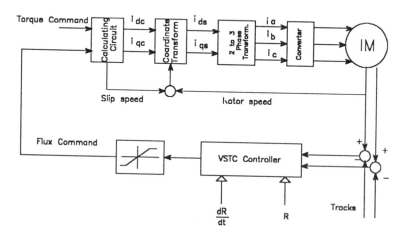

Figure 21: Main Functional Blocks of the Proposed System

The details of the VSTC block are shown in Figure 11. With VSTC tracking technique, the rotor position and speed are the only needed measurements. Both can be acquired by a digital encoder. For overcurrent protection, a current monitoring might be necessary. The control signal (electric torque or armature current) should satisfy the conditions of equations (10) and (11).

10.1 TEST SYSTEM

The test system consists of an induction motor, a mechanical load and a controller. The induction motor is a 4-pole, 3 hp, 220 V, class B machine. The rest of the motor parameters are as follows:

Stator resistance = 0.0115 pu
Rotor resistance = 0.0215 pu
Mutual inductance = 0.686 pu
Stator inductance = 0.705 pu
Rotor inductance = 0.705 pu
Mechanical time constant = 1.41 s

The induction motor is modelled by its nonlinear equations as a symmetrical machine [46]. The vector transformation blocks are modelled as described earlier, and the VSTC controller is modeled as given in Figure 11.

10.2 SIMULATION RESULTS

Two Study Cases are presented here:

Case 1: Performance of the VSTC when motor parameters vary.

Case 2: Performance of the VSTC when motor parameters and load vary.

The first case represent one of the undesirable features of the induction machines - motor parameters vary with temperature and saturation. Two parameters are considered to be continually changing: the rotor resistance (due to temperature change) and the magnetizing inductance (due to core saturation).

In the second case, the load is varied in addition to the parameter variations described in case 1. The load variation represents a real obstacle in such application as robotic and actuation - load inertia is dependent on the location of the end effector in space; and load torque may widely vary during one single movement.

The rotor resistance of the induction motor is assumed to be exponentially increasing to 10 times of its original value in 10 seconds. The per unit change in the magnetizing inductance (ΔL) due to core saturation is emulated by the following formula:

$$\Delta L = -\frac{0.5}{1 + e^{\frac{150 - V}{50}}}$$

Where V is the terminal voltage. By the above formula, the inductance is reduced in an exponential form to about 36% at 200 V.

Two types of load variations are assumed for case 2:

> *Case 2.1::* sudden reduction of the load torque and load inertia by 50% each after 5 second from starting;

> *Case 2.2:* sinusoidal changing of load torque to simulate a single joint actuator rotating 90° back and forth. The load torque changes from maximum to zero while the motor is in the forward rotation. Then increased from zero to maximum in the reverse rotation.

In all simulations, only motor and load parameters are varied. The rotor resistance and the magnetizing inductance used in the vector transformation blocks remained unchanged. This is done to demonstrate the robustness of the VSTC-based drive system.

Test results of all the above cases are shown in the following figures. The dashed lines represent the tracks and the solid lines represent the motor responses.

Figures 22 to 24 show the results of case 1. Figure 22 shows the rotor position. Figure 24 shows the control signal (electromechanical torque) generated by the VSTC block. Figure 24 shows the hyperplane.

These figures reveal a very good tracking performance of the motor with VSTC. The hyperplane $\sigma(t)$ of Figure 24 show that the switching action of the VSTC resulted in keeping the hyperplane in the vicinity of the sliding regime ($\sigma(t) = 0$). This is a key requirement for a robust controller.

Figures 25 shows the phase portrait of the rotor position and rotor speed, which is another way of displaying the results. This case corresponds to the 50% reduction of load (Case 2.1). Point (A) represents the starting operating point. The motion from (A) to (B) is the forward rotation, and from (B) to (A) is the reverse rotation. Point (B) is the holding operating point at which the load is reduced. As seen in the figure, a very good agreement is achieved between the tracks and the motor responses.

Figure 26 shows the rotor position of Case 2.2 when the load torque is sinusoidally changing.

All the above results show, from the simulation point of view, a very promising tracking capability of the proposed method, even when the motor parameters drift and the load conditions change. Indeed, a laboratory implementation is needed to verify the simulation results.

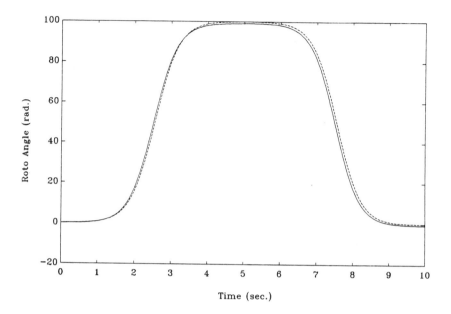

Figure 22: Rotor Position Tracking for Case 1

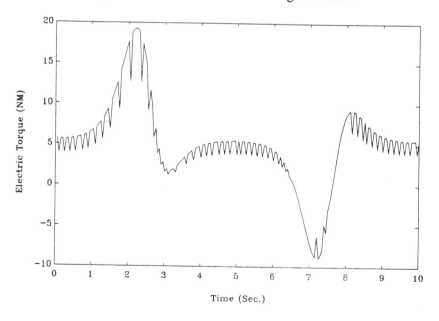

Figure 23: Electromechanical Torque of Case 1

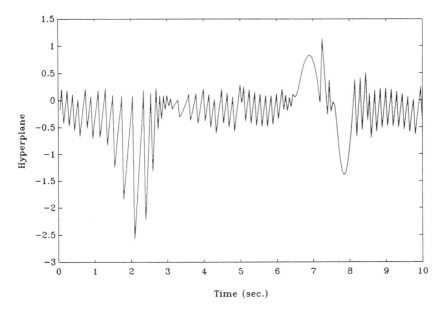

Figure 24: Hyperplane of Case 1

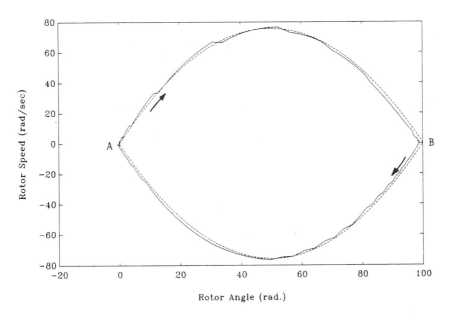

Figure 25: Phase Portrait of Position and Speed of Case 2.1

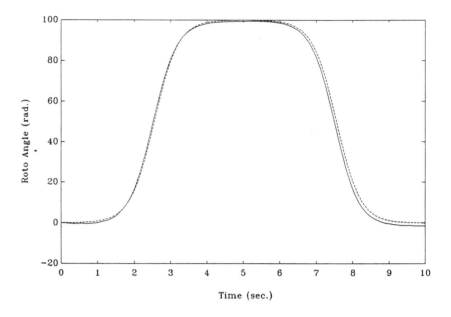

Figure 26: Rotor Position Tracking of Case 2.2

11. APPLICATIONS OF SELF-TUNING CONTROL TO DC MOTORS

A digitized model of dc motor should be developed in order to implement the Self-Tuning adaptive control technique. Under a stochastic measurement environment with uncertainties, such as random disturbances, quantization and truncation errors, an Autoregressive Moving Average Exogeneous (ARMAX) model can be used to represent the system under control [31,32]. A dc machine operating in such conditions (load changes or any other disturbance) can also be represented by the ARMAX model given in Equations(12) and (13):

An accurate ARMAX model of a dc motor depends on the proper choice of the constants n_a, n_b, n_c and k of equations (12) and (13). These constants are usually selected by using trial and error methods which may or may not lead to adequate representation of the system under control. In this section, however, a systematic method will be used to determine the plant orders and the time delay of the dc drive model. It is based on transforming the time domain model of a dc machine into a transfer function in the "z" domain, then comparing it with the transfer function of the ARMAX model of equation (12).

The z-transformation of equation (12) is given in equation (14). The first term in the right hand side of equation (14) corresponds to the transfer function of the system under control. The second term represents external disturbances, noise, etc.

When the equivalent z-transfer function of a separately excited dc motor is derived and compared to equation (14), the order of the polynomials n_a, n_b and k, can be obtained. To do this, the time domain model of the dc motor is expressed as follows:

$$K_m \, i_a(t) = J_m \, (d \, \omega_m(t)/dt) + D_m \, \omega_m(t) + T_L(t) \qquad (29)$$

$$V_t(t) = K_m \, \omega_m(t) + r_a \, i_a(t) + l_a \, (di_a(t)/dt) \qquad (30)$$

where

r_a :Armature resistance

l_a :Armature inductance

V_t :Armature voltage

i_a :Armature current

T_L :Load torque

ω_m :Rotor speed

J_m :Inertia of the system

K_m :Back emf/torque constant

D_m :Damping constant

Let the load torque be represented as a function of the speed:

$$T_L(t) = K_0(t) \, \omega_m(t)$$

where $K_0(t)$ is a time varying constant to be identified.

Transforming equations (29) and (30) to frequency domain (Laplace transformation) gives the following transfer function of dc motor:

$$\frac{\omega_m(s)}{V_t(s)} = \frac{\sigma}{(s + \beta)(s + \theta)} \tag{31}$$

where σ, β and θ are functions of the model parameters including $K_0(t)$. The z-transfer function of (31) with a Zero Order Hold (ZOH) in the forward path can be written in the form [32]:

$$\frac{\omega_m(z)}{V_t(z)} = z^{-1} \frac{b_0 + b_1 z^{-1}}{1 + a_1 z^{-1} + a_2 z^{-2}} \tag{32}$$

Comparing equation (32) to the ARMAX transfer function of equation (14) shows that the dc motor model has the following constants:

$$n_a = 2, \quad n_b = 1 \quad \text{and} \quad k = 1$$

Hence, the ARMAX model of the dc motor is:

$$A(q^{-1}) \, \omega_m(t) = q^{-1} B(q^{-1}) V_t(t) + C(q^{-1}) \, \epsilon(t)$$

where $A(q^{-1}) = 1 + a_1 q^{-1} + a_2 q^{-1}$

$$B(q^{-1}) = b_0 + b_1 q^{-1}$$

The value of n_c, which represents the order of the noise model, is usually selected to provide the best fit of the model response to the actual system. Based on laboratory tests, it is found that $n_c = 1$ is adequate, hence

$$C(q^{-1}) = 1 + c_1 q^{-1}$$

It is important to note that β and θ in (31), which are the poles of the system, are functions of the time varying $K_0(t)$. The change in $K_0(t)$ is transformed into a variation in coefficients a_1, a_2, b_0, b_1 and c_1. These coefficients are identified at each sampling time using a recursive estimation algorithm [31,32,43,44]. This is equivalent to an implicit inclusion of load torque variations in the dc motor model. The same could be said about other parameter variations such as rotor inertia, damping and random disturbances. As long as any variation has an effect on the input/output response of the system, it could be identified and included in the model polynomials.

Figure 27 shows the main functional blocks of the proposed tracking controller system for the dc motor. All blocks inside the dotted area are software blocks. The (IN) block performs the task of fetching the speed readings from a pre-specified memory location. The (OUT) block performs the task of passing the optimal control signal (U_o) generated by the controller to an output port, which is connected to the driving module of the hardware. The (TRACK DATA) block stores the trajectories (tracks) of the motor speed and/or position.

The (ESTIMATION) block identifies the controller parameters by using equations (26) and (27). The (CONTROL) block computes the optimum tracking control signal given in equation (23). The parameter vector and the optimum tracking control signal are updated at each time step.

One of the features of the setup is that the control function can be modified by simply changing the structure of the blocks in the dashed area without any change in the hardware. Hence, several control structures and strategies can be implemented in a short period of time, and with minimum effort.

The laboratory implementation, unlike computer simulations, adds another source for errors. For example, the conversion of the floating point numbers to integers, and vice versa, by the IN or OUT block introduces round off errors.

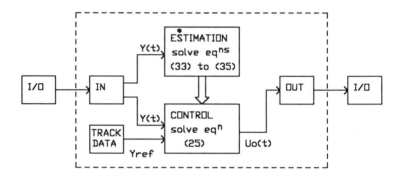

Figure 27: Functional Block Diagram Of the Controller

From [3], courtesy of IEEE, (C) IEEE 1989

11.1 EXPERIMENT RESULTS

Three test cases are presented here. In case 1 and 2, the rotor position track is composed of three segments: the first is a sigmoid to advance the rotor position by 25 revolutions (forward rotation); the second is a reverse sigmoid to return the rotor back to its starting position (reverse rotation); and the third is a constant position track to block the rotor (holding). The time for forward and reverse rotation is 7.5 s each, while the time for holding is 5 s giving a total tracking time of 20 s.

In case 3 the load torque is manually and continually disturbed to simulate an excessive form of disturbance (as high as 100%). Also, to show the versatility of the proposed controller, a different shape of track is selected. In this case, for example, the position track is composed of three segments: the first is a 15 revolution forward rotation; the second is 10 revolution forward rotation; and the third is a 25 revolution in the reverse rotation.

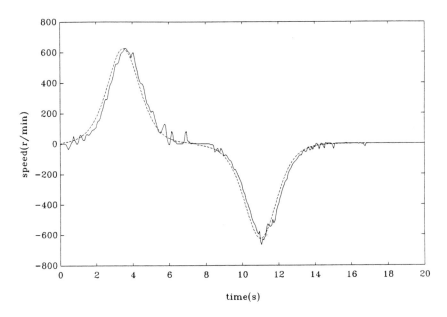

Figure 28: Measured Speed and Reference Track of case 1

From [3], courtesy of IEEE, (C) IEEE 1989

Case 1 and 2 are designed to evaluate the controller performance with respect to: 1) the initial value of the parameter vector $\theta(0)$ of equation (25); and 2) the sampling interval. Same loading profile and disturbance are applied to the electric drive system in both cases. The load is a 4.5 kg weight attached to the rotor by a pulley arrangement. The motor in forward rotation lifts the weight from rest, then returns it back in the reverse rotation, and finally suspends it in air until the end of the track. The disturbance is a 30 V dip in the field voltage for a period of 7 second, starting after 4 second into the test.

In case 1, ϴ(0) is selected as a vector of zero elements. This represent a case where no prior knowledge of the systems parameters exist. Also, the sampling time in this case is 57 ms.

In case 2, ϴ(0) is set equal to the final value of case 1. This means that apriori knowledge on the system parameters is available.

Results of case 1 are shown in Figures 28 to 30. Figures 28 and 29 show comparisons between the actual measurement and the corresponding reference tracks for speed and rotor position respectively. The dotted lines correspond to tracks while the solid lines are the actual measurement.

Figures 30 shows the estimated coefficients of polynomial A. The solid line corresponds to a_1 and the dotted line is for a_2. The coefficient variations due to the disturbance (dip in the excitation voltage) are indicated by arrows.

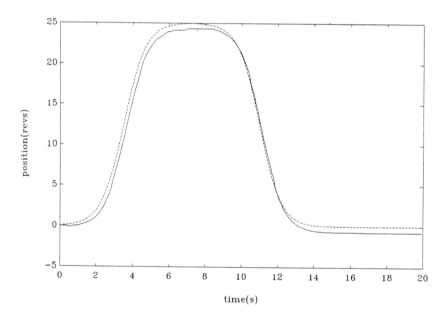

Figure 29: Measured Rotor Position and Reference Track of case 1

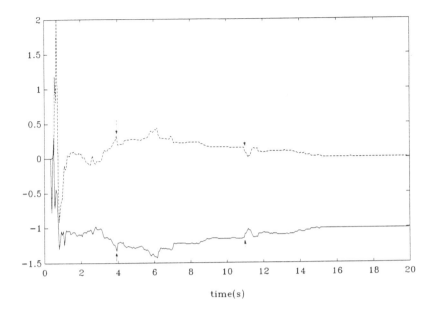

Figure 30: Coefficients of polynomial A

From [3], courtesy of IEEE, (C) IEEE 1989

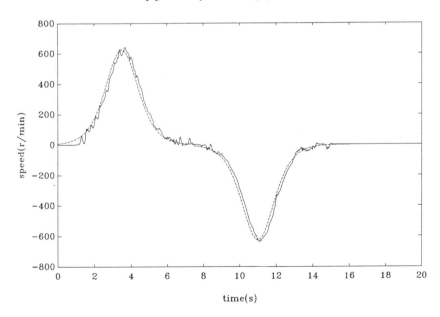

Figure 31: Measured Speed and Reference Track of case 2

From [3], courtesy of IEEE, (C) IEEE 1989

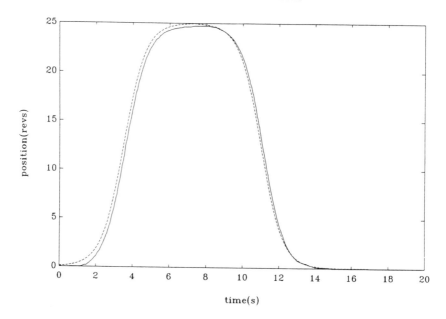

Figure 32: Measured Rotor Position and Reference Track of case 2

From [3], courtesy of IEEE, (C) IEEE 1989

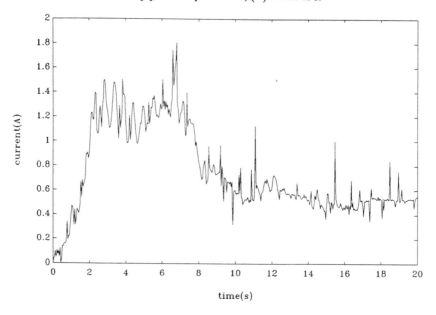

Figure 33: Armature Current of Case 2

From [3], courtesy of IEEE, (C) IEEE 1989

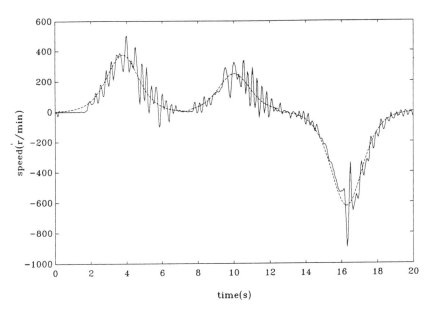

Figure 34: Measured Speed and Reference Track of case 3

From [3], courtesy of IEEE, (C) IEEE 1989

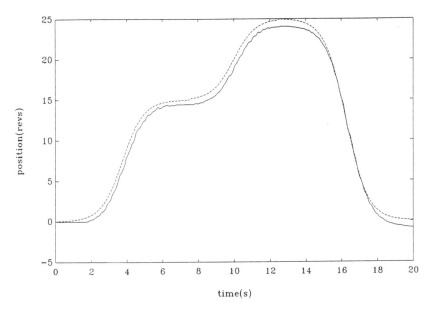

Figure 35: Measured Rotor Position and Reference Track of case 3

From [3], courtesy of IEEE, (C) IEEE 1989

The results of case 2 are shown in Figures 31 to 33. Figures 31 and 32 show the speed and position tracking respectively. If these figures are compared to Figures 28 and 29, definite improvements in the tracking accuracy can be detected for case 2. Figure 33 shows the armature current.

The speed and position tracking of the system for case 3 are shown in Figures 34. and 35 respectively. The figures show excellent tracking properties of the motor with the VSTC controller, even under excessive load variations

12. APPLICATIONS OF SELF-TUNING CONTROL TO INDUCTION MOTORS

A generalized Minimum Variance Control strategy is used in an implicit Self-Tuning Multivariable Adaptive control algorithm. As seen earlier, the generalized Minimum Variance strategy allows the minimization of the tracking error as well as the control signal itself. This prevents the control signal from overshooting which may lead to excessive mechanical and electrical stresses on the motor.

If the induction motor is treated as a single-input system, a vector control block may be needed. In this case the application of the Self-Tuning control would be similar to that presented in the previous section. Here, the induction motor is controlled without the "vector control" blocks. Hence, the discrete model of the motor has two inputs: the stator current and the electrical frequency. To implement this Self-Tuning control, a multivariable approach can be used.

The model of the induction motor without the vector transformation method can be written as

$$dx(t)/dt = A_c(x,u,t)x(t) + B_c u(t) \qquad (33)$$

$$y(t) = C_c x(t)$$

where

$$x = [\; \psi_{qr} \;\; \psi_{dr} \;\; \omega_r \;]^T$$

$$u = [\; i_{qs} \;\; i_{ds} \;\; T_L \;]^T$$

$$A_c = \begin{pmatrix} \dfrac{-\omega_b R_r}{X_{rr}} & -(\omega-\omega_r) & 0 \\[4mm] (\omega-\omega_r) & \dfrac{-\omega_b R_r}{X_{rr}} & 0 \\[4mm] \dfrac{-3P^2 X_M}{8\,\omega_b J\, X_{rr}} & \dfrac{3P^2 X_M}{8\,\omega_b J\, X_{rr}} & \dfrac{D}{J} \end{pmatrix} \;;$$

$$B_c = \begin{pmatrix} \dfrac{-\omega_b R_r X_M}{X_{rr}} & -0 & 0 \\[4mm] 0 & \dfrac{-\omega_b R_r X_M}{X_{rr}} & 0 \\[4mm] 0 & 0 & \dfrac{-P}{2J} \end{pmatrix}$$

$$C_c = [\; 0 \;\; 0 \;\; 1\;]$$

ψ_{qdr}: rotor flux linkage

i_{qds} : stator current

ω_b : base electrical frequency (rad/s)

ω : stator electrical frequency

ω_r : rotor speed

T_L : load torque

J : rotor inertia

R_r : rotor resistance per phase referred to stator

X_{rr} : rotor self reactance per phase referred to stator

X_M : mutual reactance per phase

D   :   Damping coefficient

P   :   number of poles

u in this model include the load torque (T_L) and the control vector containing (i_{qs}, i_{ds}).

The discrete model parameters (n_a, n_b, n_c and the time delay k) can be obtained by comparing the z-transform of the linearized version of the dynamic model given in (33) to the discrete model of equation (12). The parameters of the polynomials A, B and C of equation (12) are identified and updated on-line to compensate for the nonlinearity in the induction motor and for any changes in the loading condition or disturbances.

The performance index under the Generalized Minimum Variance Control is similar to that given in equation (15) with the following modifications:

$$J = E\{\alpha \, [y(t+1) - y_{ref}(t+1)]^2 + \eta \, [p(t+1) - p_{ref}(t+1)]^2$$
$$+ \, ||\mu Ru(t)||^2\}$$

where y(t) and $y_{ref}(t)$ are the rotor speed and the speed track respectively. p(t) and $p_{ref}(t)$ are the rotor position and the reference position respectively. The weighting parameters α and η are used to put more emphasis on either speed or position tracking. The control signals are also included in the cost index to limit their magnitude to avert unnecessary stresses. This also provides an avenue to introduce an integrator in the control signal path to remove steady state errors. μ is used as a weighting factor for the control signals.

12.1 SIMULATION RESULTS

A 4-pole, 3 hp 220 V, 5.8 A, 60 Hz induction machine is used in the following simulations. The load of the machine is assumed to be a fan-type. The machine has the following ratings:

R_r = 0.435 Ω

X_{rr} = 26.884 Ω

X_M = 26.130 Ω

J = 0.089 kg m^2

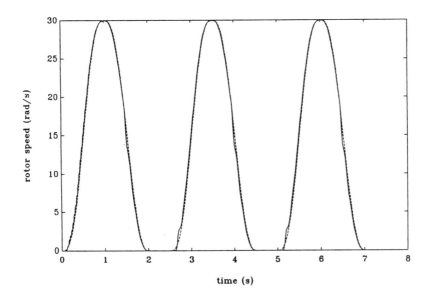

Figure 36: Speed tracking of Induction Motor Using Self-Tuning Control

Figures 36 and 37 show samples of the results. In this case, the initial values of the model parameters are set to zero. This means that the controller is operating without any a-priori knowledge of the system parameters and is learning during starting. In these cases the initial covariance matrix has to be set to large values to express the lack of confidence in the present model parameter [2,3] and therefore to support a fast adaptation.

In this test case, a fixed forgetting factor of 0.95 (equation (28)) is used during the parameter estimation.

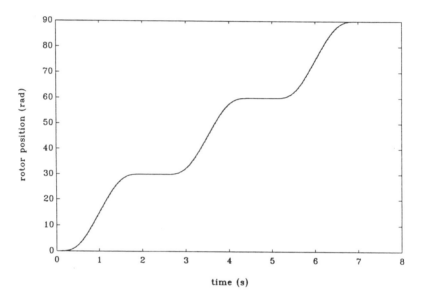

Figure 37: Position tracking of Induction Motor Using Self-Tuning Control

REFERENCES

[1]   M.A. El-Sharkawi and C. H. Huang, "Variable Structure Tracking of dc Motor for High Performance Applications," IEEE Transactions on Energy Conversion, pp. 643-650, December 1989.

[2]   S. Weerasooriya and M.A. El-Sharkawi, "Adaptive Tracking Control for High Performance dc Drives," IEEE Transactions on Energy Conversion, pp. 502-508, Sept. 1989.

[3]   M. A. El-Sharkawi and S. Weerasooriya, "Development and Implementation of Self-Tuning Tracking Controller for DC Motors," IEEE Transactions on Energy Conversion, pp. 122 - 128, March 1990.

[4]   N. A. Demerdash and T. W. Nehl, "Closed Loop Performance of a Brushless dc Motor Powered Electromechanical Actuator for Flight Control Applications," Proceedings of the IEEE National Aerospace and Electronics Conference, NAECON 80, pp. 759-767, May 1980.

[5]   M. A. El-Sharkawi, R. A. Sigelmann, Fang Shi and I. S. Mehdi, "Design and Evaluation of Electronically Commuted dc Motors," Proceedings of the IEEE National Aerospace and Electronics Conference, NAECON 84, pp. 619-625, May 1984.

[6]   M. A. El-Sharkawi, J. S. Coleman, I. S. Mehdi and D. L. Sommer "Microcomputer Control of an Electronically Commutated DC Motor," Proceedings of the IEEE National Aerospace and Electronics Conference, NAECON 86, pp. 320-325, May 1986.

[7]   N. A. Demerdash and T. W. Nehl, "Dynamic Modeling of Brushless dc Motors for Aerospace Actuation," IEEE Transactions on Aerospace and Electronic Systems, Vol. AES-16, No. 6, pp. 811-821, 1980.

[8]   Paul C. Krause and Oleg Wasynczuk, Electromechanical Motion Devices, McGraw-Hill, 1989.

[9]   C. H. Huang, M. A. El-Sharkawi and M. Chen, "Laboratory Setup for Instruction and Research in Electric Drives Control," IEEE Transactions on Power Systems, pp. 331-337, February 1990.

[10] M.A. El-Sharkawi and M. Akherraz, "Tracking Control Technique for Induction Motors," IEEE Transactions on Energy Conversion, pp. 81-87, March 1989.

[11] W. Leonard, "Microcomputer Control of High Dynamic Performance AC-Drives - A Survey," Automatica, vol.22, no.1, 1986, pp 1-19.

[12] T. A. Lipo, "Recent Progress in the Development of Solid-State AC Motor Drives," IEEE Transactions on Power Electronics, April, 1988, pp. 105-117.

[13] J.W. Brown and R.F. Dugan, "A Comparison Study of Induction Motor vs. Synchronous Motor Servo Systems," Power Conversion & Intelligent Motion, April 1987, pp.43-49.

[14] B. K. Bose, "Adjustable Speed AC Drives - A Technological Status Review" IEEE Proceedings, Feb.1982, pp.116-135.

[15] T. Matsuo and T. A. Lipo, "A Rotor Parameter Identification Scheme for Vector-controlled Induction Motor Drives," IEEE Transactions on Industry Applications, May/June 1985, pp. 624-632.

[16] R. Krishnan and F.C. Doran, "Study of Parameter Sensitivity in High Performance Inverter-Fed Induction Motor Drive Systems," Proc. of the IEEE IAS Conf. pp. 510-524, 1984.

[17] Akira Nabae, Kenichi Otsuka, Hiroshi Uchino and Ryoichi Kurosawa, "An Approach to Flux Control of Induction Motors Operated With Variable-Frequency Power Supply," IEEE Transactions on Industry Applications, vol. IA-16, No. 3, pp. 342-349, May/June 1980.

[18]   Edward Y. Y. Ho and Paresh C. Sen, "Decoupling Control of Induction Motor Drives," IEEE Trans. on Industrial Electronics, vol. 35, No. 2, pp. 253-262, 1988.

[19]   Ruppecht Gabriel, Werner Leonhard and Craiga J. Nordby, "Field-Oriented Control of a Standard AC Motor Using Microprocessors," IEEE Trans. on Industry Applications, vol. IA-16, No. 2, March/April, pp. 186-192, 1980.

[20]   R. D. Lorenz, M. O. Lucas and D. B. Lawson, "Synthesis of a State Variable Motion Controller For High Performance Field Oriented Induction Motor Drives," IEEE IAS Conf. Rec. pp. 80-85, 1986.

[21]   Satoshi Ogasawara, Hirofumi Akagi and Akira Nabae, "The Generalized Theory of Indirect Vector Control for AC Machines," IEEE Transactions on Industry Applications, vol. 24, No. 3, pp. 470-478, May/June, 1988.

[22]   L. J. Garces, "Parameter Adaptation for the Speed-Controlled Static AC Drive with a Squirrel-Cage Induction Motor," IEEE Transactions on Industry Applications, vol. IA-16, No. 2, March/April 1980.

[23]   D.P.Connors, D.A. Jarc and R.H. Daugherty, "Considerations in Applying Induction Motors With Solid-State Adjustable Frequency Controllers," IEEE Trans. on Industry Applications, Jan./Feb. 1984.

[24]   B. Courtiol and I.D. Landau "High speed adaptation systems for controlled electric drives," Automatica vol 11. pp 119-127, 1975.

[25]   H. Naitoh and S. Tadakuma "Microprocessor based adjustable speed dc motor drives using model reference adaptive control,"

IEEE Transactions on Industry Applications, pp 313-318, Mar/Apr, 1987.

[26] A Brickwedde, "Microprocessor based Adaptive Speed and Position Control for Electric drives," IEEE Transactions on Industry Applications, pp. 1154-1161, Sept/Oct, 1985.

[27] D.W. Clark and P.J. Gawthrop "Implementation and Application of Microprocessor-Based Self-Tuners", Automatica, Vol 17 No1 p 233-244,1981

[28] D.W. Clark and P.J. Gawthrop "Self-Tuning Control," Proceedings of the IEE 126 pp 633-640, 1979.

[29] O.P. Malik and G.S. Hope "Power system stabilizer based on adaptive control techniques," IEEE Transactions on Power Apparatus and Systems, pp. 1983-1988, August, 1984.

[30] P.J. Gawthrop, "Some Interpretation of the Self-Tuning Controller," Proceedings of the IEE, pp. 929-934, 1975.

[31] V.V. Chalam "Adaptive control systems; techniques and applications," Marcel Dekker Inc., 1987.

[32] C.J. Harris and S.A. Billings, " Self-Tuning and Adaptive Control: Theory and Applications," Peter Peregrinus ltd., 1985.

[33] M. A. El-Sharkawi, "Development and Implementation of High Performance Variable Structure Tracking Control for Brushless Motors," IEEE Transactions on Energy Conversion, in print.

[34] S.K. Korovin and V.I. Utkin, "Use of the Slip Mode in Problems of Static Optimization," Automatic and Remote Control, No. 4, 1972, pp. 570-579.

[35] V.I. Utkin, "Variable Structure Systems with Sliding Modes," IEEE Transactions on Automatic Control, April 1977, pp. 212-222.

[36] V.I. Utkin, "Sliding Modes and Their Application in Variable Structure Systems," MIR Publishers, Moscow, 1978.

[37] U. Itkis, "Control Systems of Variable Structure," John Wiley and Sons, 1976.

[38] K.D. Young, "Controller Design for a Manipulator Using Theory of Variable Structure Systems," IEEE Transactions on Systems, Man, and Cybernetics, February 1978, pp.101-109.

[39] B. Drazenovic, "The Invariance Conditions in Variable Structure Systems," Automatica, Vol. 5, 1969, pp. 287-295.

[40] D.B. Izosimov, V.I. Utkin, "Sliding Mode Control of Electric Motors," Proc. of the IFAC Control Science and Technology Conf., 1981, pp.2059-2066.

[41] S. Lin and S. Tsai, "A Microprocessor-Based Incremental Servo System With Variable Structure," IEEE Transactions on Industrial Electronics, Nov. 1984, pp. 313-316.

[42] F. Harashima, H. Hashimoto and S. Kondo, "MOSFET Converter-Fed Position Servo System With Sliding Mode Control," IEEE Trans. on Industrial Electronics, August 1985, pp. 238-244.

[43] The Math Works Inc. "PC - MATLAB" Version 3.2-PC," June 8, 1987.

[44] L. Ljung and T Soderstrom "Theory and Practice of Recursive Identification," MIT Press., 1985.

[45]   Edward Y. Y. Ho and Paresh C. Cen, "Digital Simulation of
       PWM Induction Motor Drives for Transient and Steady-State
       Performance," IEEE Transactions on Industrial Electronics,
       vol. IE-33, No. 1, pp. 68-77, February 1986.

[46]   P. C. Krause, C.H. Thomas, "Simulation of Symmetrical
       Induction Machinery," IEEE Transactions On Power
       Apparatus and Systems, vol. 84, No. 11. November 1965.

HIGH VOLTAGE OUTDOOR INSULATION TECHNOLOGY

R. S. GORUR
DEPARTMENT OF ELECTRICAL ENGINEERING
ARIZONA STATE UNIVERSITY
TEMPE, ARIZONA 85287-5706

SUMMARY

A review of high voltage insulation technology used for outdoor power transmission and distribution is presented. Emphasis is placed on the state-of-the art polymer materials which are finding increasing use in today's highly competitive and advanced power industry. The chapter begins with a review of the problems with porcelain and glass which have been used for many decades for outdoor insulation applications, such as, insulators, cable terminations and surge arresters. It then addresses how polymers has alleviated some of the problems, while creating a few of their own. The typical constructional features of polymer insulating devices are presented.

Next, the service experience of utilities, both worldwide and within the USA, with polymeric insulators is presented. Their performance in outdoor test stations, which is valuable for advancing the technology, is discussed. Laboratory tests which are required to obtain a better understanding of the service performance in a relatively short time, are reviewed. Popularly used accelerated tests, their advantages and shortcomings, and factors that must be considered in the development of a meaningful laboratory test which bears relevance to service experience are illustrated.

Polymers, being organic, age with time in service and it is extremely important to understand the mechanisms that are responsible for aging. This will enable in the estimation of the useful life and in the development of better materials. The current research efforts and significant findings have been reviewed. Topics which need further research are proposed.

1. INTRODUCTION

The present high technological age is to a great part due to the high degree of reliability achieved in the transmission and distribution of electric power. The consequences of power interruption in large cities, processing plants, mines, hospitals, computers, etc., is indeed frightening. All the power that is generated at the generating stations which are usually far removed from the center of civilization, to the load centers is by means of outdoor transmission lines. The reason for this is simply economical. However, the distribution of power from local substations for industries, commerce and residential use is done by both outdoor distribution lines and underground cables.

Outdoor transmission and distribution lines need to be insulated from their supporting structures for reasons of safety and to avoid short circuiting of the power source. Therefore, the reliability of electric power depends to a large extent on the insulation used with outdoor lines, which is referred to as outdoor insulation.

Porcelain has been one of the oldest, and still the most widely used material for outdoor insulation. Another material which has been used for a long time and still finds widespread use for outdoor insulation is glass. Both porcelain and glass have proven to be very reliable and cost effective materials. However, during the last twenty five years, alternative materials for porcelain and glass for outdoor insulation have emerged. The new materials are polymers, and presently they are used very widely for outdoor insulation applications. Polymers materials used for outdoor insulation have been referred to by various names, such as, polymeric, composite, non-ceramic, plastic and synthetic insulation.

Polymers are presently being used worldwide for transmission and distribution suspension insulators, line posts, terminations for underground residential distribution (URD) cables, surge arrester housings, bushings, pin type insulators, etc. In the USA, they presently account for over 30% of the total outdoor insulator market. Their share in the outdoor cable terminations for distribution is even greater (greater than 50%) than for insulators. Considering that the technology is relatively recent, it is clear that polymer based outdoor insulation has found a firm footing and is rapidly expanding.

To understand the factors that have promoted the wide spread use of polymers, it is necessary to examine some of the problems experienced with porcelain and glass, namely:

1) Porcelain and glass are brittle materials. As the insulators made from these materials shatter into pieces upon impact of a gun shot, they serve as very attractive targets for vandals. Gun shot damage is more dramatic in case of glass insulators which release their locked in stress and thus explode when struck by a bullet. Vandalism is a very serious problem experienced by many power utilities in the USA. It has been found that it is more economical to use alternatives which need not be as frequently replaced as porcelain and glass, even though the long term electrical performance of the alternative materials may not be as good.

2) Porcelain and glass are very dense materials, and thus heavy. Supporting structures must be of large dimensions to withstand the weight of the

insulators. In some urban areas, due to limitations on the right of way, this poses a serious problems for utilities. Alternative materials which are light and that can be supported by smaller towers are attractive for such applications, provided the electrical and mechanical performance is comparable to that of porcelain and glass.

3) Porcelain and glass are high surface energy materials. This simply means that the energy exerted by the surface on a drop of water is so large that it causes the water droplet to spread out into a thin film. In the presence of contamination on the surface of the insulator, water film formation leads a redistribution of the surface electric field and lowering of surface resistance. The effect of the former is to promote corona which results in radio and television interference, ozone production, where as the effect of the latter is to promote leakage current leading to dry band arcing which under certain conditions culminates in a surface flashover and hence a power outage. Such flashovers, termed as contamination flashover, usually occur on insulators located in contaminated environments, in the presence of moisture. Counter measures such as periodic hot line washing and greasing have been adopted by many utilities to reduce contamination flashover, but these are very expensive and labor intensive.

There are many sources of outdoor contamination, for example, sea salt, road salt, industrial emissions, cement dust, fly ash, automobile emissions, etc, and these contaminants get deposited on the surface of the insulators either by wind or electric forces. Moisture can be present in the environment in the form of rain, dew or fog. For voltages up to 230 kV, majority of the power outages are caused by lightning and contamination. For voltages above 230 kV, the predominant cause of power outages is contamination. Power outages are expensive and therefore should be kept to a minimum or avoided. Therefore, in contaminated environments, there is the need for alternative materials for porcelain and glass, which have a superior performance under contaminated conditions, and also have long term electrical and mechanical properties comparable to porcelain and glass.

4) The increasing use of dc for high voltage transmission links has increased the contamination flashover problem. Unlike with ac, where deposition of contamination on the insulator surface by electrostatic attraction is not significant due to the alternating polarity of the wave form, with dc, surface contamination is much greater. The leakage distance required to prevent contamination flashover must be greater for dc than with ac. A simple increase in the leakage distance with porcelain and glass may not always solve the problem. Therefore, there is a need for alternative materials to porcelain and glass which resist water film formation despite being contaminated, thus maintaining a high surface resistance which reduces the possibility of surface flashover.

Thus, it is obvious that there are several disadvantages with porcelain and glass which necessitated the search for alternative materials. Let us examine how polymers have solved the above mentioned problems.

Vandalism: Polymers which are used for outdoor insulation are predominantly elastomeric materials. Due to the non-brittle and flexible nature of these materials, they are able to absorb the energy from the projectile without shattering. Bullets usually lodge themselves into the material. This discourages vandals from further shooting. Polymer insulators shot by bullets have to be replaced eventually but not immediately.

Right of Way: Polymers are organic materials and are considerably less dense than either porcelain or glass. Polymer insulators offer almost a ten fold reduction in the weight when compared to porcelain and glass for the same voltage. Therefore supporting structures such as poles and towers can be of reduced dimensions, which can be accommodated in a limited right of way passage. The use of polymer line post insulators which eliminate the use of cross arms for the towers has become a very popular approach of further reducing the right of way requirement. Polymer line posts have as good a cantilever strength as porcelain line posts, and in addition are not as easily damaged as porcelain by shock loading caused by accidents such as impact of vehicles on poles.

Contamination: Polymers, being organic, are low surface energy materials in the virgin state, i.e., prior to exposure to electric stress. In contrast to porcelain and glass, water droplets on the surface of polymers tend to remain as discrete droplets thereby maintaining a high surface resistance even under wet conditions. The property of the polymer surface to bead up water droplets is called surface hydrophobicity, and plays a very important role in outdoor insulation, as it reduces the possibility of surface flashover and subsequent power outages.

High Voltage DC Insulation: Some types of polymers, especially silicone rubber has the unique property of retaining its hydrophobic surface even when the surface is covered with contaminants. This property has been fully exploited in the Pacific Intertie DC line between Los Angeles, California and Portland, Oregon, where the first few miles of the line is in the heavily polluted Los Angeles area. Porcelain insulators were originally used but had to be washed every 60 days to prevent contamination flashover. In 1974, the utility tried silicone rubber insulators on a trial basis have since reported a very successful experience.

While solving many of the problems commonly experienced with porcelain and glass, the use of polymers have created new problems which were not experienced with porcelain and glass. The biggest problem is material aging, which results from a loss of useful insulating properties with time. The organic nature of the polymer makes it susceptible to many elements experienced in service, such as, mechanical loading, electric stress, contamination, Ultra-Violet (UV) radiation from sunlight, automobile emissions, moisture, etc. The problem is even more complicated as most of the above factors exist simultaneously and have a synergistic effect on the aging. The resultant effect of aging can be gradual loss of mechanical strength, material degradation in the form of tracking and erosion of the weathershed, and flashover leading to power outages. Fortunately, both material and manufacturing processes have been developing at a good pace and have played an important role in the success of this new insulation technology.

2. POLYMER MATERIALS FOR OUTDOOR INSULATION

Many materials have been tried for outdoor insulation applications. These include silicone rubber, ethylene propylene rubber (EPR), epoxy, teflon, polyethylene, ethyl vinyl acetate (EVA), modified polyolefins, etc. Silicone rubber is used in two forms: as a room temperature vulcanized (RTV) sprayable coating on porcelain insulators and as a high temperature vulcanized (HTV) elastomer for weathersheds on outdoor insulation. EPR is the generic name for two types of material, a copolymer of Ethylene and Propylene Monomers known as EPM, and a terpolymer of Ethylene Propylene and Diene Monomers known as EPDM.

A majority of today's polymer insulators use either HTV silicone rubber or EPR as weathershed material. HTV Silicone rubber, EPR, EVA and modified polyolefins are used in today's polymeric cable terminations. HTV Silicone rubber, EPR and EVA are used presently for surge arrester housings.

Polymers, being organic have relatively a lower thermal stability when compared with porcelain and glass. Typically most polymers decompose at temperatures around $250^{\circ}C$. In addition, the thermal conductivity of polymers is generally lower than inorganic materials. Dry band arcing under wet and contaminated conditions can result in localized surface temperatures sufficient to cause degradation of the polymer. Polymer degradation occurs in the form of tracking or erosion. Tracking is defined as the formation of carbon deposits on the surface and erosion simply means loss of material.

In order to limit the surface temperature to a safe value, the heat generated during dry band arcing has to be quickly dissipated to the surroundings. To achieve this objective, inorganic fillers are usually added to the polymer. Two types of fillers are in use for increasing tracking and erosion resistance, namely, alumina trihydrate (ATH) and silica, of which ATH is more widely used.

Fillers are also used for economic and processing reasons. For example, the polymer by itself is not stiff enough to be molded or extruded in the final form. Therefore to aid in the processing, the base polymer is mixed with very fine silica powder (fumed silica) to about 30% by weight of the base polymer. The fumed silica is known as the reinforcing filler. Economically, the inorganic ATH filler is cheaper than the polymer. Therefore, there is an incentive to add filler to the polymer. However, there is a threshold level of ATH filler above which the material becomes so stiff that it cannot be molded or extruded easily. It is generally accepted that a filler level of 50% by weight of the polymer is the optimum in terms of cost effectiveness and ease in processing.

Other ingredients are added to the base polymer, for instance, UV stabilizers, plasticizers and oils as processing aids, coloring agents, etc.. Thus many variations are available commercially, even though the base polymer may be the same. This results in a wide difference in the electrical and mechanical properties of the end product.

3. CONSTRUCTIONAL DETAILS

(A) OUTDOOR HIGH VOLTAGE INSULATORS

Polymer insulators are also referred to as composite insulators owing to the nature of construction. They consist of a central fiber glass rod, weathersheds and metal end fittings. A typical schematic of the present day polymer insulator is shown in Fig. 3.1.

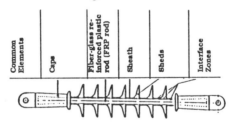

Fig. 3.1: Constructional Details of a Typical Polymeric Insulator

The fiber glass rod consists of axially aligned glass fibers (about 70-75% by weight of the rod) bonded together by means of an organic resin. The organic resin can be either polyester or epoxy and the rod can be either cast or pultruded. The latest technology is to use E-type (electrical grade) glass fibers. The fiber glass rod has to be protected from moisture, UV from sunlight, contamination, as the organic resin has very poor resistance to tracking. The weathershed provides this protection.

In today's polymer insulators, the weathershed is predominantly made up of either silicone rubber or EPR. The weathershed can be of one piece or of modular construction where the individual pieces or skirts of the weathershed are either vulcanized to each other or simply slipped on the core and the joints are filled with silicone grease. The one piece weathershed construction is obviously the best as it precludes voids that can occur when bonding or gluing individual skirts. Good bonding between the weathershed to the rod is very important for satisfactory operation in service, and is obtained by fully vulcanizing the weathershed to the core. Poor bonding gives rise to partial discharges in cavities which can lead to failure of the insulator.

The metal end fittings are cast either from aluminum or malleable iron. The attachment of the end fittings to the fiber glass rod is very important. They are either crimped, wedged or glued to the fiber glass rod. Of these techniques, the crimped or compression end fittings are superior than the rest.

(B) CABLE TERMINATIONS

Polymer cable terminations in use today can be classified into the following categories: cold shrink, heat shrink, slip-on, and roll-on. In the cold shrink type, the termination is maintained in an expanded state by means of plastic former which is then pulled out to shrink the polymer termination to the prepared cable. In the heat shrink type, the polymer is maintained in an expanded form by high energy electron radiation and on application of heat to the surface, the material regains in original size and shrinks on to the prepared cable. In the slip-on type, there are two varieties in use: one piece and modular construction, and the termination is just slipped on to the prepared cable. The roll-on termination is the latest entry in the termination market. The material is self molding and just hand pressure during installation attaches the termination on to the prepared cable.

It is clear from the above description that polymer cable terminations are very easy to install when compared to the more labor intensive installation of porcelain terminations. In fact, to satisfy the ease of application criteria, certain types of termination material need to have less inorganic filler than outdoor insulators.

Certain features are common to all types of cable terminations. These include a stress relief fitting which reduces the electric stress at the point of discontinuity of the semiconducting ground shield, and a weathershed to protect the XLPE (cross linked polyethylene) cable insulation. The stress relief fitting is a rubber of a much higher dielectric constant than the weathershed material. In some terminations a circular tube which is located internal to the weathershed is used, where in others, a short cone below the weathershed is used for stress relief. A typical polymer cable termination is shown in Fig. 3.2.

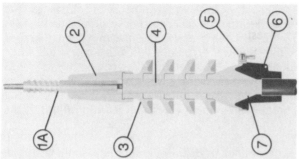

1A. Metal Connectors

1B. Cable Conductor

2. Molded Rubber Cap

3. Non-Tracking Rubber Modules

4. Cable Insulation

5. Ground Clamp

6. Ground Eye

7. Molded Stress Relief

Fig. 3.2: Constructional Details of a Typical Polymeric Cable Termination

(C) SURGE ARRESTERS

A typical polymer housed surge arrester is shown in Fig. 3.3, along with the conventional porcelain arrester. One unique feature distinguishing it from the porcelain arrester is the insulated mounting bracket. Typically, a porcelain arrester is mounted using a metal belly-band mounting bracket. The dry arcing distance (dimension A) between the arrester top end terminal and the metal bracket determines the insulation withstand capability of the arrester housing. There is also a designed-in clearance between the arrester ground terminal and metal bracket (dimension B). This clearance is designed to withstand the line-to-ground voltage that would develop should the arrester fails and remain intact following ground lead disconnector operation. If the support bracket is grounded, the ground clearance (dimension B) is sized to withstand system voltage without flashing over and locking out the system. Both of the design features contribute to the total length of the arrester.

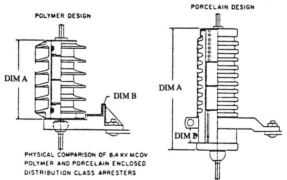

Fig. 3.3: Constructional Details of Polymeric and Porcelain Housed Surge Arresters.

The polymer housed arrester utilizes a mounting bracket attached to the ground end. The base mounting bracket is molded from an insulating material. Due to this, the ground end clearance (dimension B), instead of being designed into the arrester housing, is designed into the bracket contour, as shown. The housing insulation withstand strength is determined by the dry arcing distance (dimension A) of the entire polymer housing. Should the arrester fail and remain intact, the bracket contour is designed to withstand the power frequency voltage and prevent lockout until the arrester assembly can be replaced. Due to the insulated mounting bracket, polymer housed arrester height can be reduced to 50% of the equivalent porcelain design.

A half section view of a polymer housed surge arrester is shown in Fig. 3.4. Adjacent to each end of the disc column are a contact disc and Belleville spring washer which are compressed by applying an axial load of several hundred pounds to the end terminals. The Metal Oxide Varistor (MOV) assembly is maintained under compression by an epoxy coated fiberglass wrap which is applied to the disc assembly while the ends are under axial load. The epoxy coated assembly oven cured, yielding a completed disc module ready for assembly within the protective polymer housing. Stainless steel end plate assemblies are used on each end to compress the flexible rubber housing around the disc module and complete the arrester assembly.

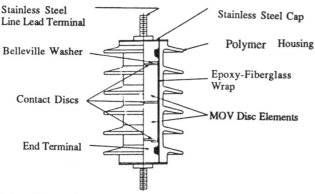

Fig. 3.4: Cross Sectional Details of a Polymer Housed Surge Arrester

The compactness of the internal construction minimizes the free gas volume in the arrester. For example, in the arrester used for 15kV class, the internal free-gas volume in the polymer housed arrester is 0.05 in^3, where as for the same rating in a porcelain housed arrester, the free gas volume is in the range 10-15 in^3. Thus polymer housed arresters are much less vulnerable to moisture ingress problems due to seal pumping action than are porcelain-enclosed units. For higher voltage ratings, the free-gas volume differences between polymer and porcelain enclosed designs is even more pronounced.

4. SERVICE EXPERIENCE

(A) INSULATORS

The first polymeric insulators were used on outdoor lines in the mid 1960's. Since then there have been many changes in the material composition and design of insulators, with the result that the actual service experience gained on a particular type of insulator is much reduced. In 1989, the CIGRE study committee 22, sub-working group 03-01, presented a report on the world wide experience with high voltage composite insulators. The salient findings are presented below:

Worldwide Service Experience

Composite insulators are being used in significant quantity in Australia, Canada, Europe, Latin America, South Africa, and the USA.

Criteria for Choosing Composite Insulators: Several reasons have been cited for choosing composite insulators over porcelain and glass. These include (in the order of importance):

Less attractive to vandalism,
Ease of handling,
Performance under pollution,
High strength to weight ratio,
Visual impact,
Lower cost, and
other reasons, such as, good behavior under conductor galloping or impact loads, line compaction, temporary structures, reduced radio and TV interference and ease of maintenance.

Number of Composite Insulators Used: Composite Insulators are used as suspension insulators, tension insulators and line post insulators. According to the survey which included data until 1986, the number of insulators used in each category worldwide is about 82,000: 19,000: and 36,000 respectively. The maximum use of composite insulators in each category is in the USA. Of the total number of composite insulators, about half of the suspension and tension insulators are for voltage levels below 200 kV, and other half is for voltage levels above 200 kV. For line post insulators, almost 80% of the total is for voltage levels below 200 kV.

Special Outdoor Locations: Composite insulators are used in special outdoor locations worldwide, such as, high altitude (>1000m), high humidity (>85%), moderate and high levels of UV exposure, severe levels of marine and industrial pollution.

Composite Insulator Failures: Table 4.1 shows the different locations of failure in composite insulators. Based on the total number of the different types of insulators listed above, the percent failure is less than 1% for the suspension insulator and much less for tension and line post insulators. It is very clear that the failure rate is extremely small, which is indicative of the high success rate of composite insulators. However, among the different types of failure, weathershed failure rate is the highest.

Table 4.1: Details of Composite Insulator Failures.

TYPE OF FAILURE	SUSPENSION	TENSION	LINE POST
Weathershed	482	196	26
Weathershed-rod interface	203	55	3
FRP Core	13	31	0
Metal Part-Rod interface (slip-out)	2	1	0
Metal part of end fitting	1	0	0

USA Service Experience

As a majority of the composite insulators that are in use (greater than 60%) presently is in the USA, a separate survey of the US utilities experience was performed by the Electric Power Research Institute (EPRI). The salient features of the survey has been included below:

Criteria for Choosing Composite Insulators: The reasons are similar to the world wide survey.

Use of Insulators at Different Voltages: Fig. 4.1 shows the distribution from which it is evident that the greatest use is for 138 kV voltage class. In addition, significant use occurs in voltage levels between 69 and 230 kV.

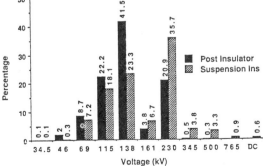

Fig. 4.1: Use of Composite Insulators at Different Voltages

Application in Contaminated Areas: Fig. 4.2 shows the distribution of composite insulators in various levels of contamination. It can noted that a majority of the usage is in locations with very light contamination, and the usage decreases with an increase in the contamination severity.

Composite Insulator Failure: The overall failure rate is less than 0.5%. Fig. 4.3 shows the different causes of failure and their distribution. It is

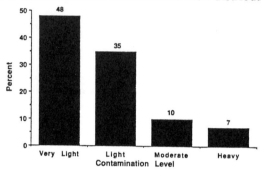

Fig. 4.2: Use of Composite Insulators in Contaminated Areas

evident that as with the world wide experience, the greatest failure rate is due to the deterioration of the weathershed.

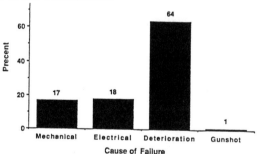

Fig. 4.3: Failure Details of Composite Insulators

Weathershed failures are in the form of corona cutting, punctures, chalking and crazing due to UV radiation from sunlight, tracking and erosion, all of which are a result of aging of the material. This suggests that the greatest improvement in the performance of composite insulators can be made through a better understanding of the aging process.

(B) CABLE TERMINATIONS

Although the use of polymeric cable termination is quite significant, there has been no publication which summarizes the service performance. However, based on the information provided by the utilities at the Working Group of Polymeric Cable Accessories of the IEEE Insulated Conductors Committee, it can be summarized that the overall experience has been very good. Very few failures have been reported that can be linked to the design or construction of the cable termination. Polymeric terminations have been in service for more than 15 years, and have been tested in outdoor test sites for over 20 years.

Unlike outdoor insulators, polymeric cable terminations are predominantly or exclusively used at distribution voltages of 15 and 25 kV. There are a few terminations used at 46 and 69 kV. The factors which are responsible for the widespread use of polymers instead of porcelain for distribution cable terminations are a combination of ease of application, better availability, cost, and good contamination performance.

(C) SURGE ARRESTER HOUSINGS

The use of polymers as housings for surge arresters is very recent, less than 3 years. At present, few utilities are using polymer housed surge arresters for distribution. The main reason for their use is that the polymer housing does not break into fragments and fly off injuring personnel, which is a common problem with porcelain. It has been reported that more than 300,000 polymer housed surge arresters are presently in service, mostly at 15-25 kV level. No field failures have been reported to date.

5. EXPERIENCE IN OUTDOOR TEST SITES

(A) INSULATORS

The service experience mentioned in the preceding section considers all types of polymer insulators as one entity. However, due to the variety of materials available, it would be of great value to know the performance of the different types of polymer insulators. This information is obtained by evaluating them in different outdoor locations which have more than the average level of contamination. The data obtained is conservative, but is useful as it suggests that a particular insulator type which does well in a more contaminated area will not fail in a less contaminated area.

The results from two outdoor test locations are summarized here. The first site is in Brighton, England. The pollution level in this station is classified as "very heavy" and it consists of salt from the sea coast and pollution from a nearby coal fired generating station. Although other types of polymer insulators were evaluated in the station, only the silicone rubber and EPR types have been considered here for discussion. The second test site is located in Anneberg, Sweden, where the main type of pollution is from salt storms.

Table 5.1 shows the different types of insulators evaluated and the Figure of Merit (FOM) which is based on the leakage current and flashover voltage data. The porcelain insulator is taken as the reference for determining the figure of merit. The FOM is defined as the mean length L_m of the control insulator divided by the length L of the insulator across which flashover had occurred. The control insulator used was a vertical string of porcelain bells. One of the silicone rubber insulator VII listed in Table 5.1 showed gross erosion in five of the weathersheds after three years of energization.

Insulator		FOM	
		Average Value	Standard Deviation %
Reference	Vertical string; Allied type 54675	1.0 -	4.9
Porcelain Controls	Vertical string; standard 132 kV disc Doulton type 8628	1.02	9.0
	SE angled string; Doulton type 9164	1.04	7.3
	SW angled string; Doulton type 9164	1.10	10.7
	NE angled string; Doulton type 9164	1.02	7.1
	NW angled string; Doulton type 9164	0.94	6.1
VII SiR	Horizontal	>1.53	-
	Vertical	>1.53	-
VI EPDM	Horizontal	‾1.28	6
	Vertical	1.12	14
V EPDM	Horizontal	1.14	12
	Vertical	1.21	11
VIII EPR	Horizontal	1.19	14
	Vertical	1.17	14

Table 5.1: Figure of Merit Data From Brighton Tests.

The test station at Anneberg has evaluated HTV and RTV silicone rubber, EPR and porcelain insulators, with ac and dc applied voltage, for several years. The leakage current data is shown in Tables 5.2.

Insul. No	Basic current		Highest pulse current	
1	ACba=	0.289 mA	AChi=	1.228 mA
2	ACba=	0.334 mA	AChi=	1.534 mA
3	ACba=	0.331 mA	AChi=	1.281 mA
4	ACba=	0.313 mA	AChi=	23.983 mA
5	ACba=	1.665 mA	AChi=	147.034 mA
6	ACba=	0.357 mA	AChi=	22.990 mA
7	ACba=	0.387 mA	AChi=	24.778 mA
8	ACba=	0.387 mA	AChi=	2.885 mA
9	ACba=	0.332 mA	AChi=	1.250 mA
10	ACba=	0.328 mA	AChi=	1.257 mA

Note: Insulators 1-3 SR, 4-6 EPDM, 7 glas, 8 porcelain, 9-10 porcelain RTV silicone compound coated.

Insul. No	Basic current		Highest pulse current	
1	DCba=	0.002 mA	DChi=	0.311 mA
2	DCba=	0.028 mA	DChi=	0.178 mA
3	DCba=	0.029 mA	DChi=	0.322 mA
4	DCba=	0.110 mA	DChi=	19.647 mA
5	DCba=	0.064 mA	DChi=	31.635 mA
6	DCba=	0.161 mA	DChi=	23.754 mA
7	DCba=	0.199 mA	DChi=	16.761 mA
8	DCba=	0.054 mA	DChi=	0.300 mA
9	DCba=	0.007 mA	DChi=	0.222 mA
10	DCba=	0.005 mA	DChi=	0.355 mA

Note: Insulators 1-3 SR, 4-7 EPDM, 8 porcelain, 9-10 porcelain
RTV silicone compound coated.

Table 5.2: Basic and Highest Pulse Currents for AC and DC Insulators in Anneberg

From the results of these two test sites, the following conclusions can be drawn:

(1) In terms of the figure of merit both silicone rubber and EPR insulators perform better than porcelain.

(2) the silicone rubber insulator evaluated gave the best performance among all types of polymer insulators, based on the figure of merit and leakage current.

(3) There are differences in the performance among the various types of EPR insulators tested which are related to the material composition, as all other factors, such as specific creepage distance (mm/kV), electric stress, contamination severity, are the same for comparison.

(4) Although the RTV coated porcelain insulators performed much better than bare porcelain and EPR insulators, there are differences in the leakage current when compared to the the HTV silicone rubber insulators. This difference could be due to the thickness of the coating and/or to the difference in material composition.

(5) Failures are predominantly related to the weathershed. Tracking, erosion, punctures, and corona cutting were some of the factors responsible for insulator failure.

It is evident both from service and outdoor test station experience, that weathershed material aging is the single most important factor that determines the performance of polymeric insulators. The subsequent portions of this chapter focuses on the laboratory research which has been performed for better understanding the aging phenomena.

6. LABORATORY RESEARCH ON ACCELERATED AGING TESTS

Laboratory tests are required for two main reasons:

1) As a screening test to filter out the undesirable material compositions during development.

2) As an evaluation test to determine the service performance in a relatively short period of time, say several hours or days.

With these goals, many tests have been developed. They can be classified into two main categories - short term tests lasting for a few hours, and long term tests lasting for a few days.

Among the short term tests are:
Inclined Plane, dust and fog, dry arc resistance, differential wet tracking tests, all of which are ASTM tests.

Among the long term tests are:
Tracking wheel, solid contaminant and fog chamber tests. None of these are "standardized".

One common factor in all the short and long term tests is that they have been developed with the aim of accelerating the aging process, which occurs in service over a period of years, in several hours or days, and therefore are commonly referred to as accelerated aging tests. Dry band arcing is considered as the single most important factor responsible for aging.

In these tests, the experimental conditions of electric stress, water conductivity and wetting rate are more severe than those normally encountered on outdoor polymeric insulators. A high leakage current is thus promoted in a very short time which gives rise to intense dry band activity. This rapidly converts the hydrophobic surface of all polymers to a hydrophillic surface and is responsible for the results to be obtained in a relatively short time. Therefore, the outcome of these tests is more dependent on the inorganic filler concentration used for increasing the tracking and erosion resistance, than on the polymer type. Generally, polymers with greater than 50% by weight of filler show high arc resistance and therefore perform well in the ASTM tests.

Although these tests are useful in evaluating the tracking and erosion resistance of materials, they must be used with caution. It is clear that the type of polymer plays a more significant role than the filler type or concentration for successful outdoor insulation applications. For example, both service and outdoor tests explained earlier have demonstrated that, in general, silicone rubber insulators perform better than any other type in contaminated areas. The filler concentration in the silicone rubber materials is relatively low (say 30% by weight) when compared to the EPR (usually greater than 50% by weight). Thus there is a contradiction between the results of earlier accelerated aging ASTM tests and service experience.

The superior performance of silicone rubber in the field is due to the hydrophobic surface which limits leakage current to very low values. This useful property is quickly destroyed in the ASTM accelerated aging tests. Thus it is clear that a meaningful test must take into account this important surface property.

The following section shows how a fog chamber can be used to assess both the ability of the surface to suppress leakage current, and tracking and erosion resistance of polymers.

The fog chamber used is shown in Fig. 6.1. It uses 4 IEC nozzles for fog generation. It is equipped with a data acquisition system, schematically shown in Fig. 6.2, for continuous monitoring of the leakage current activity. Data such as the average leakage current, peak current above a preset threshold, and cumulative charge which is the integral of current over a time period is obtained at hourly intervals.

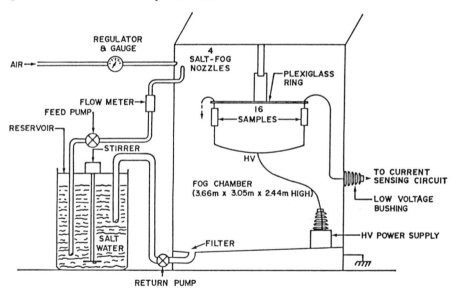

Fig. 6.1: Schematic of the Fog Chamber.

The samples evaluated are silicone rubber and EPR with differing concentrations of ATH filler, and glazed and unglazed porcelain rods for reference. The details of the samples is shown in Table 6.1. The unglazed porcelain could be considered to be equivalent to a glazed porcelain surface which has been covered with contamination.

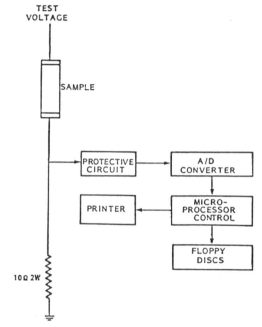

Fig. 6.2: Schematic of the Data Acquisition System

| MATERIAL | IDENTIFI-CATION | FILLER | | TIME TO FAILURE (HOURS) FOR WATER CONDUCTIVITY | |
		TYPE	LEVEL (pph)	250 μS/cm	1000 μS/cm
EPDM	EO	NONE	0	92	> 480
	E30A	ALUMINA TRIHYDRATE	30	140	165
	E120A	ATH	120	352	> 480
	E200A	$(A\ell_2O_3.3H_2O)$	200	> 480	> 480
	E250A		250	> 480	> 480
	E105A	ALUMINA $(A\ell_2O_3)$	105	300	150
	E30S		30	130	70
	E130S	SILICA	130	330	160
	E250S	(SiO_2)	250	> 480	> 480
HTV SILICONE RUBBER	S5S	SILICA	5	> 480	13
	S30S		30	> 480	15
	S120A	ATH	120	> 480	336
	S200A		200	> 480	> 480
PORCELAIN	GLAZED	NONE	0	> 480	> 480
	UNGLAZED	NONE	0	> 480	> 480

pph: PARTS PER HUNDRED OF POLYMER FORMULATION

Table 6.1: Details of Samples Evaluated.

Fog chamber testing was performed at two levels of water conductivity, 250 and 1000 µS/cm. A test duration of 30 cycles at each water conductivity was chosen. In each cycle the samples were subjected to electric stress and fog continuously for 16 hours for the first part of the cycle. For the next part of the cycle, the voltage and fog were switched off for 8 hours. The eight hour period was selected on the basis that silicone rubber recovers its surface hydrophobicity which is lost during dry band arcing. This parameter is investigated in detail later. An exposure time of 480 hours obtained in 30 cycles was found to bring out the relative performance of the materials studied. To ensure similar wetting conditions for all the samples, their position in the chamber was interchanged after every cycle.

Time to Failure

Table 6.1 also shows the time to failure in low and high conductivity fog. The following points can be noted:

a) Low Conductivity Fog (250 µS/cm).

1) There were no failures of the silicone rubber samples.

2) The EPR samples with up to 130 pph of ATH or silica filler failed by tracking.

3) At the same filler concentration, ATH, alumina and silica fillers impart similar resistance to tracking in EPR as judged by the similar times to failure.

b) High Conductivity Fog (1000 µS/cm).

1) Unfilled and filled EPR samples had longer times to failure than the correspondingly filled silicone rubber samples.

2) ATH filler imparts superior tracking and erosion resistance to EPR than does alumina or silica as judged by the longer time to failure.

Leakage Current and Cumulative Charge

a) Low Conductivity Fog

All the silicone rubber and EPR samples were hydrophobic and the porcelain samples were hydrophillic before the fog chamber exposure. The surface of the EPR samples were rendered hydrophillic due to dry band arcing in about 1 hour. But it took about 60-70 hours for the hydrophobic surface of the silicone rubber samples to be converted to a hydrophillic surface. This transition was indicated by an increase in the average leakage current from about 0.5 to 5 mA. The variation of the cumulative charge with time (cycles) is shown in Fig. 6.3. The following can be noted:

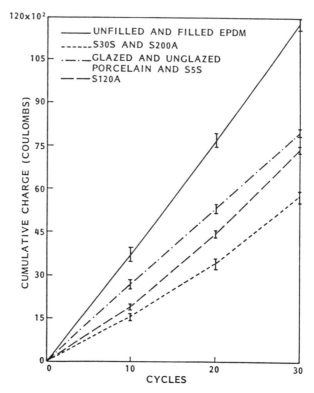

Fig. 6.3: Cumulative Charge Data in Low Conductivity Fog.

1) The EPR samples (both filled and unfilled) had substantially higher cumulative charge than porcelain or silicone rubber. The silicone rubber samples had the least cumulative charge.

2) The variation of the cumulative charge was identical both for the EPR samples that passed and failed the test. This proves that the cumulative charge is not a good indicator of material aging and failure.

3) The silicone rubber samples S30S and S200A with widely varying amount of filler had very similar cumulative charge. This suggests that there is no correlation between the filler level and leakage current in low conductivity fog.

The magnitude of the peak current and the number of current pulses above a preset threshold proved to be more effective indicators than cumulative charge of the surface condition and tracking failure of the EPR samples. The highest peak current recorded increased from about 20 mA to 50mA in the last 4 to 5 hours prior to failure. Fig. 6.4 shows that the number of current pulses exceeding 15 mA steadily increased with time of exposure and was considerably greater for the EPR samples than for silicone rubber or porcelain. This seems to be in good agreement with outdoor testing results from Anneberg reported in Table 5.2.

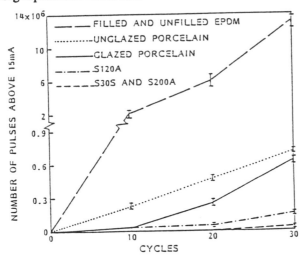

Fig. 6.4: Leakage Current Pulses Exceeding 15 mA in Low Conductivity Fog.

b) High Conductivity Fog

It was observed that the hydrophobic surface of the silicone rubber and EPR samples was converted to a hydrophillic surface within the first two hours of exposure to dry band arcing in the fog chamber. The dry band arcing was visually observed to be extremely energetic. The variation of the cumulative charge with time is shown in Fig 6.5 from which the following can be noted:

1) Filled silicone rubber and EPR samples had higher cumulative charge than porcelain.

2) The concentration of filler from 5-200 pph did not affect the magnitude of the cumulative charge.

3) The variation of cumulative charge with time was linear both for the samples that passed and failed the test and hence was not a good indicator of tracking and erosion failure.

4) Unfilled EPR sample had the lowest cumulative charge. This suggests that the presence of filler or surface contamination contributes to the increased leakage current in EPR samples.

In addition to water conductivity, other factors which have influence the results in accelerated aging tests are on/off cycle time, electric stress, wetting

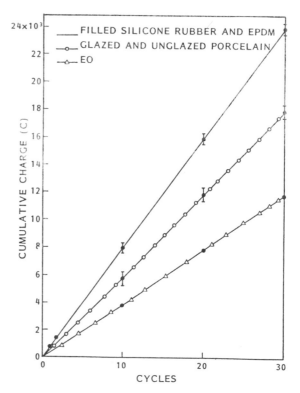

Fig. 6.5: Cumulative Charge Data in High Conductivity Fog.

rate determined by water flow rate and air pressure, etc. Some of these have been investigated with regard to silicone cable terminations. The results can be treated as fairly general as the mode of failure irrespective of the device tested in the fog chamber is either by tracking or erosion.

Table 6.2 shows the outcome of fog chamber tests performed on one type of silicone rubber cable termination under a wide range of experimental conditions. The water conductivity was varied from 250 to 22,000 μS/cm, thereby covering the range used in most laboratory tracking and erosion tests. The average electric stress, which is the ratio of the applied voltage to the leakage distance, was varied from close to the normal operating value (100% rated value) to three times the rated stress. Two values of water flow rate were chosen, 0.5 and 1.6 liters/minute. The off time was varied from 2 hours to 22 hours. The air pressure was maintained constant at 0.6 MPa.

The important points to be noted from Table 6.2 are:

1) The sample failed at 1000 μS/cm when the off time was limited to 2 hours, Test A. However, in Test G, where the off-time increased to 24

Test ID	Conductivity (μS/cm)	Voltage (%)	Flow Rate (l/min)	On/Off time	Hours to Fail
A	1000	115	1.6	22h/2h	200
B	250	300	1.6	22h/2h	NF
C	250	300	0.5	22h/2h	NF
D	600	300	1.6	22h/2h	250
E	16,000	115	1.6	22h/2h	NF
F	22,000	115	1.6	22h/2h	NF
G	1000	115	1.6	22h/24h	NF

Note: NF. No failure in 500 hours.

Table 6.2: Summary of Fog Chamber Tests on Cable Terminations.

hours, and all other experimental conditions remaining unchanged from test A, the samples did not fail. It will be shown later in this chapter that for the silicone rubber material, very little recovery of surface hydrophobicity is possible within 2 hours, and about 24 hours is needed for complete recovery. This proves that the off time has a significant bearing on the test result.

2) The samples did not fail at very low value of water conductivity, 250 μS/cm (Tests B and C), nor at very high levels of water conductivity, 16,000 and 22,000 μS/cm (Tests E and F). At the lower value, the leakage current was observed to be usually below 10 mA (peak) and therefore the energy in the dry band arc was insufficient to cause material degradation. At the very high levels, the peak current recorded was typically 100 mA. However, the duration of the individual current pulses was lower that obtained in Test A, and the interval between subsequent pulses was much longer than that recorded in Test A, with the result that the total energy of the dry band arcing (as indicated by the cumulative charge) was significantly lower than that obtained in Test A.

3) The water flow rate does not seem to have a major influence on the test results.

4) The water conductivity seems to have a greater impact on the test results than the average electric stress. In Test B, the electric stress is increased almost three fold, where as the water conductivity is reduced four fold when compared to Test A, all other conditions are unchanged. However, no failures were noted in Test B, where as the samples failed in Test A.

The results from accelerated laboratory tests lead to the following conclusions:

1) The performance of polymeric materials in accelerated aging tests is largely controlled by the experimental conditions employed. A much better correlation with service experience is obtained by studying materials at low rather than at high water conductivity, and allowing time between cycles for the materials such as silicone rubber to regain their surface hydrophobicity, which does occur in nature.

2) In order to produce material degradation, the leakage current has to be of an intermediate value, not too low nor too high. In the experimental set up used, degradation was obtained when the peak leakage current was typically 20-30 mA. No degradation was noticed when the peak current was less than 10 mA or greater than 100 mA.

3) Based on the above conclusions, it is evident that great care must be taken to analyze the test conditions before arriving at any conclusion on the electrical performance of polymers under contaminated conditions.

7. LABORATORY RESEARCH ON MATERIAL DEGRADATION

(A) ROLE OF FILLER

The type of filler and its concentration in the polymer significantly affects the resistance to tracking and erosion. Table 7.1 lists the details of the samples evaluated to understand the role of filler. The samples were cylindrical rods, 25 mm in diameter and 150 mm long, and were evaluated in the fog chamber shown earlier in Fig 6.1 with a water conductivity of 1600 μS/cm and at various levels of average electric stress. Fig. 7.1 shows the effect of the ATH filler concentration on the time to tracking or erosion failure of silicone rubber and EPR insulating materials, as determined from a salt-fog test. It is clear that the time to failure can be increased by increasing the ATH filler concentration, however for, there is an optimum value filler level of about 60% by weight of polymer, beyond which there is little improvement in the time to failure obtained by increasing the filler concentration.

The time to failure, which is an indication of the effectiveness of ATH and silica fillers in providing tracking and erosion resistance, shown in Fig. 7.1, clearly indicates that the ATH is much superior to the silica filler. The reasons for this can be better appreciated by understanding the mechanisms involved which impart improved tracking and erosion resistance. This understanding is useful in the proper selection of filler type.

MATERIAL TYPE	IDENTIFICATION	FILLER	
		TYPE	LEVEL (pph)
	EO	NONE	0
	EA	$Al_2O_3 \cdot 3H_2O$	30
	EB	ALUMINA TRIHYDRATE (ATH)	60
	EC		80
	ED		105
EPR	EE		130
	EF		250
	EDA	Al_2O_3 (ALUMINA)	105
	EAS	SiO_2 (SILICA)	30
	ECS		80
	EES		130
	EFS		250
EPOXY	XA	$Al_2O_3 \cdot 3H_2O$	220
	XB		350
HTV SILICONE RUBBER	SA		30
	SB		60
	SC	$Al_2O_3 \cdot 3H_2O$	80
	SD		105
	SE		130

pph: PARTS PER HUNDRED OF POLYMER

Table 7.1: Details of Samples Evaluated to Understand Role of Filler.

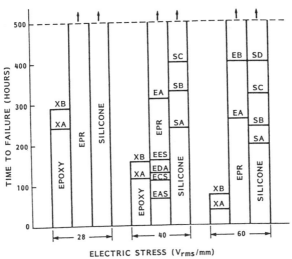

Fig. 7.1: Time of Failure of Samples Shown in Table 7.1.

Initially it was believed that both physical and chemical processes were involved. Physical mechanisms include the removal of carbon tracks by sputtering which occurs during the release of the water of hydration from hydrated fillers, and volume effects in which, very simply fewer organic molecules are exposed to the heat of dry band arcing. Chemical mechanisms include the removal of free carbon before tracks are created through the formation of CO and CO_2 gases. Although these mechanisms were suggested many years ago, it was only recent that the dominant mechanisms involved were discovered.

Sputtering or the physical cleaning action of the hydrated filler occurs when the temperature of the filler particles reach about 200^oC. The chemical mechanism of carbon removal occurs only when the temperature of the filler on the surface of the polymer exceed 600^oC. It has been established with the aid of temperature sensitive paints that change their color irreversibly at set temperatures, that during dry band arcing, the surface temperature is in the order of $200\text{-}300^oC$. In addition, measurement of released gases during dry band arcing using a mass spectrometer showed that there was no additional CO or CO_2 generated over the ambient level. This proves that it is the physical, and not the chemical mechanism, which is dominant and is chiefly responsible for imparting tracking and erosion resistance to the polymer.

Table 7.2 lists the physical properties related to heat transfer of ATH and silica fillers, silicone rubber and EPR.

MATERIAL	THERMAL CONDUCTIVITY W/cm^oC	THERMAL DIFFUSIVITY cm^2/s	DENSITY g/cm^3
SILICONE RUBBER	19×10^{-4}	15×10^{-4}	1.074
EPDM	19×10^{-4}	15×10^{-4}	0.99
ATH	2135×10^{-4}	675×10^{-4}	2.42
SILICA	150×10^{-4}	80×10^{-4}	2.65

Table 7.2: Physical Constants of Rubber and Filler Materials.

If x_1 is the volume fraction of the rubber and x_2 ($x_2=1\text{-}x_1$) is the volume fraction of the filler, the effective thermal conductivity of a filled polymer sample is calculated from the following formula:

$$K = \frac{K_1 K_2}{K_1 x_2 + K_2 x_1}$$

where K_1 and K_2 are the thermal conductivities of rubber and filler respectively.
If p_1 and p_2 are the densities of the polymer and filler respectively, Cp_1 and Cp_2 are the specific heat capacities of the polymer and filler respectively, it can be shown that the effective thermal diffusivity is given by,

$$\alpha = \frac{K}{\rho_1 x_1 C_{p_1} + \rho_2 x_2 C_{p_2}}$$

Fig 7.2 shows the effective thermal conductivity and thermal diffusivity of the polymer filled material as a function of the filler type and concentration. It can be observed that although the thermal conductivity of the ATH filler is significantly higher than that of silica, the resultant thermal conductivity of the filled polymer is not very different for the same amount of filler. However, experimental tests of the time to failure in the fog chamber under identical experimental conditions have indicated that the ATH filled polymer samples have a significantly higher time to failure than the silica filled polymer. This is a clear proof that the physical mechanism which is the most dominant in imparting tracking and erosion resistance is the sputtering of the hydrated filler which occurs only in the ATH filled polymer.

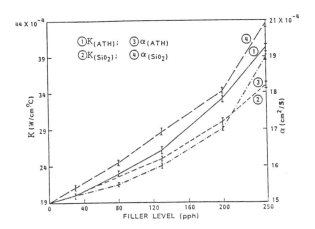

Fig. 7.2: Effective Thermal Conductivity and Diffusivity for Rubber Samples as a Function of Filler Type and Concentration. The Error Bars are an Indication of the Difference in Values for Silicone Rubber and EPR.

In addition to the choosing the right amount and type of filler, another important factor that must be considered for satisfactory operation is the dispersion of the filler in the polymer. Molded polymer materials can be expected to have some dispersion in filler uniformity. The fact that dispersion plays an important role in the initiation of tracking and erosion has been proven experimentally. Fig. 7.3 illustrates tracking and erosion of polymer surge arrester housing due to gross non-uniformity of filler dispersion.

Fig. 7.3: Material Degradation along Mold Join Due to Non-Uniform Filler Dispersion.

Fig 7.4 shows a study of the filler dispersion made using the Energy Dispersive X-Ray Analysis (EDX) technique. This is a standard attachment of a scanning electron microscope. In this technique small specimens about 1cm square and 1mm thick, are cut from the sample in the region of interest. They are coated with a conductive layer and subjected to high energy electron beam, which interacts with the sample. X-Rays from the metallic elements, Aluminum from ATH and silicon from Silica fillers, are liberated. The count of X-Rays is proportional to the amount of filler in a particular location. The dispersion ratio is defined as the ratio of the maximum X-Ray count obtained on the specimen to the X-Ray count obtained in a particular location.

Generally improved uniformity of the filler distribution is obtained by increasing filler concentration. However, mold join lines are still prone to non-uniform filler distribution due to the difference in the mobility of the polymer and filler particles. During the injection molding process, the relatively high mobile polymer is pushed outward, leaving the filler behind. Regions with low filler that are sandwiched between regions with more filler are the most likely locations of tracking and erosion initiation. This is due to the fact that the hygroscopic nature of the filler will cause the high filler locations to retain moisture for a longer time and low filler locations to dry up faster. Therefore, dry band arcing repeatedly occurs across the low filler locations initiating tracking and erosion. Preferential tracking and erosion along mold joins have been noticed on some insulators in the field also.

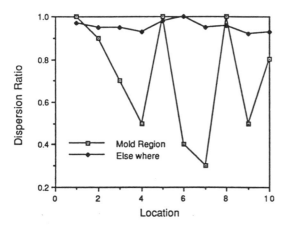

Fig 7.4: Filler Dispersion of Sample Shown in Fig. 7.3.

B. SURFACE HYDROPHOBICITY OF POLYMERS

Polymers are inherently low surface energy materials, where as porcelain and glass are high surface energy materials. The low surface energy is responsible for the initial hydrophobic surface of polymers. This is a very significant difference which has an important bearing on the electrical performance of polymeric insulators, and is one of the main factors that contribute to their superior contamination performance in the initial stages when compared to porcelain and glass. However, this condition is not permanent. Dry band arcing causes the surface to lose its hydrophobicity. However, the hydrophobicity is recovered if sufficient time is elapsed between bouts of dry band arcing. It has been found both from service experience and laboratory research that silicone rubber maintains its initial hydrophobicity for a much longer time, and also has the ability to recover its hydrophobicity faster than does EPR. Considerable research has been performed in order to understand the mechanisms that are responsible for the loss and recovery of surface hydrophobicity. Some of the chief findings are explained in this section.

The important aspects evaluated are:

1) Understand the mechanisms involved in the loss of and subsequent recovery of surface hydrophobicity of silicone rubber and ethylene propylene diene monomer (EPR) polymers.

2) Investigate various experimental techniques which could be used for characterizing surface hydrophobicity.

3) Theoretical calculations which add to the value of the experimental results.

As the technology of polymeric outdoor insulation is still undergoing development, there are variations in the material composition of silicone rubber and EPR that are presently available. These variations can and have resulted in significant differences in the electrical performance of the materials under contaminated conditions. Having more than one technique to evaluate the surface hydrophobicity would be beneficial until it can be demonstrated that a particular method is better than the rest. This was the motivation for examining Electron Spectroscopy for Chemical Analysis (ESCA) and Cross Over Voltage (COV) in detail as possible tools to study the surface.

One of the major problem with the failure analysis of polymeric outdoor insulation is the dynamic nature of the material when compared to porcelain. Although there have been flashovers of polymer insulators in the field due to the wettable surface caused by dry band arcing, by the time the surface can be analyzed in the laboratory, sufficient time has been elapsed for the material to regain their original surface characteristics. This is especially true for silicone rubber. Thus there is a need for a simple, yet reliable, measurement that can be performed quickly in the field, which gives an indication of the surface condition. This was the motivation for examining the contact angle and relating it to the results obtained from more sophisticated techniques such as ESCA and COV.

EXPERIMENTAL

(a) Electrical Testing

A fog chamber shown in Fig. 6.1, used to subject the polymer samples to a simulated outdoor contamination. The samples were cylindrical rods 25mm in diameter and 150 mm long, and were initially hydrophobic. The experimental conditions of electric stress and water conductivity were chosen such that the peak leakage current on the samples was in the range of 25-50 mA, which corresponds to that measured on outdoor insulators located in contaminated areas. This order of current was obtained after the samples were completely wettable, which occurred in about 20 hours of continuous exposure to dry band arcing in the fog chamber.

Although the fog chamber exposure causes the surface to become hydrophillic in a relatively short time when compared to service, this is still substantial if several compositions are to be quickly evaluated for their resistance to water filming and recovery characteristics. Thus it becomes essential to have a more simple technique which can cause the transition from a hydrophobic to a hydrophillic surface in a few minutes. However, the mechanisms by which the surface transition is caused should be similar to that obtained from well established methods for accelerated testing. This was the reason for examining the Tesla coil as a bench top tool.

A Tesla coil, which is commonly used to detect leaks in glass vacuum systems, was used to subject the specimens to corona discharges, and hence cause the surface transition. The specimens were thin slices cut from the sample. They were mounted on a metallic plate and with the tip of the Tesla coil about 5mm away from the specimen, the high frequency corona discharges caused the surface transition in 3 to 5 minutes.

(b) Surface Analysis Techniques

The transition of the surface from hydrophobic to hydrophillic, and its recovery back to a hydrophobic surface is limited to the top monolayers. Therefore, the analytical tools must be capable of analyzing only the top monolayers of the surface in order to give meaningful results. In addition, the loss and recovery of hydrophobicity is a dynamic surface property. Thus, it is desirable that sample preparation for the analysis be minimized. It is well known from surface chemistry that the ESCA technique analyzes only the top 5 to 50 Angstroms of the surface, and detects all elements except hydrogen. The samples to be analyzed with this technique do not need any conductive coating, hence delays are minimized. Therefore, this technique was believed to yield meaningful results on polymer surface transition.

Electron Spectroscopy for Chemical Analysis (ESCA): The ESCA measurements were done on a KRATOS X800 SAM machine which utilized an Al X-Ray source. The test specimens were 10mmX10mmX1mm thick slices cut from the surface of the samples. The surface composition was monitored before fog chamber exposure, and periodically after the surface became hydrophillic following fog chamber exposure. A broad scan was done on all the specimens in order to determine the surface composition. A selected number of specimens were subjected to a narrow scan, which gave information on the bonding arrangement of the surface molecules. Analysis of each specimen took about one hour.

Characterization of polymer surfaces using Cross Over Voltage measurement is a very recent development. It is performed in a scanning electron microscope using a low energy (<1kV) electron beam for probing the surface. The specimens can be examined without any conductive coating. Although this technique relies on the visual ability of the operator to precisely detect the cross over, it has been found that good repeatability of the results can be obtained after a few trials. The advantage of this technique is that it takes only about 15 minutes for analysis when compared to an hour required for ESCA measurements.

An additional advantage of the COV measurement is that the SEM is not as sensitive as ESCA to the type of material. ESCA measurements cannot be performed until the specimen chamber has attained a vacuum of 10^{-9} torr. A few specimens of both silicone rubber and EPDM used here started to degas, thereby preventing the required vacuum needed to start the measurement. However, no such problems were encountered with the SEM, as the vacuum required is of the order of 10^{-6} torr.

Cross Over Voltage (COV): A scanning electron microscope (SEM) used to determine the COV was a JEOL JXA-840 machine. The parameter monitored was the "Cross Over Voltage" (COV) which is defined as the energy of the incident electron beam which causes the ratio of the back scattered plus secondary electrons to the incident electrons to be unity. If this ratio is greater than unity, the surface gets positively charged and is indicated by white image on the screen. When this ratio is less than unity, the surface gets negatively charged and is indicated by a black image on the screen. The energy of the incident beam is varied gradually until the cross over occurs.

Contact Angle Measurement: This technique was used because of its simplicity. The contact angle was measured on the samples before fog chamber exposure and periodically after the samples had lost their surface hydrophobicity. A simple set up was built which involved dropping a 100 µl water droplet and measuring the advancing angle of contact between the droplet and the sample surface with a graduated reticule and a 10X magnifying glass. The specimens were either the entire rod samples or small sections from the sample.

RESULTS AND DISCUSSION

The terms EPR and EPDM have been used interchangeably in many places in this chapter.

(a) ESCA Studies

Tables 7.3 shows the surface composition obtained of a virgin sample which is hydrophobic (before fog chamber exposure) and a sample which was rendered wettable from fog chamber exposure. The time elapsed between the deenergization of the fog chamber and the start of the ESCA analysis was about one hour. It can be seen that there is a large increase in the oxygen concentration on the wettable specimens in comparison with the virgin specimens, for both silicone rubber and EPDM.

A narrow scan was obtained for the wettable samples and compared with that for the virgin samples. For silicone rubber, it was observed that the oxygen, carbon and silicon peaks had shifted by 1.4, 1.7 and 0.7 eV. By comparison with standard ESCA Tables, it was determined that the shift was due to the hydroxyl (OH) group formation on the surface of the wettable samples. On further examination, it was noted that the hydroxyl groups were associated with the carbon atom. This indicates that in silicone rubber, the side chains composed mainly of the methyl groups have reacted to form the hydroxyl groups.The hydroxyl groups could have been generated by two possible actions: (1) due to interaction of the dry band discharges, moisture and the polymer surface and (2) due to interaction of the dry band discharges, polymer and the alumina trihydrate filler, which was added to the samples for improving tracking resistance.

Material Type and Condition	Surface Composition(%)		
	O	C	Si
Silicone Rubber, not wettable	26.0	48.0	26.0
EPDM,not-wettable	16.0	83.0	<1
Silicone Rubber, wettable	62.0	15.0	23.0
EPDM, wettable	51.0	48.0	<1

Table 7.3. Surface Composition of Samples Before Fog Chamber Exposure (non-wettable) and Immediately After Loss of Hydrophobicity in the Fog Chamber (wettable).

To determine the source of the hydroxyl species, silicone rubber and EPDM samples without any hydrated fillers were also evaluated and a similar increase in the hydroxyl concentration was measured. Thus a positive proof, that the loss of surface hydrophobicity is due to hydroxyl groups caused by the interaction of the dry band arc with the polymer surface and moisture, has been established. The above results also indicate that the backbone of the silicone rubber polymer is not affected by the dry band arcing which causes the surface transition.

The hydroxyl groups associated with the carbon atom were also detected on the wettable samples of EPDM. However, it was not possible to discriminate whether these hydroxyls originated from the side chains or from the backbone of the polymer, due to the fact that the EPDM material is mostly composed of carbon and hydrogen.

Table 7.4 shows the surface composition measured periodically after the samples were removed from the fog chamber. It can be seen that the initial surface composition is regained in about 24 hours and 48 hours, respectively for the silicone rubber and EPR material.

Table 7.5 shows the surface composition for silicone rubber and EPDM respectively, as obtained by ESCA on specimens which have been rendered hydrophillic by fog chamber and Tesla coil treatments. The close similarity in the surface composition produced by the two methods demonstrates that the

Tesla coil, which is an inexpensive, off-the-shelf instrument, can be reliably
used for the surface studies of these polymers.

	Hours After	Surface Composition (%)		
Material	Removal	O	C	Si
	1	62.0	15.0	23.0
Silicone	5	59.0	27.0	14.0
Rubber	10	54.0	30.0	16.0
	24	27.0	50.0	23.0
	48	26.0	49.0	25.0
	1	51.0	48.0	<1
	5	47.0	52.0	<1
	10	43.0	56.0	<1
EPDM	24	29.0	70.0	<1
	29	20.0	79.0	<1
	34	19.0	78.0	<1
	48	18.0	81.0	<1

*Table 7.4: Periodic ESCA Analysis on Samples removed from
the fog chamber which were Converted to a Hydrophillic
Surface.*

Material	Surface condition	Surface Composition(%)		
	and method of treatment	O	C	Si
SR	Wettable, fog chamber (FC) treated	62	15	23
SR	Wettable, Tesla Coil (TC) treated	60	18	22
SR	24h recovered after FC treatment	26	49	25
SR	24h recovered after TC treatment	29	53	18
EPDM	Wettable, FC treated	51	48	<1
EPDM	Wettable, TC treated	55	44	<1
EPDM	48h recovered after FC treatment	18	81	<1
EPDM	48h recovered after TC treatment	20	79	<1

*Table 7.5: ESCA Data of Silicone Rubber (SR) and EPDM
Samples after Fog Chamber Dry Band Arcing and Tesla Coil
Corona Treatments.*

The recovery of the hydrophobicity is due to the diffusion of the low
molecular weight (LMW) polymer chains, which are in a fluid form, from
the bulk to the surface of the material. The mobility of the short polymer
chains which are in the fluid form is much higher than long polymer chains.

The mobility is also related to the surface tension, which is among the lowest for silicone rubber. All these factors are responsible for the faster recovery of surface hydrophobicity of silicone rubber when compared to EPR.

In order to demonstrate that the diffusion in silicone rubber occurs even in the presence of surface contamination, virgin samples which were coated with a 1 μm thick layer of carbon and fog chamber exposed samples which had accumulated contamination were analyzed with ESCA. Table 7. 6 shows the results of the surface analysis performed at various time intervals. It is evident that the surface composition after a certain time period is very similar to the virgin uncoated samples. This proves that the mobile LMW chains have diffused through the surface contamination, thus restoring the surface hydrophobicity. This observation is consistent with service experience.

MATERIAL	COATING MATERIAL AND THICKNESS	SURFACE COMPOSITION (%)					TIME LAPSED AFTER COATING (HOURS)
		0	Si	C	Ca	Mg	
SILICONE RUBBER (S120A)	NONE (VIRGIN)	26.60	21.10	52.30	0	0	--
	CARBON ≈1 μm	12.63	10.26	77.11	0	0	1
		13.24	13.34	73.42	0	0	12
		14.76	13.94	71.30	0	0	18
		16.29	16.90	66.81	0	0	24
		25.00	20.49	54.51	0	0	50
	WATER SCALES ≈0.335mm	30.90	13.52	40.10	2.32	13.16	200
EPDM (E120A)	WATER SCALES ≈0.26mm	41.52	2.11	30.22	4.53	21.61	200

Table 7.6: ESCA Analysis of Silicone Rubber Samples with Surface Contamination.

(b) COV Studies

Table 7.7 illustrates the results of the COV measurements. The transition from a hydrophobic to a hydrophillic surface can be explained as being caused when the polymer chains on the surface are broken by the dry band arcs and the free radicals subsequently interact with the moisture to form hyroxyl groups. Bond breaking generates free electrons. It is reasonable to expect that a material which generates a large number of free electrons can become hydrophillic in a shorter time than a material which generates a smaller quantity of free electrons, if the magnitude of the discharge is the same. This is equivalent to saying that greater the COV, the more resistant will the material be towards water filming. The Tesla coil was used to simulate the effect of dry band arcing in a shorter time.

R. S. GORUR

Material	Virgin	COV (V) after corona				
	COV(V)	discharge for minutes				
		1	2	3	4	15
SR	900	700	600	500	200	200
EPDM	600	500	400	200	200	200

Table 7.7: COV of Silicone Rubber (SR) and EPDM samples after Corona Discharge from a Tesla Coil.

The results in Table 7.7 indicate that the virgin silicone rubber has a COV which is greater than EPDM by 50% (900 vs. 600 V). This in agreement with the observations made from the service experience of these outdoor insulators, that silicone rubber resists the formation of a water film better than EPDM. In addition, the COV is reduced for samples which have been treated with corona discharge from the Tesla Coil. This is also in agreement with the observation that the resistance to water filming reduces as the duration of the corona discharge increases. It can also be noted that the COV for silicone rubber and EPDM is the same (200 V) after a prolonged exposure to discharge from the Tesla coil. This explains the observation that silicone rubber and EPDM wet out very quickly in the ASTM and other accelerated aging tests, which promote intense dry band activity from a very early stage of the test. These results leads us to speculate that the dry band discharges which occur during service are mostly of a low energy, which makes the silicone rubber to be more resistant to water filming than EPDM.

Table 7.8 shows the COV measured at various time time intervals in the recovery process. It can be observed that the time required for the COV to increase back to the initial value is the same as obtained from ESCA. Thus there is good correlation between ESCA and COV measurements.

(c) Contact Angle Studies

Fig. 7.5 shows the results of contact angle measurements on silicone rubber and EPDM. The samples were initially cleaned with methanol to remove any

Sample	COV (V) after hours of removal							
	from fog chamber							
	0.5	1	2	4	6	8	24	48
SR	200	200	400	500	600	700	900	900
EPDM	200	200	200	200	300	300	400	600

Table 7.8: COV as a Function of Recovery Time.

mold release agent which could affect the measurement. The contact anglebefore fog chamber exposure was 100° and 85° on the silicone rubber and EPDM samples, respectively. They were subjected to dry band arcing in the fog chamber and were removed after their surfaces were completely wettable. The samples were dried and the contact angle was measured periodically. It can be seen from the figure that the initial contact angle, or the surface hydrophobicity, is regained in 24 and 48 hours of removal from the fog chamber, respectively for the silicone rubber and EPDM materials. The recovery of the contact angle follows a straight line with respect to the Logarithmic of time in minutes.

The time for the recovery of surface hydrophobicity is seen to be similar from three widely differing types of measurements. This is encouraging as the contact angle measurements can be easily performed on insulating devices in the field, and the instrument can be made simple and portable.

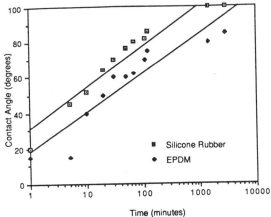

Fig. 7.5: Recovery of the contact angle with time for silicone rubber and EPDM after their surfaces were wettable due to fog chamber exposure.

In the EPDM material, fluids such as oils and plasticizers are used during processing. The exact composition of the LMW chains were not known. However, in a silicone rubber polymer, the LMW chains exist in the form of silicone oil. If this is responsible for the hydrophobicity, then a sample from which the silicone oil has been extracted should take a significantly longer time to recover its hydrophobicity. To demonstrate this, thin slices (5mmX5mmX1mm thick) of silicone rubber were dipped in hexane, which dissolves silicone oil. Specimens which were dipped in hexane for 12, 24, 48 and 96 hours were measured for weight loss using an analytical balance (Mettler M3, Model SNR A 98384) with a sensitivity of 1 µg in a class 100 semiconductor clean room. They were then treated with corona discharges from the Tesla coil for 3 minutes, which rendered the surface completely wettable. The contact angle was measured periodically on these specimens. The results shown in Fig. 7.6 indicate that due to the loss of silicone oil, the recovery of hydrophobicity has been significantly retarded. To avoid cluttering of the points, results of the samples dipped for 6, 24 and 96 hours are not shown.

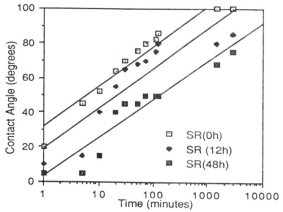

Fig. 7.6: Recovery of Contact Angle with Time for 1mm thick Silicone Rubber (SR) samples Discharged with a Tesla Coil after being dipped in Hexane for 12 and 48 h. SR (0h) is the sample not dipped in Hexane.

(d) Theoretical Considerations

The diffusion of the LMW chains can be further demonstrated with the aid of the following equation, based on the diffusion of a substance from a solid:

$$\frac{M_t}{M_0} = 4 \left(\frac{D}{\ell^2}\right)^{1/2} t^{1/2} \left[\frac{1}{\pi^{1/2}} + 2\sum_{n=1}^{\infty} (-1)^n \, erf\frac{n\ell}{2(Dt)^{1/2}}\right]$$

where M_t = change in mass after time t
M_0 = initial mass
t = time in s
D = diffusion coefficient in m^2/s
l = sample thickness
erf = error factor, obtained from standard Tables.

For values of time between 0 to 50 hours, the error factor approaches zero, and the above expression can be simplified as:

$$M_t / M_0 = 4(Dt / \pi l^2)^{1/2}$$

The average diffusion coefficient D for silicone oil has been reported to be about 3.5×10^{-12} m²/s. The above equation suggests that if the weight loss is plotted as a function of $t^{1/2}$, the graph should be a straight line, as all the other parameters in the equation are constant. It can be observed from Fig. 7.7 that this is indeed the case.

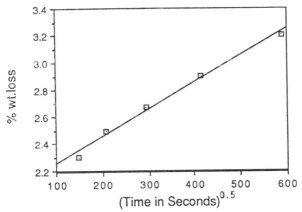

Fig. 7.7: Variation of Weight Loss with Time for 1mm thick Silicone Rubber samples dipped in Hexane.

These results show that due to the presence of silicone oil, the silicone rubber has a faster recovery than EPDM. It would be of great practical interest to determine if the hydrophobicity of materials such as EPDM and epoxies, which are also widely used for outdoor insulation applications, can be significantly improved with the addition of silicone oil.

The following are the conclusions on the surface hydrophobicity of polymers:

1) The loss of surface hydrophobicity is due to the formation of hydroxyl species as a result of the interaction of the polymer and moisture caused by the dry band discharges.

2) The recovery of hydrophobicity is due to the diffusion of low molecular weight polymer chains. The recovery process can be been studied using more than one experimental technique.

3) The measurement of Cross Over Voltage seems to explain several commonly made observations from the service experience and laboratory testing of silicone rubber and EPDM rubber insulators. In addition, from the point of view of testing time and repeatability of results, this technique seems to be the most promising. However, contact angle measurements are simple to perform and they correlate well with ESCA and COV data.

(C) AGING OF WEATHERSHED POLYMER

It is clear from service experience, outdoor test sites and accelerated laboratory tests that the polymer weathershed materials undergo both temporary and permanent changes with time. The temporary change is the change in surface hydrophobicity, which has been dealt in detail in the preceding section. The permanent changes are due to material aging which ultimately decides the useful service life. Aging is dependent on the type of material and material composition, all other factors being the same. Whereas aging in EPR is apparent, in silicone rubber, it is not so due to the recovery of surface hydrophobicity. In fact, there is a general feeling in the industry that even if the insulator surface is contaminated, as long as it is hydrophobic, the electrical performance of the insulator has not significantly changed with time in service.

This section describes a hypothesis for aging in silicone rubber, which has been experimentally proven both for RTV and HTV varieties. It is believed that the hypothesis would be extremely useful in predicting service life, in the development of better materials for outdoor insulation, and could later be used to explain aging in other types of polymer materials used. The effort was motivated by the following observations which indicate that the electrical performance of silicone rubber insulators in service for several years, whose surface is still hydrophobic although contaminated, has reduced with time in service:

1) Outdoor tests in Brighton, UK, have demonstrated that silicone rubber insulators exhibit erosion after several years, despite the fact that the flashover voltage was consistently higher due to the hydrophobic surface, than other polymeric and porcelain insulators located in the same site.

2) A recent paper has reported that the surface resistance under wet conditions of field aged RTV coated insulator surfaces, was high and similar to that of the virgin material in the initial stages of measurement. However, after several minutes of wetting, the surface resistance of the field aged material was considerably lower than that of the virgin RTV material, but higher than that of porcelain. A similar result was reported on a new, rubbed RTV surface which was allowed to recover its hydrophobicity for six days.

3) Another recent paper which evaluated RTV coatings for bushings has described that artificially contaminated RTV coated bushings repeatedly flashed over under certain conditions despite the fact that the surface was hydrophobic at the start of the test, as judged by the beading up of water droplets.

4) Fog chamber tests on RTV coatings and silicone elastomer rods have demonstrated that the material undergoes erosion, despite the fact that the

on/off time of voltage and fog application are sufficiently long for the hydrophobicity to recover, as judged by ESCA and contact angle measurements.

The above findings suggest that the surface hydrophobicity could be a superficial property, limited to the top few monolayers of the surface, which masks the permanent changes occurring deeper in the material. Techniques such as ESCA and COV measurement are extremely surface sensitive and analyze only the top 5-50 Angstroms of the surface. It requires a water film thickness of several hundred Angstroms on the surface to promote leakage current of few mA. Therefore, it is possible that although the top surface monolayers (about 50 Angstroms thick) of the new and aged silicone rubber material show little difference in their chemical composition or in the temporary water beading behavior, the electrical performance under contaminated conditions, which is determined by the underlying material in addition to the top monolayers, can show significant differences in the new and aged condition.

HYPOTHESIS FOR MATERIAL AGING

The aspect of aging considered in this paper is that resulting from dry band arcing. The hypothesis will first provide the reasons for above field and laboratory observations. The hypothesis was developed in conjunction with experiments described later. However, the hypothesis and the experimental proofs are presented separately for improved clarity of the subject.

The development of leakage current on an aged material and the events which lead to material degradation are explained as follows:

The LMW polymer chains on the surface are responsible for the hydrophobicity. There is a reduction in the quantity of LMW polymer chains on the surface of an aged material when compared to a virgin material. Dry band arcing causes scission of the long polymer chains into short chains. (Based on results from gel permeation chromatography, long polymer chains are those which have a molecular weight of greater than 1000, where as short chains are those which have a molecular weight of less than 1000). During the process of chain scission, the amount of free radicals (i. e., elements which have an excess or deficiency in the number of electrons in their outer orbit) is increased.

The free radicals are very reactive. The siloxane bonds are attacked hydrolytically to form smaller cyclic or silanol-functional molecules. This has been schematically illustrated in Fig. 7.8. As the molecules become smaller in size, the silanol-functional species become more water soluble and the cyclic species more volatile. Eventually the silanol products are completely miscible in water and hence cause a depletion in the LMW polymer chains. Thus there is a difference in the thickness of the continuous

water film formed on new and aged silicone rubber material. This explains
the difference in the surface resistance and leakage current in a new, field
aged RTV and porcelain (no LMW chains). The reduction in the quantity of
the LMW polymer chains on the surface of silicone rubber is a permanent
change caused by the dry band arcing.

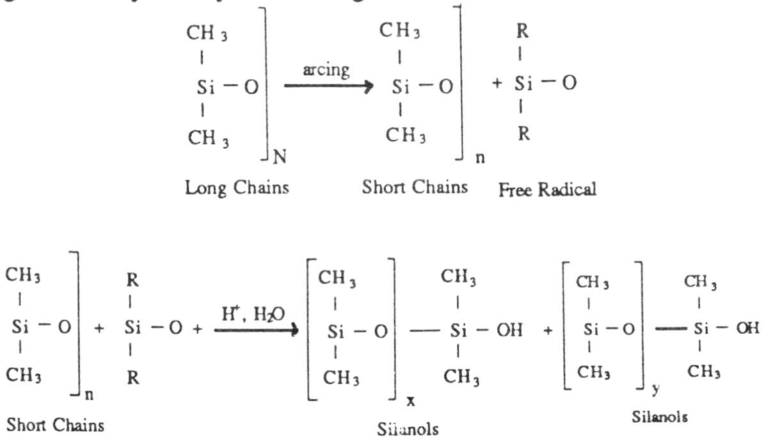

$$
\begin{bmatrix} \begin{array}{c} CH_3 \\ | \\ Si-O \\ | \\ CH_3 \end{array} \end{bmatrix}_N \xrightarrow{\text{arcing}} \begin{bmatrix} \begin{array}{c} CH_3 \\ | \\ Si-O \\ | \\ CH_3 \end{array} \end{bmatrix}_n + \begin{array}{c} R \\ | \\ Si-O \\ | \\ R \end{array}
$$

Long Chains   Short Chains   Free Radical

$$
\begin{bmatrix} \begin{array}{c} CH_3 \\ | \\ Si-O \\ | \\ CH_3 \end{array} \end{bmatrix}_n + \begin{array}{c} R \\ | \\ Si-O \\ | \\ R \end{array} + \xrightarrow{H^+,\ H_2O} \begin{bmatrix} \begin{array}{c} CH_3 \\ | \\ Si-O \\ | \\ CH_3 \end{array} \end{bmatrix}_x \begin{array}{c} CH_3 \\ | \\ Si-OH \\ | \\ CH_3 \end{array} + \begin{bmatrix} \begin{array}{c} CH_3 \\ | \\ Si-O \\ | \\ CH_3 \end{array} \end{bmatrix}_y \begin{array}{c} CH_3 \\ | \\ Si-OH \\ | \\ CH_3 \end{array}
$$

Short Chains   Silanols   Silanols

Fig. 7.8: Schematic of Silanol and Cyclic Polymer Formation.

The LMW polymer chains encapsulate the contamination and filler particles
on the surface of silicone rubber material. As there is a depletion in the LMW
polymer chains on the surface caused by dry band arcing and subsequent
wetting, the hygroscopic contamination and filler particles absorb water and
form a thicker water film. This also results in a higher leakage current on a
aged silicone rubber material when compared to a virgin material.

Another permanent change due to aging is an increase in the surface
roughness. This provides an increase in the surface wetting area and
generally enhances the leakage current.

The above processes explain the build up of leakage current in the aged
material. The events which lead to material degradation are the following:

A permanent change causing degradation is depolymerization of the surface.
It is possible that depolymerization of the surface increases with aging. This
means that the filler particles on the aged surface are not as tightly bound to
the polymer as they are on a virgin surface. The filler particles now act as
individual entities and promote a thicker water film due to their hygroscopic
nature. In addition, as the bonding to the polymer on the surface is
weakened, the interaction with the polymer is reduced. This leads to a
reduction in the erosion resistance. There is a further reduction in the erosion
resistance as it is takes lesser energy from the arc to erode shorter chains than

longer chains. Thus a cumulative process of degradation occurs from depolymerization of the surface.

An additional consequence of the depletion of the LMW chains and depolymerization on the surface is the clustering of the filler particles, i. e., the conglomerates of filler appear on the surface in place of smaller filler particles. Then will again lead to a reduction in the erosion resistance as the interaction of the polymer-filler is reduced due to increased size of the filler. There is published data to demonstrate that the erosion resistance decreases as the particle size of the filler increases.

The polymer in a virgin silicone rubber is amorphous, which means that there is no well defined structure within the material. The amorphous structure can also be thought of as a structure in which the degree of crystallinity is low. This permits a fair amount of interaction between the polymer and filler for the mechanisms which impart increased erosion resistance to be operative. In addition, the amorphous structure permits easier diffusion of the LMW polymer chains from the bulk to the surface. With aging, there is a physical restructuring in the material brought about by an increase in the crystallinity of the polymer. The polymer molecules are now confined to definite locations in the lattice and not as freely available as in the amorphous virgin material. Hence, the interaction between the polymer and filler is decreased, thereby causing a reduction in the erosion resistance.

Material degradation follows on a layer by layer basis. As the top surface layers are rendered less erosion resistant by the above processes, they are removed by the dry band arcs. The layer beneath the top layer will then undergo the same processes.

Summarizing, it is hypothesized that the aging mechanism involves the following processes:

1) Dry band arcing resulting in loss of surface hydrophobicity,

2) Reduction in the quantity of Low Molecular Weight polymer chains on the surface resulting in increased leakage current,

3) Increased surface roughness which generally enhances the wetting and leakage current,

4) Depolymerization of the top surface layers resulting in further depletion of LMW polymer chains on the surface and decreased tracking and erosion resistance,

5) Changes in physical structure brought about by crystallization of the polymer, and clustering of filler particles on the surface, resulting in a further decrease in the erosion resistance, and

6) Ultimate tracking and/or erosion of the material when the energy input from the dry band arcs exceeds the rate of removal of degradation by products.

The first and the last aspect has been proven in the preceding sections. The following section provides experimental evidence for the remaining aspects of aging, which are indicated by permanent changes.

EXPERIMENTAL

In order to prove the hypothesis, a detailed experimental investigation of new and aged silicone rubber material was performed. A fog chamber, shown in Fig. 6.1 was used for accelerated aging of the silicone rubber samples.

The samples for the fog chamber aging were: (1) RTV silicone rubber coatings over porcelain rods, and (2) silicone rubber elastomer rods. All the rods were 25 mm diameter and 25 cm long. The thickness of the RTV coating on the porcelain rods was about 0.5 mm. The samples were subjected to a dc electric stress of 40 V/mm and conductivity of 1000 μS/cm/cm. DC was used to obtain material degradation in a short time.

The duration of the test was sufficient to provide samples with various degrees of aging. For example, some samples were subjected to the test for about 100 hours so that their hydrophobic surface was temporarily converted to a hydrophillic surface, but no degradation was visible. Other samples were exposed to time durations varying from 200 to 300 hours to provide samples with increasing amount of erosion.

All the samples analyzed for permanent changes were allowed to recover their temporarily lost hydrophobicity due to dry band arcing. The recovery was determined by the use of ESCA, Cross Over Voltage (COV) and contact angle measurements. Details of the equipment used have been provided earlier.

The techniques used for the determination of permanent changes are:

Energy Dispersive X-Ray (EDX) Analysis: This technique (also popularly referred to as EDAX) was used to demonstrate the depletion of low molecular weight polymer chains on the surface of the aged material.

The experiments were performed on a Scanning Electron Microscope (Model JEOL JXA 840), equipped with a EDX probe (From Tracor-Northern). The samples were small pieces 3mmX3mmX0.5mm thick, cut from the surface of the new and aged samples. They were carbon coated and the thickness of the coating is estimated at 100 Angstroms. A 'Flextran' software was employed to obtain the magnitude of the X-Ray signals emanated from the various elements in the material.

X-Rays in silicone rubber are obtained from the silicon in the polymer and reinforcing filler, and aluminum from the ATH filler. The depth of the surface analyzed depends on the accelerating voltage of the electron beam, and is calculated using the following formula:

$$R = 0.033(E^{1.7} - E_c^{1.7}) \frac{A}{\rho Z}$$

where
E=electron beam energy, keV,
E_c=critical energy required to excite the X-Ray line of interest,
A, Z and p=atomic weight, atomic number and density of the material, respectively,
R=depth of penetration.

The count of X-Rays per minute is dependent on the concentration of the particular element.

Fourier Transform Infra-Red (FT-IR) Spectroscopy: This technique was used to demonstrate that there is an increase in the depolymerization of the surface with aging.

The samples were analyzed using a Nicolet 205 FT-IR spectrometer. The technique used was the Attenuated Total Internal Reflection (ATR) method employing a KRS-5 crystal. The spectrum obtained is a plot of the percent transmittance of the IR light through the sample. vs. wavelength (expressed as wave numbers) of the various chemical groups in the material. Each chemical group in the material has a distinct frequency of transmittance (or absorbance) of the IR signal based on the type of elements and their bonding. The intensity of the transmitted signal is indicative of the concentration of the chemical group in the material. Zero transmittance at a particular wavelength means that all the light is absorbed and indicates abundance of a particular group in the material.

The depth of surface analyzed is obtained from the following equation:

$$dp = \frac{\lambda}{2\pi \eta_C [Sin^2\theta - (\eta_S/\eta_C)^2]^{1/2}}$$

where
λ=wavelength in cm
η_S=refractive index of sample (silicone rubber=1.43)
η_C=refractive index of crystal (KRS-5=2.38)

θ=incident angle, 45 degrees
dp=depth of penetration in microns

For silicone rubber, the depth analyzed was calculated to be ranging from 4,000 Angstroms at a wave number of 4000 cm^{-1} to 40,000 Angstroms at a wave number of 400 cm^{-1} (wave number=1/wavelength in cm).

The samples were about 10mmX50mmX0.5mm thick pieces. Two such pieces were used on either side of the crystal for increased sensitivity of the measurement. The samples were not cleaned in any manner prior to the measurement and were handled with plastic gloves to prevent any oil film deposition on the surface from fingers.

X-Ray Diffraction (XRD): This technique was used to demonstrate that there is a change in the physical structure of the polymer with aging. In addition, the clustering of the filler particles on the surface with aging was determined by this technique.

The instrument used was a Rigaku D/Max IIB automated powder diffraction system with a diffracted beam curved graphite crystal monochromator for collecting the X-Ray data. A copper anode sealed X-Ray tube operated at 50 kV/30 mA provided the X-Ray source and a scintillation counter was used for detection. The diffraction data on each polymer sample was obtained from 3 to 90 degree 2-theta, with a sampling interval of 0.02 degrees and a scan rate of 2 degrees per minute. Data reduction was done using the software provided by the manufacturer. This included removal of the K-Alpha component of the copper radiation prior to peak identification.

The samples were 10mmX10mmX0.5mm thick sections. The diffraction pattern obtained is a plot of the intensity vs. 2 theta angle. The measured intensity arises from an estimated sample depth of 10,000 to 60,000 Angstroms, the depth increasing with 2-theta angle.

Surface Roughness: This measurement was used to prove that the surface of the aged material is much rougher than a new material, and to provide quantitative data for comparing the roughness of an aged surface with a virgin surface.

The instrument used was a mechanical instrument (Rank Taylor Hobson, Model Talysurf 10) which uses a diamond stylus of width 0.0025 mm. The samples used were 25mmX6mmX0.5mm thick pieces mounted on a glass slides. The parameters monitored were the height of the various peaks of the surface aberrations, the distance between the peaks and the count of peaks per inch of the sample.

RESULTS AND DISCUSSION

The samples chosen for the determination of the permanent changes are described in Table 7.9. Two types of RTV coating and one type of elastomeric rod were used. They were analyzed by the various techniques at three different stages: in the virgin state (samples 1, 2 and 3), after 100 hours of exposure to dry band arcing in the fog chamber, which provided partially aged surfaces, and after 200 hours of exposure to dry band arcing in the fog chamber resulting in erosion of various degrees, thereby providing considerably aged surfaces. The term "partially aged surface" is subjective and has been used to characterize samples which permitted a significantly higher leakage current than a virgin sample, but which showed no visible signs of degradation. The considerably aged samples 7, 8 and 9 were obtained from portions close to the eroded parts of the sample.

All the samples in Table 1 were analyzed for permanent changes, and very similar results were obtained for samples 1, 2 and 3, samples 4, 5 and 6, and samples 7, 8 and 9. Therefore, only typical results are presented to prove the hypothesis.

All the aged samples were allowed to recover their hydrophobicity prior to the determination of the permanent changes. The complete recovery of hydrophobicity was confirmed through the use of ESCA, COV and contact angle measurements, which were performed after 24 hours of removal from the fog chamber.

Sample ID	Description	Hours in Fog Chamber	Comments
1	RTV Coating A	0	Virgin Material
2	RTV Coating B	0	Virgin Material
3	Elastomer Rod	0	Virgin Material
4	RTV Coating A	100	No Degradation, partially aged surface
5	RTV Coating B	100	No Degradation, partially aged surface
6	Elastomer Rod	100	No Degradation, partially aged surface
7	RTV Coating A	200	Erosion in parts, considerably aged sufrace
8	RTV Coating B	200	Erosion in Parts, considerably aged sufrace
9	Elastomer Rod	200	Erosion in Parts, considerably aged sufrace

Table 7.9: Details of Samples Evaluated.

The following results were obtained on all samples in Table 7.9: ESCA measurements showed that the oxygen, silicon and carbon composition were 25±2%, 25±3%, and 50±5%, respectively. The COV was 0.9±0.1 keV. The contact angle was about 95±5°. The very similar results on the new and aged samples prove that the aged samples in Table 7.9 had recovered their temporarily lost hydrophobicity.

EDX Results: Table 7.10 shows the results of the EDX analysis. The ratio of silicon to aluminum has been determined at various depths of the surface. The surface is made of LMW polymer chains due to its low surface energy. It is evident in the virgin samples 1 and 2, that on the surface, i. e, when the

R. S. GORUR

depth of penetration of the electron beam is the least (at 3 keV), there is an abundance of LMW polymer chains, as indicated by the large ratio of silicon to aluminum. However, on an aged sample, the quantity of silicon is significantly reduced at the same depth. This trend continues for until a depth of penetration of about 6,000 Angstroms.

This proves that there is a depletion in the amount of LMW polymer chains on the top surface layers, which is the proof of the hypothesis. Beyond a depth of 6,000 Angstroms (1μm), there appears to be little change in the silicon to aluminum ratio. This indicates that the chemical composition of the deeper layers of the aged material is similar to that of the virgin material.

Beam Voltage	Penetration Depth (A^o)	Si:Al			
		Sample 1	Sample 7	Sample 2	Sample 8
3	1,290	>100:1	4:1	>100:1	13.69:1
4	2,630	4:1	3:1	10:1	4.13:1
5	4,240	3:1	2:1	8:1	4:1
6	6,090	2:1	2:1	4:1	4:1
8	10,470	2.5:1	2:1	3:1	4:1
9	12,900	2.5:1	2:1	2:1	2:1

Table 7.10: Typical EDX Results Illustrating the Si:Al Ratio on Samples Described in Table 7.9.

FT-IR Results: For brevity, the spectrum from the present samples is not included, but only the results which prove the hypothesis are presented here. Fig. 7.9 shows the variation of Absorbance $A\{A=-\log_{10}(T/100)$, where $T=$ % Transmittance$\}$ of the relevant chemical groups in the virgin sample 1, a partially aged sample 4 and a considerably aged sample 7. The regions of interest in the spectrum are at wave numbers of 2960 cm^{-1}, corresponding to the CH$_3$ chemical group in the side chain of the silicone polymer, at 1260 cm^{-1}, corresponding to the Si-CH$_3$ chemical group involving the side chain and the back bone of the polymer, and at 477cm^{-1}, corresponding to the Si-O-Si chemical group of the back bone of the polymer.

It can be observed from Fig. 7.9, that there is a reduction in all the chemical groups of the silicone polymer due to aging. This reduction indicates that depolymerization of the top surface layers increases with aging, and this is the experimental evidence of the hypothesis.

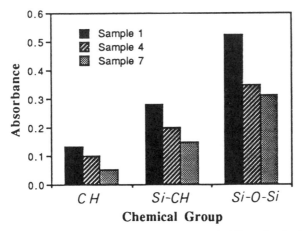

Fig. 7.9: Reduction in the Absorbance of Chemical Groups in Silicone Rubber due to Aging.

In order to prove that the above changes are limited to the top few microns on the surface, the surface of the aged sample 7 was mechanically scraped with a knife, and subsequently analyzed with the FT-IR technique. With each scraping, it was observed that the absorbance of the above functional groups was closer to the values obtained for the virgin material. After 3 scrapings, the absorbance data of the aged sample was almost identical to that of the virgin sample. Although it was not possible to exactly determine the thickness removed from each scraping, it is estimated to be of the order of a few microns.

XRD Results: Fig. 7.10 (A) shows a typical wide angle scan of the virgin silicone rubber material. The very broad low intensity hump between 10-15 degrees indicates that the silicone polymer is mostly amorphous. The other peaks observed are due to the ATH filler in the polymer.

Fig. 7.10 (B) illustrates the diffraction pattern of the aged sample, which shows a sharpening of the amorphous hump indicating that the polymer has become more crystalline. A similar effect is noted for the peaks due to the filler. This is better seen in Fig. 7.11 for one of the ATH diffraction peaks. The smaller width associated with the aged sample (curve b) suggests that the filler particles are present in larger crystallites than in the virgin sample (curve a).

Since X-Ray diffraction occurs from the top several thousand Angstroms of the material, the bulk of the material in addition to a thin surface layer, contributes to the diffraction process. The observed data indicates that:

(a) the silicone polymer becomes more crystalline with aging, and

(b) the crystallite size of the filler material also increases, supporting the hypothesis that dry band arcing causes permanent changes in the structure of the material.

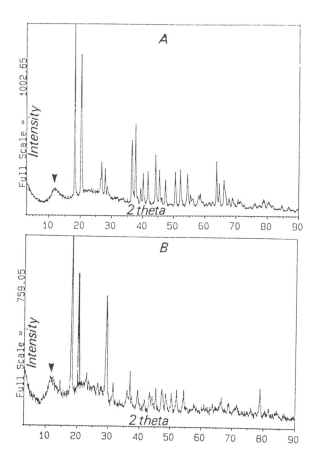

Fig 7.10: Typical Wide Angle Scan of Virgin Sample 1 (A) and Considerably Aged Sample 7 (B).

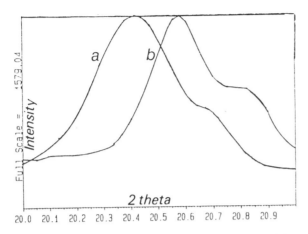

Fig 7.11: Narrow Scan Corresponding to ATH Filler of Virgin Sample 1 (Curve a) and Considerably Aged Sample 7 (Curve b).

Surface Roughness Results: The surface roughness of a new and aged sample is shown in Fig. 7.12.

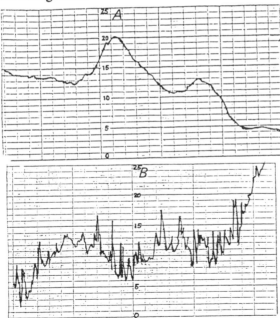

Fig. 7.12. Surface Roughness of the Virgin Sample 1 (A) and Aged Sample 4 (B).

Table 7.11 shows the data which help in quantifying the surface roughness. The last column in the Table gives the number of peaks which have a greater depth than the average value of the virgin sample, and has been included to provide more information on the change in roughness of an aged sample in comparison with the virgin sample. It is evident from the Table that the aged surface has a significantly greater roughness than a virgin surface, which is the experimental evidence for the hypothesis.

Sample ID	Maximum Peak (mm)	Average Peak (mm)	Number of Peaks > 0.004 mm/cm Length of Sample
1	0.020	0.004	4
4	0.025	0.014	32

Table 7.11: Typical Surface Roughness of Virgin Sample 1 and Aged Sample 4.

The following are the conclusions on the aging of silicone rubber weathershed material.

1) For the first time, a hypothesis for aging of silicone rubber material used for outdoor insulation has been proposed and experimentally proven.

2) Aging of the material is indicated by permanent changes, such as, depletion of the low molecular weight polymer chains on the surface, increase in depolymerization, increase in surface roughness, changes in the surface physical structure brought about by increase in the degree of crystallization of the polymer and conglomeration of small filler particles into larger ones.

3) Analytical techniques, such as Fourier Transform Infra-Red Spectroscopy, Energy Dispersive X-Ray Analysis, and surface roughness measurements are useful tools for the measurement of permanent changes. A new technique, namely, X-Ray Diffraction has been demonstrated to be a useful tool for the analysis of polymer insulation aging.

8. TOPICS FOR FURTHER RESEARCH

Polymer based outdoor insulation has been a topic which has received attention both from industry and academia, and a significant number of publications are available. Yet, there a number of areas which need more research. This will enable manufacturers to develop more cost effective products without sacrificing the desirable electrical and mechanical properties, and users to specify and select the right material and design of product for optimal performance. Some of them are:

1) The relationship between the magnitude of dry band arcing activity and material degradation is not yet quantified. It is important to establish the energy required to cause material degradation in terms of a relatively simple measurable quantity such as leakage current. This will enable in the selection of the correct type and composition of the material, and the design for the end product.

2) Estimation of life expectancy is another area which needs a lot of research. At present, there are methods to predict the life of the fiber glass rod with various types of mechanical loading, but none for the useful life of the polymer weathershed. Research in this area requires a detailed knowledge of the aging phenomena, and the relationship between electrical discharge activity and rate of material degradation.

3) Correlation between the results of accelerated laboratory tests and service experience at this stage is only qualitative. For a laboratory test to be universally acceptable, it is essential to relate the life in laboratory test to that in service. A "STANDARD" method of testing is of great benefit to the industry.

4) It has been assumed both in laboratory test development and in the aging hypothesis, that insulation aging is predominantly determined by dry band activity. Yet, it is well known that other factors such as heat, UV, moisture ingress, chemicals, all affect aging. The synergistic effect of all these parameters ultimately decide the useful life of the polymer insulating device. The ultimate aging model and a meaningful accelerated test has to take into account this synergism.

9. BIBLIOGRAPHY

The bibliography has been classified into several sections. Quite frequently, the same publication may provide information of more than one aspect. In such cases, they have been included only in the section representing the major focus of the paper. Only Transactions papers are listed for the simple reason that they can be more easily found in most libraries. The list is by no means complete.

A. EFFECT OF CONTAMINATION ON OUTDOOR INSULATION

This is probably the one of the most extensively researched topic with more than 250 transaction papers. A few good review papers on this topic are:

F.A.M. Rizk, "Mathematical Models for Pollution Flashover", Electra, Vol. 78, pp 71-103, 1981.

D.C. Jolly, "Contamination Flashover Theory and Insulator Design", Journal of Franklin Institute, Vol. 294, pp 483-500, Dec 1972.

E. Nasser, "Flashover of Contaminated Surfaces", ETZ-A, Bd-83, H-6, pp 321-325, 1972.

P.J. Lambeth, "Effect of Pollution on High Voltage Outdoor Insulators", Proc. IEE (London), IEE Reviews, Vol. 118, pp 1107-1130, 1971.

B. SERVICE EXPERIENCE OF POLYMER INSULATION

CIGRE Committee 22, Subworking Group 03-01, "Worldwide Service Experience with HV Composite Insulators, Electra No. 130, 1989.

H.M. Schneider et al., "Non-Ceramic Insulators for Transmission Lines", IEEE Trans. Power Delivery, Vol. 4, No. 4, pp 2214-2221, October 1989.

J.N. Edgar and E. A. Cherney, "Canadian Utility Operating Experience with Polymer Insulators for Transmission Lines", Canadian Electrical Association (CEA) Report, ST-276, May 1989.

R.G. Houlgate et al., "Field Experience and Laboratory Research on Composite Insulators for Overhead Lines", Paper 15-12, CIGRE 1986.

H. Dietz et al., "Latest Developments and Experience with Composite Long Rod Insulators", Paper 15-09, CIGRE 1986.

C. de Tourreil, P. Burdon and J. Lanteigne, "Aging of Composite Insulators Caused by Service and Simulated Service Conditions", Paper 15-01, CIGRE 1986.

H.D. Chandler and G.R. Reynders, "Electrochemical Damage to Composite Insulators", Paper No. 33-08, CIGRE 1984.

L. Vo Van, C. de Tourreil and P.J. Lambeth, "735 kV Composite Insulators: Observations During 6 Years of Service", Paper Presented at the CEA Spring Meeting, Toronto, 1984.

A. Bradwell and J.C.G. Wheeler, "Evaluation of Plastic Insulators for Use on British Railways 25 kV Overhead Line Electrification, IEE, Proc 129B, pp 101-110, 1982.

E. Bauer et al., "Service Experience with the German Long Rod Insulator with Silicone Rubber Weathersheds since 1967", Paper 22-11, CIGRE, 1980.

M. Cojan et al., "Polymeric Transmission Insulators, Their Application in France, Italy and the U.K", Paper 22-10, CIGRE 1980.

Vayaweiheh, Macey and Reynders, "Field Experience and Testing of New Insulator Types in South Africa", Paper 22-43, CIGRE 1980.

R.J.T. Clabburn et al., "The Outdoor Performance of Plastic Materials used as Cable Accessories", Trans IEEE, Vol. PAS-92, pp 1833-1842, 1973.

C. OUTDOOR TESTING

R.G. Houlgate and D.A. Swift, "Composite Rod Insulators for AC Power Lines: Electrical Performance of Various Designs at a Coastal Testing Station", Trans. IEEE, Power Delivery, Vol. 5, No. 4, pp 1944-1955, Oct 1990.

A.E. Vlastos and E. Sherif, "Experience from Insulators with RTV Silicone Rubber Sheds and Coatings", IEEE Trans. Power Delivery, Vol. 5, No. 4, pp 2030-2038, Oct. 1990.

S.M. Gubanski and A.E. Vlastos, "Wettability of Naturally Aged Silicone Rubber and EPDM Insulators", IEEE Trans. Power Delivery, Vol. 5, No. 3, pp 1527-1535, July 1990.

A.E. Vlastos and E. Sherif, "Natural Aging of EPDM Composite Insulators", IEEE Trans. Power Delivery, Vol. 5, No. 1, pp 406-414, Jan 1990.

A.E. Vlastos and S.M. Gubanski, "Surface Structural Changes of Naturally Aged and EPDM Composite Insulators", IEEE Trans. Power Delivery, Paper No. 90 SM 424-2 PWRD.

R.E. Carberry and H.M. Schneider, "Evaluation of RTV Coating for Station Insulators Subjected to Coastal Contamination", IEEE Trans. Power Delivery, Vol. 4, No. 1, pp 577-585, Jan 1989.

T. Orbeck and J. Hall, "Evaluation of a New RTV Protective Coating for Porcelain Insulators", IEEE Trans. Vol. PAS-101, pp 4689-4696, 1982.

C. Aubin et al., "Natural and Artificial Weathering of EPDM Compounds used in Outdoor High Voltage Insulation", IEEE Trans., Electrical Insulation (EI), pp 290-296, 1981.

E.A. Cherney and D.J. Stonkus, "Non-Ceramic Insulators for Contaminated Environments", IEEE Trans. Vol. PAS-100, pp 131-142, 1981.

D. LABORATORY TESTING

S.H. Kim, E.A. Cherney and R. Hackam, "Leakage Current Suppression Mechanisms in RTV Coatings", IEEE Trans. Power Delivery, Paper No 91 WM 115-5 PWRD.

R.S. Gorur, G.G. Karady, A. Jagota, M. Shah and A.M. Yates, "Aging of Silicone Rubber for Outdoor Insulation", IEEE Trans. Power Delivery, Paper No. 91 WM 115-6 PWRD.

R.S. Gorur, E.A. Cherney and R. Hackam, "Effect of Environmental Degrading Factors on the Electrical Performance of Polymeric Insulating Materials", Paper No. MS 203-860, International Journal on Energy Systems, 1991.

H.M. Schneider and A.E. Lux, "Mechanism of HVDC Wall Bushing Flashover in Non-Uniform Rain", IEEE Trans. Power Delivery, Vol. 6, No. 1, pp 448-455, Jan 1991.

R.S. Gorur, V. Chaudhry, M. Dyer and R.S. Thallam, "Electrical Performance of Polymer Housed Surge Arrestors Under Contaminated Conditions", IEEE Trans. on Power Delivery, Paper No. 90 SM 295-6 PWRD, 1990.

G.G. Karady and R.S. Gorur, "Use of RTV Silicone Rubber Coatings for Station Apparatus", CIGRE Letter, 1990.

R.S. Gorur, E.A. Cherney and R. Hackam, "Electrical Performance of Polymeric Insulators Under Contaminated Conditions", International Journal of Energy Systems, Vol. 10, No. 3, pp 136-139, 1990.

R.S. Gorur, L.A. Johnson and H. Hervig, "Contamination Performance of Silicone Rubber Cable Terminations", Paper No. 90 WM 074-5 PWRD, IEEE Trans. Power Delivery, 1990.

R.S. Gorur, J. Chang and O.G. Amburgey, "Surface Hydrophobicity of Polymeric Materials Used for Outdoor Insulation Applications", IEEE Trans. Power Delivery, Vol. 5, No. 4, pp 1923-1933, Oct 1990.

C. de Tourreil and P.J. Lambeth, "Aging of Composite Insulators: Simulation by Electrical Tests", IEEE Trans. Power Delivery, Vol. 5, No. 3, pp 1558-1567, July 1990.

S.H. Kim, E.A. Cherney and R. Hackam, "The Loss and Recovery of Hydrophobicity of RTV Silicone Rubber Insulator Coatings", IEEE Trans. Power Delivery, Vol. 3, pp 1491-1499, July 1990.

D. Dumora, D. Feldmann and M. Gaudry, "Mechanical Behavior of Flexurally Stressed Composite Insulators", IEEE Trans. Power Delivery, Vol. 5, No. 2, pp 1066-1073, April 1990.

S.M. de Oliveria and C. de Tourreil, "Aging of Distribution Composite Insulators Under Environmental and Electrical Stresses", IEEE Trans. Power Delivery, Vol. 5, No. 2, pp 1074-1077, April 1990.

R.S. Gorur, E.A. Cherney and R. Hackam, "Polymer Insulator Profiles Evaluated in a Fog Chamber", IEEE Trans. Power Delivery, Vol. 5, pp 1078-1085, April 1990.

H.M. Schneider et al., "Rain and Contamination Tests on HVDC Wall Bushings with and without RTV Coatings", IEEE Trans. Power Delivery, Paper No. 90 SM 393-9 PWRD.

R.S. Gorur, S. Sundhara Rajan and O.G. Amburgey, "Contamination Performance of Polymeric Materials for Outdoor Insulation Applications", IEEE Trans. Electrical Insulation, Vol. EI-24, pp 713-716, August 1989.

R.S. Gorur, E.A. Cherney and R. Hackam, "Performance of Polymeric Cable Terminators in Salt-fog", IEEE Trans. Power Delivery, Vol-4, no. 2, pp 842-849, April 1989.

S. P. Nunes et al., "Tracking Degradation and Pyrolysis of EPDM Insulators", IEEE Trans. Electiical Insulation, Vol. EI-24, No. 1, pp 99-105, Feb 1989.

R.S. Gorur, E.A. Cherney and R. Hackam, "The ac and dc Performance of Polymeric Insulating Materials Under Accelerated Aging in a Fog Chamber", IEEE Trans. Power Delivery, Vol.3, no. 4, pp 1892-1902, October 1988.

E.A. Cherney, "Long-Term Mechanical Life Testing of Polymeric Post Insulators for Distribution and a Comparison to Porcelain", IEEE Trans. Power Delivery, Vol. 3, No. 3, pp 1141-1145, July 1988.

R.S. Gorur, E.A. Cherney, R. Hackam and T. Orbeck, "The Electrical Performance of Polymeric Insulating Materials Under Accelerated Aging in a Fog Chamber", IEEE Trans. Power Delivery, Vol. 3, No.3, pp 1157-1164, July 1988.

D.W. Lenk, F.R. Stockum and D.E. Grimes, "A New Approach to Distribution Arrester Design", IEEE Trans. Power Delivery, Vol. 3, No. 2, pp 584-590, April 1988.

R.S. Gorur, E.A. Cherney and R. Hackam, "Performance of Polymeric Insulating Materials in Salt-fog, IEEE Trans. Power Delivery, Vol. 2, no. 2, pp 486-492, April 1987.

T. Ishikara et al., "Aging Degradation of the Mechanical Properties of Composite Insulators and its Analytical Approaches", IEEE Trans. Power Delivery, Paper No. 86 SM 423-428.

M. Ieda et al., "Testing of High Polymer Insulation for Outdoor Application Review, Analysis and Development", Paper 15-11, CIGRE 1986.

R. S. Gorur, E. A. Cherney and R. Hackam, "A Comparative Study of Polymer Insulating Materials Under Salt-fog Conditions", IEEE Trans. Electrical Insulation, Vol. EI-21, no. 2, pp 175-182, April 1986.

R. S. Gorur, E. A. Cherney and R. Hackam, "A Comparative Study of Polymeric Insulating Materials Under Contaminated Conditions", Canadian Electrical Association, Transactions on Engineering and Operation Division, vol-25, 1985-86.

C. de Tourreil, R. Roberge and P. Bourdon, "Long Term Mechanical Properties of High Voltage Composite Insulators", IEEE Trans. PAS, Paper No. 85 WM 192-0.

IEEE Insulated Conductors Committee WG 12-39, "Proposed Extended Time Artificial Pollution Tests for Polymeric Cable Accessories", IEEE Trans. PAS, Paper Number 84 SM 503-9.

E.A. Cherney et al., "Evaluation and Application of Dead-End Type Composite Insulators for Distribution", IEEE Trans. PAS, Vol. PAS-102, pp 121-132, 1984.

G.R. Mitchell, "Present Status of ASTM Tracking Test Methods", Journal of Testing and Evaluation, Vol. 2, No. 1, pp 121-132, 1984.

"Test Method for Evaluating Resistance for Tracking and Erosion of Electrical Insulating Materials used under Severe Ambient Conditions", IEC 587, 1984.

E.A. Cherney et al., "The AC Clean-Fog Test for Contaminated Insulators, IEEE Trans. PAS, Vol. PAS-102, pp 604-612, Mar 1983.

D.C. Jolly, "A Quantitative Method for Determining the Resistance of Polymers to Surface Discharges", IEEE Trans. Electrical Insulation, Vol. EI-17, pp 293-299, 1982.

"VDE Guide for Power Cable Accessories with Rated Voltage to 30 kV", VDE Guide 0278, October 1982.

K.J.L Paciorek et al., "Moist Tracking Investigations of Organic Insulating Materials", IEEE Trans. Electrical Insulation, Vol. EI-17, pp 423-428, 1982.

E.A. Cherney et al, "Minimum Test Requirements for Non-Ceramic Insulators", IEEE Trans. PAS, Vol. PAS-100, pp 882-890, 1981.

"Standard Test Method for Liquid-Contaminant , Inclined Plane Tracking and Erosion of Insulating Materials", Annual Book of ASTM Standards, Vol. 39, pp 548-560, 1981.

Working Group 04, Study Committee 33, "A Critical Comparison of Artificial Pollution Test Methods for HV Insulators", Electra, pp 117-136, May 1979.

J. M. Mason, "Discharges", IEEE Trans. Electrical Insulation, Vol. EI-13, pp 211-238, 1978.

T. Tanaka, K. Naito and J. Kitagawa, "A Basic Study on Outdoor Insulators of Organic Materials", IEEE Trans. Electrical Insulation, Vol. EI-13, pp 184-193, 1978.

G.G. Karady, N. Souchereay and R. Vinet, "New Test Methods for Synthetic Insulators", Paper 22-15, CIGRE 1976.

"Artificial Pollution Tests on High Voltage Insulators to be used on AC Systems", IEC Report 507, 1975.

G. Karady, "The Effect of Fog Parameters on the Testing of Contaminated Insulators", IEEE Trans. PAS, Vol. PAS-75, pp 378-387, 1975.

R.M. Scarisbrick, "Electromechanical Erosion of Epoxy Resin Outdoor Insulators", Trans. IEE, Vol.21, pp 779-783, 1974.

R.G. Niemi and T. Orbeck, "Test Methods Useful in Determining the Wet Voltage Capability of Polymeric Insulator Systems after Time Related Outdoor Exposures", IEEE Trans. Electrical Insulation, Vol. EI-9, No. 3, pp 102-108, 1974.

T.W. Dakin, "Application of Epoxy Resins in Electrical Apparatus", IEEE Trans. Electrical Insulation, Vol. EI-19, pp 121-128, 1974.

M.J. Billings, A. Smith and R. Wilkins, "Tracking in Polymeric Materials", IEEE Trans. Electrical Insulation, Vol. EI-21, pp 175-182, 1972.

R.G. Niemi and T. Orbeck, "High Surface Resistance Protective Coatings for High Voltage Insulators", IEEE Trans. PAS, Vol. PAS-91, pp 2263-2270, 1972.

M.J. Billings, L. Warren and R. Wilkins, "Thermal Erosion of Electrical Insulating Materials", IEEE Trans. Electrical Insulation, Vol. EI-6, pp 82-90, 1971.

M. Kurtz, "Comparison of Tracking Test Methods", IEEE Trans. Electrical Insulation, Vol. EI-6, pp 76-81, 1971.

R.W. Wilkins and M.J. Billings, "Effect of Discharges Between Electrodes on the Surface of Organic Insulation", Trans. IEE, Vol-116, pp 1777-1784, 1969.

D.J. Parr and R.M. Scarisbrick, "Performance of Synthetic Insulating Materials Under Polluted Conditions", Trans. IEE, Vol-112, pp 1625-1632, 1965.

N. Parkman, "Electrical Breakdown by Tracking", Trans. IEE, 109B, Suppl. 22, 1962.

R.S. Norman and A.A. Kessel, "Internal Tracking Mechanism for Non-Tracking Organic Insulators", Trans. AIEE, 77, pp 532-536, 1958.

E. BOOKS AND BOOK CHAPTERS RELATED TO POLYMER MATERIALS AND DIAGNOSTIC TECHNIQUES

J.I. Kroschwitz (Editor), "Encylcopedia of Polymer Science and Engineering", vol-15, John Wiley and Sons, 1989.

J.D. Andrade (Editor), "Polymer Surface Dynamics", Plenum Publishing Corporation, 1988.

S.M. Sze (Editor), "VLSI Technology", McGraw Hill, 1988.

R.J.H. Clark and R.E. Hester (Editors), "Spectroscopy of Surfaces", Chister: New York: Wiley, 1988.

J.D. Andrade (Editor), "Surface and Interfacial Aspects of Biomedical Polymers", Vol-1, Plenum Press, 1985.

Souheng Wu, "Polymer Interface and Adhesion", Marcel Dekker Inc, New York and Basel, 1982.

C. Hall, "Polymer Materials", The Macmillan Press Ltd., 1981.

G.E. Muilenberg (Editor), "Handbook of X-Ray Photoelectron Spectroscopy", Perkin Elmer Corporation, 1978.

B.D. Cullity, "Elements of X-Ray Diffraction", Addison Wesley, 1978.

J.S. Mattson, H.B. Mark Jr., and H.C. McDonald Jr. (Editors), "Infra-Red, Correlation and Fourier Transform Spectroscopy", Marcel Dekker, 1977.

A.W. Adamson, "Physical Chemistry of Surfaces", John Wiley and Sons, 1976.

J.I. Goldstein and R.K. Yakowitz, "Practical Scanning Electron Microscopy and Ion Microprobe Analysis", Plenum Press, 1975.

J. Crank and G.S. Park, "Diffusion in Polymers", Academic Press, 1968.

A.J. Barry and H.N. Beck, in the book "Inorganic Polymers", Edited by W.A.G. Graham and F.G.A. Stone, Academic Press, pp 189-320, 1962.

F.M. Clark, "Insulating Materials for Design and Engineering Practice", John Wiley and Sons, 1962.

POWER SYSTEM GENERATION EXPANSION PLANNING USING THE MAXIMUM PRINCIPLE AND ANALYTICAL PRODUCTION COST MODEL

KWANG Y. LEE

Department of Electrical and Computer Engineering
The Pennsylvania State University
University Park, Pennsylavania 16802

YOUNG MOON PARK

Department of Electrical Engineering
Seoul National University
Seoul 151, Korea

I.   INTRODUCTION

Historically, the electric utility demand in most countries has increased rapidly, with a doubling of approximately 10 years in the case of developing countries. In order to meet this growth in demand, the planners of expansion policies were concerned with obtaining expansion plans which dictate what new generation facilities to add and when to add them.

However, the practical planning problem is extremely difficult and complex, and required many hours of the planner's time even though the alternatives examined were extremely limited. In this connection, increased motivation for more sophisticated techniques of evaluating utility expansion policies has been developed during the past decade.

CONTROL AND DYNAMIC SYSTEMS, VOL. 44

Among them, the long-range generation expansion planning is to select the most economical and reliable generation expansion plans in order to meet future power demand over a long period of time subject to a multitude of technical, economical, and social constraints.

But it involves generally the following difficult and complex problems to overcome:

i) What kind of optimization technique is to be utilized?

ii) How can the uncertainties such as future load growth, fuel cost and availability, hydrological quantities, availability of facilities, economic fluctuations, etc. be predicted?

iii) How can the social constraints such as environmental, legal and political factors, social impact by supply shortage, etc. be quantified?

As far as the optimization techniques are concerned, their validity of resulting solution in terms of the realistic planning problem is only as good as how well the model realistically approximates the planning problem.

Much work has been done in applying the formal optimization techniques to the planning problems such as linear programming, mixed-integer programming, integer programming, dynamic programming, nonlinear programming, etc.

There are a number of linear programming formulations of the generation planning problem [1-4], but the linear programming models, in general, increase problem size, and result in a poor measure on the operating costs.

The nonlinear programming formulation [5] of the generation planning problem, which makes it easier to represent the operating cost more accurately and also to reduce the number of constraints, has been proved to be not efficient in computation as well as to face with difficulties in obtaining optimum values.

Furthermore, another shortcoming of both the linear and nonlinear formulations is that they cannot recognize the fact that the investment cost of each year depends on the decisions made in the preceding years. As a consequence, considerable work has been done to improve the cost model as a multi-stage decision making problem within the dynamic programming framework [6]. There are several variations of this formulation [7-9]. For example, the WASP (Wien Automatic System Planning) package [7] which utilizes the dynamic programming, is designed to find the optimal generation expansion policy for an electric utility system with a probabilistic simulation model incorporated. However, most of the reported work in dynamic programming is for small problems since the dimensionality of the problem highly restricts its use.

Another recent approach to the dynamic expansion problem is to use the optimal control theory. Its typical work can be found in the MNI (Model of National Investment) package developed by the EDF (Electricite de France) in France [10-12]. Here, the system reliability constraints are replaced by the supply shortage cost which is included in the cost function, and the optimal control problem is solved by using the gradient method.

Recently, Park, Lee and Youn [23] developed a new approach to the optimal generation expansion planning. It also uses the discrete optimal control theory which the MNI introduced, but differs from that in many respects. The probabilistic simulation model for operation is approximated by Gaussian random distributions for load curves (rather than load duration curves) resulting in an analytical production costing model. The optimization problem for each year is solved by using a modified gradient projection method rather than the gradient method.

II.   PROBLEM STATEMENT

A.   Objectives

The problem is to determine the most economical and reliable generation and expansion plan in order to meet forecasted load over a long-range horizon (usually five to thirty years), subject to a multitude of technical, economical and social constraints.

B.   Long-Term Generation Expansion Planning

The optimal long-term generation expansion planning (OLGEP) problem is now formally defined.

Given the following information:

- the load forecast over the planning horizon;
- the existing system description;
- hydrological data;
- economic data such as investment prices, money escalation rates, price discount rates, useful life for each type of generating unit;
- possibilities on alternative energy sources;
- description of uncertainties in energy sources;
- failure rates and maintenance rates of generating units;
- other technical and economical constraints.

Then the OLGEP is to find:

- The annual capacity additions and their investment costs, annual operation costs, reliability indices or failure costs, expected annual or seasonal energy to be produced by each type of plant;
- the expected generation schedules and maintenance schedule for each plant type;
- the operational reserves for each plant type;
- the marginal production costs and other economical data usable for rationally determining electric rates and load management strategies;
- cash flow in operation and investment costs.

To solve the OLGEP problem, the following simplifying, but realistic, assumptions will be made:

- the random load at a given point in time (rather than the load duration curve commonly used) is represented by a Gaussian random variable where mean is available by load forecast;
- the generating plants are grouped into several types, such as nuclear, base oil, intermediate oil, base coal, intermediate coal, peaking gas turbine, hydro, pumped storage, etc. All units of the same type are assumed to have the same economic and technical characteristics.

The first assumption makes it possible to develop a closed-form analytical production costing model and reliability indices rather than simulation models. This will result in a considerable reduction in computation time. The second assumption makes it possible to represent each plant type with a number of identical single units of the same size, which also contributes to the development of the closed-form analytical production costing model.

III.   PRODUCTION COST MODEL

During the past decade this problem has received considerable attention, and the motivation for more efficient and more sophisticated techniques of evaluating utility expansion policies has been increased [1,2,4,8,13,14]. Also, it has undergone significant changes during the past decade. Since capacity expansion planning is to be carried out in an environment of relative certainty, tools are becoming more complicated and sophisticated. However, in these studies there has always been an issue of trade-offs between computational effort and accuracy of the solution. Here, presents a new analytical approach is presented for the production cost model, which is the basis for the generation expansion planning. The model provides a comparable accuracy with much less computational effort than the conventional one.

A.   Load Modeling

The main trends for production cost model are to use re-ordered load duration curves rather than the original (chronological) load variation curves so that the cumulant or inverted load duration curves may be used in order to apply convolution and/or deconvolution techniques. The applied techniques in general, are less analytical, time consuming and recursive, thus yielding unacceptable erroneous results; particularly in the case of the loss-of-load probability (LOLP) calculation.

The efficiency-accuracy trade-offs become particularly pronounced when the hydro and other limited energy plants are part of the generating system, and must be optimally dispatched by taking into account thermal unit forced outages. This requires repeated loading and unloading of thermal units (via. convolutions and deconvolutions). It is also common to use only one load curve to represent a yearly load demand. If different load curves for each season of a year is used, the number of convolution processes would increase as well, resulting in a big increase of computational time.

1.   Analytical Approach for Production Cost Modeling

The basic idea of the analytic approach is similar to the probabilistic production model of Booth and Baleriaux [13,14], where the load and generating units are modelled as random variables. However, the new analytical production cost model is developed under the assumption of Gaussian probability distribution for random load fluctuations and plant outages.

One of the primary features of this approach is the direct use of load curves rather than the use of inverted load duration curves. Thus, not only is the computational requirement reduced in the order of magnitudes due to the absence of numerical convolution processes, but also greater accuracy is achieved due to the analytical nature of the production cost model. Even in the case where hydro and other limited energy plants are included in the generating system, the new analytical approach eliminates the convolution and deconvolution processes.

The Gaussian assumption for load fluctuations may seem to be inadequate and produce much error. This becomes an issue when one models the skewed distribution of load duration curves. However, when the original (chronological) load curves are used rather than the load duration curves, the Gaussian assumption never poses a problem. Instead of the total accumulated load for a cycle, the (chronological) load curve is divided into several time bands, each with few hours of duration. Thus the variance is very small in each time bands and the error is negligible in spite of the Gaussian assumption. If this is viewed as a problem for someone the higher order moments can be included in the analysis. However, it is more practical to reduce the duration of each time band than to include the higher moments.

Another important feature is that several load curves can be used to represent an annual load demand without increasing the computational effort too much; which is not the case in conventional method. Thus a representation of annual demand becomes more realistic and accurate.

2. Representation of Random Load Fluctuations

The load, at a particular time of the day of the week in a given season, fluctuates randomly and it is reasonable to assume that it behaves with Gaussian distribution since its value is obtained from a large number of previous historical data.

Any kind or length of load curve can be used for the generation expansion planning purpose. It is, however, common to use a weekly load curve because it represents more accurately the behaviour of the load demand for a season. When a weekly load curve is used there are several options to represent it as follows:

1) use 7 daily load cycles; or

2) use one equivalent weekday load cycle for 5 days, and one equivalent weekend load cycle for 2 days; or

3) use one equivalent load cycle for 7 days.

In a planning problem, although the overall pattern of the load curve is predictable, it is difficult to make hourly predictions. For this reason in the conventional method, the time-dependent load curve is converted into a load duration curve. When planning with limited energy and storage plants, this procedure ignores the inherent chronological nature of storage operations. In the new analytical production cost model, depending upon the planning purpose the load curve can be either time-dependent or time-independent. For example the option 1) mentioned above is time-dependent and options 2) and 3) are time-independent. The options 2) and 3) would be more suitable for long range generation expansion planning purpose.

The statistical means and variances for the past years can be easily computed following the standard statistical analysis. However, they can not be computed for the future years and must be forecasted on the basis of past values as well as future economic and social considerations.

Given the statistical values of future loads on basis of load forecasting, the expected annual energy demand for year i, $\overline{D}_i$, can be computed as

$$\overline{D}_i = \sum_{s=1}^{S} \sum_{t=1}^{T} n_s \tau_t \overline{L}_{i,s,t} = \sum_{s=1}^{S} \sum_{t=1}^{T} n_s \tau_t \overline{L}_\Delta, \qquad (1)$$

where

$\Delta = (i, s, t)$: index of the year, season and time-band, respectively,

$\overline{L}_\Delta$: statistical mean of random load L_Δ for year i, season s, and time-band t [MW],

τ_t : length of time-band t [hrs.],

n_s : number of load cycles in a season s,

T : number of time-bands in a load cycle,

S : number of seasons in a year.

B.   Generation Modeling

1.   Available Generation Capacities

The generating plant are grouped in several types, such as nuclear, base oil, intermediate oil, base coal, intermediate coal, peaking gas turbine, etc. All units which are the same type are assumed to have the same economic and technical characteristics. This makes it possible to represent each type with a number of identical units of same size. The unit capacity, of course, will differ from type to type.

The probabilistic nature of plant outages of units in each plant type j makes the available generating capacity $\overline{y}_\Delta^j$ as a random variable. Each unit is considered to have two stages of operation: full-capacity operation with availability p, and full-outage with failure rate (1-p). It is assumed that each plant type has many such units which are mutually independent. Then, according to the law of large numbers the total available generating capacity y_Δ^j for type j can be represented by normal distribution with its statistical mean $\overline{y}_\Delta^j$ and variance $\sigma_\Delta^{j^2}$, which are derived as [2]

$$\overline{y}_\Delta^j = p^j(1 - \gamma_{i,s}^j)\beta_\Delta^j(x_i^j + \alpha^j u_i^j) \tag{2}$$

$$\sigma_\Delta^{j^2} = p^j(1 - p^j)(1 - \gamma_{i,s}^j)\beta_\Delta^j uc^j(x_i^j + \alpha^j u_i^j), \tag{3}$$

where

p^j : availability of plant type j [p.u.],

uc^j: unit capacity of plant type j [MW],

x_i^j : previously installed capacity of plant type j in year i [MW],

u_i^j : newly added capacity of plant type j in year i [MW],

α^j : teething factor for units in type j [p.u.],

β_Δ^j : energy resource distribution factor of plant type j in time-band t of season

s

  of year i [p.u.],

$\gamma_{i.s}^j$: maintenance rate of plant type j in season s of year i [p.u].

The teething factor is defined to be the discount factor to reflect the nature of newly installed unit which cannot operate at its fully capacity during the first year of operation. The energy resource distribution factors β_Δ^j are fixed to one for the non-hydro plant, and are determined optimally for hydro plants in order to find the optimal use of hydro resources. However, if there are situations where the available generating capacity is limited by other factors, the value of β_Δ^j less than one can be used for thermal plant. For example it can be used to represent the availability of fuel for thermal plants.

Note that the formulae for available generating capacities contain three control (decision) variables: newly added plant capacity u_i^j, energy resource distribution factor β_Δ^j, and maintenance rate $\gamma_{i,s}^j$. These variables may later be determined optimally in long-range investment, hydro, and maintenance optimization problems, respectively.

2.   Expected Plant Output and Annual Energy Generation

Comparing forecasted load and available generating capacities, the expected plant output can be determined optimally in the production scheduling. If the generating plant types are confined to non-hydros, the optimal solution for this operating problem is the conventional loading order concept. The loading order is defined as the economic merit order in which the plant is loaded in the order of increasing operating costs. Although the generating units use the same fuel type, they may be grouped into different plant types according to their fuel efficiency.

a. Expected Power Output. Let j=1,...,J be the indices of plant types already in the order of increasing operating costs. The total power output from plant type 1 to j, $^jP_\Delta$, in a time-band t does not exceed either the load or the total available generating capacity of the system. Thus it can be expressed as

$$^jP_\Delta = min(L_\Delta, \sum_{k=1}^{j} y_\Delta^k) \tag{4}$$

where

$L_\Delta = L_{i,s,t}$: load at time-band t of season s in year i [MW],

y_Δ^k : available generating capacity of plant type k at time-band t of season s in year i [MW].

Since the load and available generating capacities have the Gaussian probability distributions, the total output $^jP_\Delta$ is also a random variable with Gaussian distribution, thus it can be represented by its mean $^j\overline{P}_\Delta$ and variance $^j\sigma_\Delta^{p^2}$, derived as [2]

$$^j\overline{P}_\Delta = \overline{L}_\Delta - {}^j\overline{V}_\Delta[0.5 + erf({}^j\overline{V}_\Delta/{}^j\sigma_\Delta^p)] - ({}^j\sigma_\Delta^p/\sqrt{2\pi})exp(-0.5\,{}^j\overline{V}_\Delta^2/{}^j\sigma_\Delta^{p^2}) \tag{5}$$

$$^j\sigma_\Delta^{p^2} = \sigma_\Delta^{L^2} + {}^j\sigma^{y^2}\Delta = \sigma_\Delta^{L^2} + \sum_{k=1}^{j} \sigma_\Delta^{k^2} \tag{6}$$

where

$$^j\overline{V}_\Delta = \overline{L}_\Delta - {}^j\overline{y}_\Delta = \overline{L}_\Delta - \sum_{k=1}^{j} \overline{y}_\Delta^k. \tag{7}$$

Here $^j\overline{V}_\Delta$ represents the difference between the load and the sum of the available generating capacities from type 1 to j, and $^j\sigma_\Delta^{y^2}$ represents the variance of the total available generating capacity from type 1 up type j.

The expected power output of each plant can be computed simply by

$$\overline{P}_\Delta^j = \begin{cases} {}^j\overline{P}_\Delta - {}^{j-1}\overline{P}_\Delta, & \text{if } j = 2,3,..,J; \\[2mm] {}^1\overline{P}_\Delta, & \text{if } j = 1, \end{cases} \tag{8}$$

where

$^{j}\overline{P}_{\Delta}$:expected total power output of all plants from type 1 up to type j [MW],

$\overline{P}^{j}_{\Delta}$:expected power output of plant j alone [MW].

b. Expected annual energy. From Eq. (8) the expected annual energy generated by the plant type j, $\overline{E}^{j}_{i}$, is obtained by integrating the expected power output Eq. (8), i.e.,

$$\overline{E}^{j}_{i} = \sum_{s=1}^{S} \sum_{t=1}^{T} n_s \tau_t \overline{P}^{j}_{\Delta}, \tag{9}$$

where

S : number of seasons in a year,

T : number of time-bands in a load cycle,

n_s : number of load cycles in season s,

τ_t: length of time-band t [hrs.].

The expected total energy in year i is, simply by adding for all plant types,

$$\overline{E}_{i} = \sum_{j=1}^{J} \overline{E}^{j}_{i}. \tag{10}$$

Note that the power output formulae, Eqs. (5) - (8), contain simple error and exponential functions. Together with Eqs. (2) and (3), these analytical formulae are extremely simple to solve; taking only a few seconds. As one can see, there is no need of convolution or deconvolution throughout the procedure, and consequently computation time is very small.

C.   Reliability Measures

The expected unserved energy and the loss-of-load probability (LOLP) are the common reliability measures. These reliability measures can also be obtained analytically, mainly using the results in the production cost model.

1.   Expected Annual Unserved Energy

The expected unserved power $\overline{PN}_\Delta$ is the excess of the expected load compared to the expected total power output from all types, i.e., by setting j=J in Eq. (5),

$$\overline{PN}_\Delta = \overline{L}_\Delta - {}^J\overline{P}_\Delta. \tag{11}$$

Thus the expected annual unserved energy, $\overline{EN}_i$, is simply

$$\overline{EN}_i = \sum_{s=1}^{S} \sum_{t=1}^{T} n_s \tau_t \overline{PN}_\Delta. \tag{12}$$

2.   The Annual Loss-of-Load Probability

The positive value of random variable ${}^J V_\Delta$ defined in Eq. (7) is a shortage of generation. Its probability density function is Gaussian since the load and available generation capacity are Gaussian. Thus by integrating the probability density function the loss-of-load probability in each time band is computed as [2]

$$LOLP_\Delta = 0.5 + erf({}^J\overline{V}_\Delta/{}^J\sigma_\Delta^p), \tag{13}$$

where the mean and variance are defined by Eqn. (6) and (7) with j=J.

The annual loss-of-load probability, $LOLP_i$ is simply obtained by integrating over a year as

$$LOLP_i = (\sum_{s=1}^{S} \sum_{t=1}^{T} n_s \tau_t LOLP_\Delta)/n_y \tag{14}$$

where

n_y: total hour in a year [hrs].

Note that the reliability measures are given explicitly as simple analytical formulae. Again, there is no convolution or deconvolution throughout the formulae, and computation is extremely simple.

D.   Production Cost

1.   Expected Annual Cost

Having computed the generation capacities and the expected energy generated, the expected annual cost can be computed in a straightforward manner. The annual cost G_i consists of three terms: capital cost C_i, fuel cost F_i and maintenance cost M_i. The annual capital cost is calculated based on the newly added capacities determined optimally each year. A credit may be given as salvage value for the unused portion of the plant life.

Computing an accurate value for annual fuel cost is often very difficult in the conventional method since its computation is based on load curve and numerical convolution processes. Especially, the error in the values of energy generated by each plant is more pronounced in generating plants of high loading order. In this production costing model the expected energy generated by each plant is calculated by Eqs. (5) - (8). The analytical nature of of the production costing model provides an accurate value of the total annual fuel cost with much less computational effort. The new added capacity of each plant is included in its installed capacity to compute the maintenance cost. In summary, the expected annual cost can be expressed as

$$G_i(x_i^j, u_i^j, \beta_\Delta^j, \gamma_{i,s}^j) = C_i + F_i + M_i$$
$$= \sum_{j=1}^{J} [pc_i^j c_i^j r_i^j u_i^j + pf_i^j f_i^j \overline{E}_i^j + pm_i^j m_i^j (x_i^j + u_i^j)] \qquad (15)$$

where

pc_i^j, pf_i^j, pm_i^j : factors to levelize the resultant value unit capacity cost, unit fuel

price and non-fuel maintenance cost, respectively, of plant type j in year i [p.u.],

c_i^j, f_i^j, m_i^j : unit capacity cost [\$/MW], unit fuel price [\$/MWH], and unit

non-fuel maintenance cost [\$/MW], respectively, of plant type j in year i,

r_i^j : factor for the salvage value of capital cost for type j installed in year i [p.u.],

x_i^j : previously installed capacity of plant type j in year i [MW],

u_i^j : newly added capacity of plant type j in year i [MW],

$\overline{E}_i^j$:expected total energy generation of plant type j in year i [MWH].

This total annual cost is expressed as a function of three decision variables: u_i^j, β_Δ^j, and $\gamma_{i,s}^j$. Thus it can be used as an objective function for optimal investment, hydro operation, or maintenance scheduling problems, respectively. For the long-range expansion planning, the yearly cost function is summed up for all years over the planning horizon [2], and the maximum principle can be applied to solve for the yearly investment. This approach is similar to the transmission planning problem, and the VAR planning problem [15].

2.   Expected Marginal Values

When the cost function in Eq.(15), or its summation for several years, is to be optimized it is required to know its marginal value to find the optimal direction and the sensitivities with respect to decision variables. In conventional methods, the marginal values are obtained numerically after repeated simulation. However, since our cost function is given in analytical formulae, its marginal values can be obtained analytically. Similarly, marginal values for the reliability measures can also be obtained analytically.

The expected marginal values are computed analytically by taking the partial derivatives of the reliability measure $\overline{EN}_i$ in Eq. (12) and the expected annual cost G_i in Eq. (15) with respect to the previously installed capacity x_i^j and newly added capacity u_i^j. They are summarized as follows:

$$\frac{\partial \overline{EN}_i}{\partial u_i^k} = \alpha^k \sum_{s=1}^{S} \sum_{t=1}^{T} n_s \tau_t [-p^k (1 - \gamma_{i,s}^k) \beta_\Delta^k [0.5 + erf(^J\overline{V}_\Delta / ^J\sigma_\Delta^p)]$$

$$+ [p^k (1 - p^k)(1 - \gamma_{i,s}^k) \beta_\Delta^k u c^k / (2\sqrt{2\pi} ^J\sigma_\Delta^p) exp(-0.5 ^J\overline{V}_\Delta / ^J\sigma_\Delta^p)] \quad (16)$$

$$\frac{\partial G_i}{\partial x_i^k} = \frac{\partial F_i}{\partial x_i^k} + \frac{\partial M_i}{\partial x_i^k}$$

$$= m_i^k + \sum_{j=1}^{J} f_i^j \sum_{s=1}^{S} \sum_{t=1}^{T} n_s \tau_t [p^k (1 - \gamma_{i,s}^k) \beta_\Delta^k (h(j - k) [0.5$$

$$+ erf(^j\overline{V}_\Delta/^j\sigma^P{}_\Delta)]) - h(j-1-k)[0.5 + erf(^{j-1}\overline{V}_\Delta/^{j-1}\sigma^P{}_\Delta)]$$
$$+ [p^k(1-p^k)(1-\gamma^k_{i,s})\beta^k_\Delta uc^k/(2\sqrt{2\pi}^j\sigma^P_\Delta)] - h(j-k)exp(-0.5^j\overline{V}^2_\Delta/^j\sigma^{P^2}_\Delta)$$
$$+ h(j-1-k)exp(-0.5^{j-1}\overline{V}^2_\Delta/^{j-1}\sigma^{P^2}_\Delta) \tag{17}$$

$$\frac{\partial G_i}{\partial u^k_i} = \frac{\partial C_i}{\partial u^k_i} + \frac{\partial F_i}{\partial u^k_i} + \frac{\partial M_i}{\partial u^k_i}$$
$$= c^k_i r^k_i + m^k_i + uc^k \frac{\partial F_i}{\partial x^k_i} \tag{18}$$

where

$$h(x) = \begin{cases} 1, & \text{if } x \geq 0; \\ 0, & \text{if } x < 0. \end{cases} \tag{19}$$

Note that the expected marginal values of annual costs and reliability measures are given in analytical formulae. Thus they can be readily computed for optimization problems. These marginal values are also useful managerial information.

E. Simulation Results For Production Cost Model

The load cycle of Scenerio B and the full scale system of Scenerio D from EPRI Synthetic Utility Systems [16] are used as load and generation system data in order to test the accuracy of the analytical formulae. The result is then compared with the conventional method, which is based on the use of equivalent load duration curves and their recursive convolutions. System data and load cycle are shown in Tables I and II, respectively. The sample system has 174 units in 11 plant types. Simulation period, peak load, and minimum load are 728[hrs.], 23,889 [MW], and 11,575[MW], respectively.

The result of using the conventional method was obtained from EPRI Report EA-1411 [17]. It is shown in Table III and compared with the result computed using the formulae in the analytical production cost model.

The load cycle in Table II is a chronological load curve. Unlike the conventional methods, we do not convert it into an equivalent load duration curve. The load cycle

Table I.   System Data

Plant Type	Unit Capacity [MW]	Total Capacity [MW]	Availa- bility	Fuel Cost [$/MWH]
1.Nuclear	1,200	7,200	0.850	7.90
2.Nuclear	800	800	0.850	7.90
3.Coal	800	800	0.760	17.50
4.Coal	600	1,800	0.790	17.81
5.Coal	400	2,000	0.870	18.13
6.Coal	200	6,600	0.920	19.37
7.Oil	800	800	0.760	35.06
8.Oil	600	1,800	0.790	36.45
9.Oil	400	800	0.870	36.78
10.Oil	200	4,600	0.926	39.68
11.Combu. Turbine	50	4,800	0.760	61.46

Table II.   Load Data

Time-Band	Load [GW]	Variance [GW^2]
1	16.4860430	0.80398500
2	15.4261680	0.15315281
3	14.7908398	0.17671687
4	14.5754102	0.10683937
5	14.8881562	0.33597462
6	16.0490547	1.51785700
7	18.0770547	4.41450000
8	20.4079570	6.32913200
9	22.2255391	5.04400600
10	23.3756680	3.69544000
11	23.8780234	3.22498700
12	23.8887969	2.87046400
13	23.8327109	3.24225800
14	23.6607383	3.38979800
15	23.3923828	3.36281100
16	23.3427969	3.57017500
17	23.5860820	3.23661800
18	23.5550703	2.73120600
19	23.1721250	2.48781000
20	22.7585391	2.25092500
21	22.7460977	1.92058100
22	22.3393828	1.47737100
23	20.4621836	0.83491650
24	18.1576562	0.99275250

Table III.   Comparison Between Conventional and
Analytical Methods

Plant Type	Availa- bility	Conventional Method		Analytical Method	
		Energy Generated [GWH]	Capacity Factor [%]	Energy Generated [GWH]	Capacity Factor [%]
1.Nuclear	0.850	4,455.36	85.00	4,455.36	85.00
2.Nuclear	0.850	495.04	85.00	495.04	85.00
3.Coal	0.760	442.16	75.92	442.62	76.00
4.Coal	0.790	1,028.11	78.46	1,035.21	79.00
5.Coal	0.870	1,222.92	83.99	1,266.65	87.00
6.Coal	0.920	3,492.61	72.69	4,127.18	85.90
7.Oil	0.760	214.00	53.92	334.93	57.51
8.Oil	0.790	635.52	48.50	733.36	55.96
9.Oil	0.870	298.41	51.24	333.64	57.29
10.Oil	0.926	1,597.30	47.70	1,465.83	43.77
11.Combu. Turbine	0.760	298.19	8.53	309.33	8.85

a) Total Energy Generated [GWH]	14,279.62		14,999.12
b) Unserved Energy [GWH]	14.11		18.08
c) Total (a + b)	14,293.73		15,017.22
d) Total Load Demand in Both Cases [GWH]		15,017.24	
e) % Error in Energy Balance	4.82		0.00
f) LOLP [p.u.]	0.0241		0.0256
g) Total Fuel Cost [$]	281,836		296,118

is broken into 24 time-bands, each with the length of 1 hr. The same load cycle is repeated for 30 days (728 [hrs]). Thus in the analytical formulae we have $\tau_t = 1$, T=24, and $n_s = 30$. This gives the total load demand of 15,017.24 [GWH] from Eq. (1).

The expected energy generated for each plant type is computed using Eqs. (5)-(8), and its capacity factor is computed; in columns 5 and 6, respectively, in Table III. As we move down the table, these values are becoming more different than the ones obtained from the conventional method (columns 3 and 4).

According to the given data, the total installed capacity up to 5th plant type is 12,600 [MW]. Since it is less than the minimum load, the capacity factor of each plant type, from 1 to 5, should be the same as its availability. The result by the conventional method, however, as shown in Table III, exhibits inconsistencies as the number of plant types increases. For example for the 5th plant type, the capacity factor is 0.8399 instead of 0.870. This is due to recursive convolutions which cause accumulated error. It can be observed from Table III that up to the 5th plant type each capacity factor obtained by using the analytical formulae is in perfect match with its corresponding availability. For example, for the 5th plant type, both capacity factor and availability are 0.870.

The overall accuracy was also evaluated by comparing the energy balance in both methods. The sum of the total generated energy and unserved energy in each case (item c) in Table III) is compared with the total load demand of 15,017.24 [GWH] found directly from the load curve. The result from the conventional method is 14,293.74 [GWH] or 4.82% error in energy balance, while the result from the analytical formulae is 15,017.20 [GWH] or 0.00% error. This is also reflected in the total fuel cost in Table III.

The LOLP calculated by the conventional method is normally inaccurate since it is obtained from the final equivalent load duration curve after all units (great in number) are convolved. Thus in view of the above discussions, the LOLP calculated by simple analytical equation (14), is believed to be more accurate. If the LOLP is to be calculated for one year instead of 30 days, the difference between the two methods will be significant.

IV.   LONG-TERM GENERATION PLANNING: HYDRO-THERMAL SYSTEM

A.   The Master Problem For Optimal Planning

The long range problem for a system of generating plants is to minimize the present value of total costs subject to reliability constraints, and is now stated formally as follows:

$$\min_{u_i^j} G = \sum_{i=1}^{I} G_i(x_i^j, u_i^j) \tag{20}$$

such that

$$x_{i+1}^j = x_i^j + u_i^j \tag{21}$$

subject to

$$R_i(x_i^j, u_i^j) \leq \varepsilon_i \tag{22}$$

$$u_i^j \geq 0 \tag{23}$$

where

i=1,2,...,I: index of years in planning horizon,

j=1,2,...,J: index for plant types,

x_i^j : previously installed capacity of plant type j in year i [MW] (state variables),

u_i^j : newly added capacity of plant type j in year i [MW] (decision variable),

G_i : present value expected annual cost [\$] defined in (15),

R_i : expected annual unserved energy [MWH],

ε_i : desired reliability level of annual unserved energy [MWH] .

Equation (21) is the dynamic state equation describing the state of installed capacities each year, that is, the sum of previously installed capacity and newly added capacity is the installed capacity for the next year. Considering the newly added capacities u_i^j as decision or control variables, the master problem (20)-(23) is now formulated as an optimal control problem.

According to the Pontryagin's maximum principle of modern optimal control theory, the optimal control must satisfy the necessary conditions:

$$x_{i+1}^j = x_i^j + u_i^j, \qquad x_1^j = \text{specified, (state equation)} \tag{24}$$

$$\lambda_i^j = \frac{\partial G_i}{\partial x_i^j} + \lambda_{i+1}^j; \qquad \lambda_{I+1}^j = 0, \qquad \text{(costate equation)} \tag{25}$$

and u_i^j minimizes the Hamilitonian

$$H_i(x_i^j, u_i^j, \lambda_{i+1}^j) \triangleq G_i(x_i^j, u_i^j) + \sum_{j=1}^J \lambda_{i+1}^j (x_i^j + u_i^j), \tag{26}$$

subject to

$$R_i(x_i^j, u_i^j) \leq \varepsilon_i, \qquad \text{(reliability constraint)}, \tag{27}$$

$$u_i^j \geq 0, \qquad \text{(control constraint)} \tag{28}$$

where λ_i^j is the costate (adjoint) of the plant type j in year i.

Note that the minimization of the Hamiltonian H_i is to be done one year at a time while the minimization of the total cost G in Eq.(20) needs to be performed for all years in the planning horizon simultaneously. This reduces the number of decision variables u_i^j to J from JxI in the minimization algorithm.

The minimization of Hamiltonian (26) subject to inequality constraints (27)-(28) is a typical nonlinear programming problem. It is solved by using a version of the Gradient Projection Method (GPM) [2], which has been developed specially for this purpose. The results from several sample studies show remarkable advantages of this method in computational efficiency and reliability.

The reliability constraint was linearized using the gradients or marginal values $\partial R_i / \partial u_i^k$. The marginal values of expected annual costs $\partial G_i / \partial x_i^k$ and $\partial G_i / \partial x_i^k$

with respect to marginal changes in the previously installed capacity, and the newly installed capacity, respectively, are also required in the GPM. These marginal values are given in Eqs. (16)-(18) and are not only required to find the optimal solution, but also provide a planner useful technical and managerial informations.

B.   Hydro Subproblem

In the master problem, the Hamilton (26) contains the expected annual cost G_i which consists of capital cost C_i, fuel cost F_i, and maintenance cost M_i as defined in Eq. (15). The hydro subproblem is to determine an optimal distribution of hydro resources so that the fuel cost F_i is minimized. This can be achieved by optimally selecting the energy resource distribution factor β, which can be formulated as follows:

$$\min_{\beta_\Delta^j} F_i = \sum_{j=1}^{J} f_i^j \overline{E}_i^j, \tag{29}$$

subject to

$$\sum_{s=1}^{S} \sum_{t=1}^{T} n_s \tau_t \overline{P}_\Delta^j = \sum_{s=1}^{S} \overline{E}_{i,s}^j \leq \overline{W}_i^j, \quad \text{for hydro plants,} \quad j = 1, ..., JH \tag{30}$$

$$\overline{E}_{i,s}^j \leq \overline{EM}_{i,s}^j, \tag{31}$$

$$0 \leq \beta_\Delta^j \leq 1, \tag{32}$$

where

$\overline{E}_{i,s}^j$: expected total energy generated by plant j in season s of year i [MWH],

$\overline{W}_i^j$: available hydro energy of plant type j in year i [MWH],

$\overline{EM}_{i,s}^j$: Maximum available hydro energy of plant type j in season s of year i [MWH],

JH : total number of hydro plants.

Equation (30) is the hydro balance equation describing the sum of energies generated by hydro plant type j in season s during year i is equal to its annual available hydro energy. Equation (31) is the seasonal hydro energy constraint to prevent the case that the amount of distributed energy from the annual available hydro energy to each plant exceeds its available hydro resources. The hydro sub-problem is a typical nonlinear programming problem and the gradient projection method is suitable for this problem.

Often, a portion of hydro capacity is available for peak-shaving due to run-of-the river hydro capacity requirement. Thus two types of hydro capacity, continuous-on-stream and peak shaving, are represented as two-piece unit. The base piece capacity equaling the run-of-river capacity is considered the same as thermal in production costing model.

C.   Simulation Results For Hydro-Thermal System

1.   Accuracy Test for Computational Algorithm of Hydro Subproblem

The same load and generating system of 'Accuracy Test' in Section III, E, the load cycle of scenario B and the full scale system of scenario D from EPRI Synthetic Utility System [16] are used in order to compare the results obtained without and with hydro plants in the generating system. Three hydro plant types are considered to be available for peak-shaving operations. The data for hydro plants are summarized in Table IV.

The result is shown in Table V. In part A of Table V, the case of hydro plant is compared to the case without hydro plant plant [2]. The total fuel cost decreased 16.51 % after the peak-plant operation. Also, the unserved energy and LOLP decreased significantly. This is due to the use of very reliable hydro plants in the study case. Normally the inclusion of hydro plants in the generating system makes the system become more reliable because they substitute the thermal plants of high loading order with low availability rate.

The part B of Table V shows the optimal values of β found for each hydro plant in each time-band. All available generating capacities of hydro plants are used in time-bands 7-24 (high load demands) for peak shaving operation, i.e., $\beta_\Delta = 1$, $j = 1, ..., JH$. In time-band 4 (lowest load demand) no hydro energy is assigned, i.e., $\beta_\Delta = 0$, $j = 1, ..., JH$. In time-band 3 all three hydro plant have limited in production, i.e., $0 < \beta < 1$. In other time-bands, it occurs as a combination of three

Table IV.   Data For Hydro Plants For Accuracy Test

Type	Unit Capacity[MW]	Total Capacity[MW]	Availa- bility	Available Energy[GWH]
1	300	900	0.99	56×10^4
2	200	600	0.99	36×10^4
3	200	400	0.985	25×10^4

Table V.   Results From Peak-Shaving Operation
of the Hydro Subproblem

A. Computed Energies, LOLP, and Total Cost

Plant Type	without Hydro Plants Energy Generated [GWH]	Including Hydro Plants Energy Generated [GHW]
1.Hydro		559.81
2.Hydro		359.70
3.Hydro		249.95
4.Nuclear	4,455.36	4,455.36
5.Nuclear	495.04	495.04
6.Coal	442.62	442.62
7.Coal	1,035.21	1,035.21
8.Coal	1,266.65	1,266.63
9.Coal	4,127.18	3,981.33
10.Oil	334.93	299.24
11.Oil	733.36	653.28
12.Oil	333.64	285.78
13.Oil	1,465.83	890.24
14.Combu. Turbine	309.33	42.88

a) Total Energy Generated [GWH]	14,999.12	15,017.06
b) Unseved Energy [GWH]	18.08	0.18
c) Total (a + b)	15,017.22	15,017.24
d) LOLP [p.u.]	0.0256	0.0005
e) Total Fuel Cost [$]	296,118	247,223

Table V.   (continued)

B. Computed Optimal Energy Resource Distribution Factor β

Plant Type Time-Band	Optimal β		
	1	2	3
1	1.0000	1.0000	1.0000
2	0.6584	0.1273	0.4784
3	0.0000	0.0453	0.4784
4	0.0000	0.0000	0.0000
5	0.0546	0.1273	0.4784
6	1.0000	0.6635	0.4784
7	1.0000	1.0000	1.0000
8	1.0000	1.0000	1.0000
9	1.0000	1.0000	1.0000
10	1.0000	1.0000	1.0000
11	1.0000	1.0000	1.0000
12	1.0000	1.0000	1.0000
13	1.0000	1.0000	1.0000
14	1.0000	1.0000	1.0000
15	1.0000	1.0000	1.0000
16	1.0000	1.0000	1.0000
17	1.0000	1.0000	1.0000
18	1.0000	1.0000	1.0000
19	1.0000	1.0000	1.0000
20	1.0000	1.0000	1.0000
21	1.0000	1.0000	1.0000
22	1.0000	1.0000	1.0000
23	1.0000	1.0000	1.0000
24	1.0000	1.0000	1.0000

cases mentioned above. It is concluded that the optimal distribution of available hydro resource is determined accurately.

The CPU time requirements for this example is 3.30 sec in the AS/N 9000 computer system. This compares with 1.30 sec. when no hydro plant was included.

2.   Result of Generation Expansion Planning Without Hydro Plants

The results in reference [2] is summarized here, where the generation planning problem is based on full scale system of scenerio D from EPRI Synthetic Utility Systems [16]. The system has 174 existing units of total 32,000 [MW] capacity.

All three weekly load cycles of Scenerio D corresponding to spring (or fall), summer and winter, respectively, are used in order to represent more realistic energy demand in a year since the analytical production costing model allows for the use of different load curves in each season.

The planning horizon (I) of 15 years beginning in 1988, with 4 seasons (S) for each year, 4 time-bands (T) for each day, and 1980 as base year for costs and loads, are used in planning, and the growth in energy demand in subsequent years is set at an annual growth rate of of 5.3 %. Peak load and energy in the base year are taken to be 19,774 [MW] and 129,142 [GWH], respectively. The reliability standard, expected unserved energy, is set at 0.5 % of the energy demand. Future cost escalated at annual rate of 10%. A discount rate of 15% is used in computing present values. The costs are based on projections for 1980 escalated to the beginning of planning horizon. The energy resource distribution factor β is 1 for all non-hydro units. Five types of non-hydro units are considered to be available for installation in each year, as described in Table VI. The teething factor for each plant type is 0.8

The computed optimal solution is shown in Table VII. The 3rd plant type (Coal II) has more added capacity during the planning period than the 2nd plant type (coal I). This is because, although coal I units have higher availability and lower maintenance rate, but have the same fuel cost. By varying the planning horizon, it was noticed that each run gives almost identical capacity additions compared to 15 year run. The most of the combustion units were installed during the last years of planning period. This is a reasonable solution path since the fuel savings associated

Table VI. Data For Canditate Non-Hydro Plants

Plant Type	Capacity [MW]	Availa-bility	Capital Cost [$/KW]	Fuel Cost [$/MWll]	Maintenance Cost [$/MW]	Maintenance Rate
1. Nuclear	1,200	0.851	1,157	2.41	210	0.1342
2. Coal I	400	0.876	735	4.21	450	0.0959
3. Coal II	200	0.917	695	4.21	450	0.0575
4. Oil	400	0.876	341	11.30	195	0.0575
5. Combu. Turbine	50	0.895	152	12.16	235	0.0384

with the base nuclear units are less pronounced during the last years of the planning period. If a salvage value assigned to the capital stock along with the associated costs is not considered, there will be a tendency of installing more capacities to the lower capital cost units as the year increases.

In this scenario the variation of capital costs showed to give a sensitive change to a configuration of added capacities in Table VII. For example, an increase of 10 % in capital cost of nuclear units caused 36 % reduction in their total added capacity while most of this reduced capacity was dislocated to the coal I and II units. The grand total of the added capacities was decreased by 3.2 % since the units with higher availability and lower maintenance rate need less capacity to be installed in order to meet the reliability constraints and the load demands. The total cost, although the capital cost was increased for nuclear units, decreased by 0.7 % due to the reduction in the total installed capacity.

3.   Result of Generation Expansion Planning With Hydro Plants

Beside five thermal types of alternative units, three types of hydro are scheduled to be installed during the planning period. The data for the hydro are shown in Table VIII. As described in Table VIII, the hydro plants are augmented in different years due to its nature of being limited energy plants constrained by local available hydro resources. Table IX shows the result of savings in annual investment costs due to the optimal peak-shaving operation of hydro plants. The expected annual cost given in the first column are obtained from the generating system without hydro plants [2]. The other columns show the annual total costs as each hydro plant is constructed in the respective planned year. For example, the second column shows the new annual total costs of the case that only first hydro plant is installed during the whole planning period. As more hydro plants are constructed, the new total costs are shown in the next column. Thus from Table VII, it is possible to compute the saving in each year due to the optimal peak-shaving operation of each hydro-plant. The total amount of saving for the whole planning horizon is the difference between the total costs of first and fourth columns, i.e., 2.0406 x 10^9\$. The optimization algorithm in the master routine was iterated in the average of 9 times and average CPU time was 12.53 sec. in the AS 9000/N computer system.

Table VII.   Result of the 15-Year Generation Expansion
                    Planning Without Hydro Plants

plant type / year	New Added Capacities[MW]				
	1	2	3	4	5
1988	1,021		6,898		
1989	1,681				637
1990	1,559				98
1991	1,721	897			
1992	1,942		107		
1993	1,968		130		
1994	1,984				254
1995	2,085		312		
1996	2,633	144	207		
1997	2,820	203	275		
1998	3,037	172	233		
1999	3,118	219	290		
2000	3,156	215	414		
2001	3,329	200	454		
2002		137	1,664		1,824

Total[MW]:   32,054   2,287   10,984   0   2,813
Grand total[MW]:   48,138

Table VIII. Data For Future Hydro Plants.

Type	Unit Capacity [MW]	Total Capacity [MW]	Availa-bility	Installing Year	Annual Avail. Energy [GWH]	Maint. Rate	Capital Cost [$/KW]	Maint. Cost [$/MW]
1	300	1,200	0.990	1991	8,500	0.005	670	270
2	200	800	0.985	1994	5,500	0.010	730	235
3	250	1,000	0.985	1996	7,000	0.010	750	250

Table IX.   Annual Investment Cost With Respect to the
             Number of Hydro Plants Installed

Planning year	Annual cost w/o hydro plants [$\times 10^9$ $]	Annual cost w/ 1 hydro plant [$\times 10^9$ $]	Annual cost w/ 2 hydro plants [$\times 10^9$ $]	Annual cost w/ 3 hydro plants [$\times 10^9$ $]
1988	4.9703	4.9703	4.9703	4.9703
1989	2.4100	2.4100	2.4100	2.4100
1990	2.4416	2.4416	2.4416	2.4416
1991	3.9681	3.8421	3.8421	3.8421
1992	2.7375	2.6230	2.6230	2.6230
1993	2.2500	2.1420	2.1420	2.1420
1994	2.2517	2.1486	2.0840	2.0840
1995	2.3293	2.2342	2.1750	2.1750
1996	2.1650	2.0769	2.0220	1.9560
1997	2.3646	2.2834	2.2327	2.1720
1998	2.2498	2.1750	2.1283	2.0710
1999	1.8982	1.8284	1.7847	1.7333
2000	1.9279	1.8627	1.8218	1.7735
2001	1.8163	1.7552	1.7168	1.6714
2002	1.7043	1.6458	1.6089	1.5638
Total Cost [$\times 10^9$ $]	37.4846	36.4932	36.0032	35.6248

This shows a considerable advantage on speed due to the analytical production costing model. In EGEAS [1], the reported CPU times are 191 sec., 969 sec., and 2100 sec., respectively, for linear programming (LP), generalized Bender (GB), and dynamic programming (DP) options.

V. LONG-TERM GENERATION PLANNING: PUMPED-STORAGE SYSTEM

A. Pumped-Storage Subproblem

One of the primary features of our analytical production costing model is the direct use of load curves rather than load duration curves. This analytical approach can be extended to model pumped-storage plant operation on the basis of random load distribution.

Let's denote X_p and X_s as the pumping and peak shaving level of the pumped-storage unit, respectively. When the load L_Δ is below the pumping level the unit will be in a pumping mode, thereby the effective load will be increased to $L_\Delta + y_p$, where y_p is the pumping power. On the other hand when load is above the peak-shaving level, the unit will be in a peak-shaving mode and the effective load will be reduced to $L_\Delta - y_s$, where y_s is the peak-shaving generation.

The variable y_p and y_s can also be assumed to be Gaussian random variable since the load was assumed to be Gaussian. Since the load is altered due to the pumping and peak-shaving operations, the total power output (4) needs to be modified as

$$^{j}P\Delta = \min(L_\curlyvee, \, ^{j}y_\Delta), \tag{33}$$

where the _effective load_ $L_\curlyvee$ is defined by

$$L_\curlyvee = \begin{cases} L_p = L_\Delta + y_p & \text{for} & -\infty \leq L_\Delta \leq X_p - y_p, \\ X_p & \text{for} & X_p - y_p \leq L_\Delta \leq X_p, \\ L_\Delta & \text{for} & X_p \leq L_\Delta \leq X_s, \\ X_s & \text{for} & X_s \leq L_\Delta \leq X_s + y_s, \\ L_s = L_\Delta - y_s & \text{for} & X_s + y_s \leq L_\Delta \leq \infty. \end{cases}$$

$$\tag{34}$$

Consequently, the statistical mean of the total power output $,^j\overline{P}_\Delta$, in (5) will be modified and expressed as a function of the pumping and peak-shaving levels, i.e.,

$$^j\overline{P}_\Delta =^j \overline{P}_\Delta(X_p, X_s).\tag{35}$$

Once the total power output (35) is computed the power output for each plant type, $\overline{P}_\Delta^j$, is computed by (8), and the expected total energy is computed by Eq. (9)-(10). The the total expected annual cost can be computed by (15). Since the total power output (35) is a function of the pumping and peak-shaving levels X_p and X_s, the pumping-storage subproblem is defined to find the optimal X_p and X_s for each load cycle.

Therefore, the pumped-storage subproblem for each season is to minimize the total expected seasonal operation cost, which can be formulated as following

$$\min_{X_p, X_s} \overline{f}_T = \sum_{t=1}^T \tau_t \sum_{j=1}^J f_i^j \overline{P}_\Delta^j,\tag{36}$$

subject to

$$g(X_p, X_s) \stackrel{\Delta}{=} \eta_p \overline{E}_p - \overline{E}_s = 0,\tag{37}$$

$$\overline{E}_p \le E_p^{max},$$

where

$\overline{E}_s$: expected peak-shaving energy [MWH],

$\overline{E}_p$: expected pumping energy [MWH],

E_p^{max} : maximum allowable pumping energy [MWH],

η_p : cycle efficiency [p.u.].

Equation (37) is the energy balance equation for pumped storage operation. The pumped storage subproblem is a simple mathematical programming problem. It can easily be solved by augmenting the constraint (37) to the cost (36) and applying the Kuhn-Tucker optimality condition. It can also be solved directly by using the gradient projection method.

B.   Supply-Shortage Cost

The reliability in generation planning is usually measured with probability indices such as the unserved energy (12) or the loss of load probability (14). However such indices are rather vague and based on the supply side only. The customer's position due to supply-shortage should also be evaluated and this can best be done with socio-economic considerations [43-46]

It is proposed to develop an analytical supply-shortage cost model from the system's marginal cost concept [18]. When the conventional loading-order concept is employed, the system's marginal cost increases monotonically with the load increase. When there is a supply-shortage the marginal cost, I_Δ, is likely to increase as a function of the unserved power, and is represented in a quadratic form

$$I_\Delta = A + BZ_\Delta + CZ_\Delta^2, \quad [\$/\text{MWH}] \tag{38}$$

where, A, B, and C are constants to be defined for a given power system, and Z_Δ is the random variable defined by the difference between the load and available generation capacity:

$$Z_\Delta = L_\Delta - {}^J y_\Delta, \quad [\text{MW}] \tag{39}$$

whose expected value is defined by (7) with j=J being omitted.

Then the supply-shortage cost per unit time is

$$q_\Delta = I_\Delta Z_\Delta, \quad [\$] \tag{40}$$

which is a random variable since Z_Δ is a random variable. By taking the mathematical expectation, the expected supply-shortage cost can be derived as

$$\overline{q}_\Delta = (A + B\overline{Z}_\Delta + C\overline{Z}_\Delta^2)\overline{PN}_\Delta + (B + C\overline{Z}_\Delta^2)\sigma_\Delta^2 . \text{LOLP}_\Delta + 2C\sigma_\Delta^2 \overline{PN}_\Delta, \quad [\$] \tag{41}$$

where the variance σ_Δ^2, expected unserved power $\overline{PN}_\Delta$, and the loss-of-load probability LOLP_Δ are defined in Eqs. (6), (11), and (13), respectively.

Consequently, the annual supply-shortage cost can be integrated over a year as

$$Q_i = \sum_{s=1}^{S} n_s \sum_{t=1}^{T} \tau_t \overline{q}_\Delta. \quad [\$] \tag{42}$$

This supply-shortage cost can be added to the total annual cost (4.15) to find the optimal generation planning policy.

C.   Simulation Results For Pumped-Storage System

1.   Accuracy Test for Computational Algorithm of Pumped-Storage Subproblem

The Accuracy of the new algorithm for thermal and hydro-thermal systems were already proved in Sections III,E, and IV,E. Therefore, it will be appropriate to present only the accuracy test of newly developed pumped-storage algorithm as it is used in the production costing model.

Two different tests were performed. First, the algorithm is compared with EPRI's conventional method [1], and then it is also compared with the "Le Modele National d'Investissment (MNI)" of EDF in France [18]. For the first study the load cycle of the EPRI's Synthetic System I is used, and for the second study a real data taken from Korea Electric Power Company (KEPCO) is used. Table X summarizes loads for both cases. The load cycle is broken into 7 time-steps (T), each with different time duration (τ_t). The load variance (σ_λ^2) are taken to be 5% of the mean ($\overline{L}_\lambda$) for KEPCO, and calculated from the load duration curves for EPRI data.

The result of the first study for one month operation is shown in Table XI. The first two columns show the comparison between the proposed algorithm and the conventional method of EPRI for thermal plants only. The third column shows the case of the proposed algorithm when the pumped-storage is introduced.

The result of EPRI is based on probabilistic simulation(cumulant method) with the equivalent load duration curves and a number of single generating units. On the other hand the proposed method uses a discrete load duration curves and group generations, one group for each plant type. The accuracy of the production model is again demonstrated because of the close comparison as in the earlier studies. When the sum of the total energy supplied (row 1) and energy not supplied (row 2) is compared with the area under the load duration curve (13,988.94 GWH), the proposed method shows a much smaller energy mismatch (0.0011 compared to

Table X.   Load Data.

Season \ Time-step			1	2	3	4	5	6	7
KEPCO LOADS [MW]	BASE YEAR I	I	7,637.02	7,325.82	6,802.71	6,658.44	5,727.29	5,354.16	5,283.46
		II	8,539.03	8,398.44	8,166.21	8,148.31	6,817.31	6,349.3	6,308.8
		III	8,518.91	8,169.29	7,702.92	7,636.19	6,439.91	6,025.94	5,950.75
		IV	8,896.65	8,896.65	8,107.52	7,877.67	6,653.84	6,239.03	6,164.7
	BASE YEAR II	I	11,983.08	11,690.68	11,133.98	10,378.00	8,654.04	8,065.89	7,913.29
		II	13,553.28	13,333.28	12,878.38	12,209.98	10,093.3	9,203.37	9,027.37
		III	13,157.33	12,836.46	12,476.1	11,584.0	9,562.2	8,918.02	8,7559.12
		IV	13,775.98	13,445.78	13,114.2	11,998.0	9,981.1	9,254.22	9,097.05
	BASE YEAR III	I	18,253.64	17,809.63	16,964.93	15,942.01	13,178.6	12,285.4	12,053.2
		II	20,957.48	20,619.18	20,138.85	18,846.5	15,575.3	14,204.9	13,933.4
		III	20,126.8	19,482.98	19,090.74	17,569.5	14,508.8	13,533.8	13,292.6
		IV	20,892.0	20,392.5	20,018.08	18,187.0	15,040.3	14,036.0	13,797.7
EPRI LOADS [MW]			23,255.102	22,036.898	20,740.602	18,923.301	15,809.898	14,778.398	14,415.301
Time Duration [h]	KEPCO		1	3	5	7	5	2	1
	EPRI		1	8	3	5	2	3	2

Table XI. Accuracy Test For EPRI System

(Period=728 [h])

	Proposed Method (Pure thermal)	EPRI (Pure thermal)	Proposed Method (Pumping)	Remarks
Total Supplied Energy [GWh]	13,943.4	13,902.0	14,176.8	
Not Supplied Energy [GWh]	45.38	21.93	24.3	
LOLP [p.u.]	0.0374	0.03333	0.0218	
Supplied Energy by Group [GWh] — Nuclear	4,950.34	4,950.34	4,950.34	8,000[MW]
Supplied Energy by Group [GWh] — Coal	6,656.00	6,431.45	6,917.56	11,200[MW]
Supplied Energy by Group [GWh] — Oil	2,091.36	2,240.28	1,993.22	8,000[MW]
Supplied Energy by Group [GWh] — Gas Turbine	245.70	279.88	167.08	4,800[MW]
Supplied Energy by Group [GWh] — Pumping	-	-	148.64	
Total Operating Cost [k$]	252,633	255,868	249,140	
Energy Mismatch[%]	-0.0011	-0.465		AREA of LDC = 13,988.94 [GWh]

0.465). This also shows that the proposed method is more accurate than the other method.

When the pumped-storage is added for operation, the low-cost plant energy is increased, while the energy from the high-cost plants (oil and gas turbine) is decreased. The results are reduced operating cost and improved reliability, and economic justification of pumped-storage is clearly demonstrated.

The second study with KEPCO data is shown in Table XII. The base year in Table X is chosen for the load cycle. The proposed method and MNI show a very close comparison in general. We note that the nuclear group produces the same energy for both cases. The MNI shows slightly lower generation cost due to the use of more low-cost coal plants and pumped-storage plants. The proposed method shows a slight improvement in reliability. Since the proposed method does not involve the convolution and deconvolution procedures, it is believed that it gives a more accurate result.

2.   Result of Generation Expansion Planning With Pumped-Storage

The results of generation expansion for thermal and hydro-thermal systems were already reported in [2], and in Section IV,E, respectively. Therefore it will be appropriate to present only the results of generation planning with additional pumped-storage option. The base year I of KEPCO in Table X is chosen as base year in the planning study. Two different cases were studied to see the effect of planning interval, one for a 10-year planning interval and another for a 15-year planning interval.

Table XIII shows the summary of a 10-year planning problem for optimal investment. The proposed method and MNI are compared each year. The results are very close to one another, except at the end of the planning interval where the proposed method invests more coal type plant while MNI invests more pumped-storage plant.

Table XIV summarizes the results for the a 15-year planning problem. Here, a major difference is noticed in the investment of pumped-storage plants. The MNI invests a total of 2,518 MW] by the year 2000, while the proposed method

Table XII. Accuracy Test For KEPCO System

Period / Method / Item	Season 1 Proposed	Season 1 MNI	Season 2 Proposed	Season 2 MNI	Season 3 Proposed	Season 3 MNI	Season 4 Proposed	Season 4 MNI
Total Operating Cost [k$]	293,905	292,021	440,761	440,425	388,506	387,467	418,465	417,933
Total Supplied Energy [GWh]	14,326.3	14,333.7	16,991.5	16,996.2	15,944.7	15,954.4	16,387.7	16,395
Not Supplied Energy [GWh]	5.4×10^{-2}	2.7×10^{-5}	0.016	0.048	5.4×10^{-3}	0.014	0.032	0.083
Pumping Cost [k$]	2,754	2,221	6.3	104	80.48	358	17.40	178
LOLP [p.u.]	9.9×10^{-8}	7.6×10^{-8}	3.38×10^{-5}	1.14×10^{-4}	1.17×10^{-5}	3.73×10^{-3}	6.3×10^{-5}	1.53×10^{-2}
Supplied Energy by Group [GWh] — Nuclear	6,978.37	6,978.37	6,978.10	6,978.37	6,901.59	6,902.52	6,826.10	6,826.67
Coal 1	5,052.28	5,136.13	5,194.37	5,193.50	5,114.67	5,134.13	5,078.71	5,080.00
Coal 2	1,114.34	1,067.75	1,422.15	1,436.52	1,316.4	1,343.45	1,347.99	1,380.37
O i l	1,126.20	1,091.19	3,384.78	3,383.17	2,606.52	2,563.79	3,123.59	3,100.28
L N G	-	-	-	-	-	-	-	-
Combined	0.015	0.019	11.63	1.87	4.01	0.86	10.59	2.94
Gas Turbine	-	0.0004	0.38	0.052	0.1	0.026	0.42	0.095
Pumped-Storage	55.09	60.29	0.93	2.68	1.38	9.66	0.26	4.52

Table XIII. The 10-Year Optimal Investment

Year	Demand [MW]	Reserve [%]		Nuclear [MW]		Coal [MW]		Pumped-storage [MW]		Present Worth Cost [k$]		Remarks
		Proposed	MNI	Proposed	MNI	Proposed	MNI	Proposed	MNI	Proposed	MNI	
Initial				3816		2120		400	–			Coal=850 [MW] Oil=5791 Combined =920 Gas Turbine =100
1986	10656	55.48	55.48	900	900	1000	1000	–	–	3655550	3649179	
1987	11679	54.82	56.0	900	900	613	705	–	–	2931590	3001431	
1988	12749	51.41	53.65	1000	1000	222	266	–	–	2480390	2490945	
1989	14014	44.88	47.39	1000	1000	–	–	–	–	2136640	2120781	
1990	15340	45.39	48.09	2000	2000	–	–	–	–	2981090	2969619	
1991	16796	44.1	47.53	2000	2000	–	–	–	–	2685280	2677230	
1992	18317	43.05	46.89	2000	2000	–	–	–	–	2433770	2427665	
1993	19977	41.18	45.33	2000	2000	–	–	–	–	2221570	2217750	
1994	21779	38.84	42.83	2000	2000	35	34	–	41	2056880	2063465	
1995	23744	38.14	41.09	2000	2000	561	284	–	108	2122430	2026108	
1996	25881	26.73	29.44	–	–	–	–	–	–	6788270	6702434	
Total Present Worth Cost [k$]										32,493,456	32,346,607	

233

Table XIV. The 15-Year Optimal Investment

Year	Demand [MW]	Reserve [%] Proposed	Reserve [%] MNI	Nuclear [MW] Proposed	Nuclear [MW] MNI	Coal [MW] Proposed	Coal [MW] MNI	Pumped-storage [MW] Proposed	Pumped-storage [MW] MNI	Present Worth Cost [k$] Proposed	Present Worth Cost [k$] MNI
Initial											
1986	10656	58.24	59.05	900	900	914	1000	-	-	3577140	3649336
1987	11670	56.83	58.39	900	900	555	650	-	-	2892010	2955486
1988	12749	53.20	54.79	1000	1000	214	236	-	-	2488050	2472826
1989	14014	46.50	47.95	1000	1000	-	-	-	-	2150050	2128781
1990	15340	46.88	48.20	2000	2000	-	-	-	-	2991470	2975783
1991	16796	45.80	47.26	1957	2000	-	-	-	-	2655580	2682288
1992	18317	44.14	45.95	1914	2000	-	-	-	-	2374990	2432065
1993	19977	41.91	44.15	1947	2000	-	-	-	63	2198380	2235132
1994	21779	39.35	42.38	2000	2000	-	-	-	212	2058480	2083159
1995	23744	36.24	40.46	2000	2000	-	-	-	342	1903060	1949174
1996	25881	32.72	38.29	2000	2000	-	42	-	398	1771510	1839755
1997	27984	31.69	38.61	2000	2000	504	613	-	385	1826810	1909215
1998	30260	31.21	38.77	2000	2000	798	826	54	377	1797120	1837632
1999	32723	33.00	39.13	2000	2000	1309	1170	506	366	1870970	1807641
2000	35384	35.99	40.17	2000	2000	2000	1696	600	375	1926030	1821679
2001	38200	25.97	29.84	-	-	-	-	-	-	5990660	5730482
Total Present Worth Cost [k$]										40,472,304	40,510,384

invests only 1,160 [MW]. It appears to be rather impractical to invest too much for pumped-storage plants in the case of MNI. The effect of planning interval from 1986 to 1995 is compared in both Tables XIII and XIV, the investments of coal plants and pumped-storage are quite different in later years.

In both cases a heavy committment of nuclear plants is evident. This reflects the unique Korean economic policy of high dependency on nuclear energy. The total present worth costs for investment are comparable each other, showing 32,393,456 [k$] in the proposed method and 32,346,607 [k$] in MNI for the 10-year planning case, and 40,472,304 [k$] and 40,510,384 [k$], respectively, for the 15-year planning case. This shows that the proposed method performs realistically. The most important advantage of the proposed method is in the computation time. Both methods were simulated in eclipse MV-8000 computer, which is compatible with IBM 370. It was shown that the proposed method is more favorable by 1/4 than MNI for 10 year planning case and 1/8 for the 15-year planning case. A 10-year planning problem by the proposed method was run in about 4 minutes. This shows a considerable advantage on the speed due to analytical production costing model and the new optimization scheme.

VI.   CONCLUSIONS

An analytical technique is developed for optimization of long-term generation expansion planning problem. Under probabilistic assumptions made on the load and plant outages, a closed-form analytical production costing model is developed. This eliminates the use of convolutions and results in a significant reduction in computation time while yielding much accurate results. The maximum principle is used to reduce the size of optimization for long-term problems. This approach results in the Hamiltonian minimization problem and the size of the problem becomes invariant to the number of years in the planning horizon.

The development presents an analytic algorithm which will ensure more exact and realistic solutions in economic and technical planning. It can reflect future uncertainties in load, system failures, fuel supply, maintenance, and economic conditions. It introduces the concepts of system failure cost and marginal costs in operation and planning. It can also reflect load management, financial constraints, and energy storage facilities.

ACKNOWLEDGMENTS

This work is supported jointly in part by the National Science Foundation under Grant INT-8617329, the Korea Science and Engineering Foundation, Allegheny Power System, and the Korea Electric Power Corporation. The authors wish to thank former Ph.D. students L. T. O. Youn and B. Y. Lee for their significant contribution to this work.

REFERENCES

1. EPRI, " Electrical Generation Expansion Analysis System (EGEAS)," Project RP 1529, Final Report EL 2561. EPRI, Palo Alto, 1982.
2. Y. M. Park, K. Y . Lee, and L. T. O. Youn, *IEEE Trans. Power Apparatus and Syst.*, *PAS-104*, 390-397 (1985).
3. D. G. Luenberger, "Introduction to Linear and Nonlinear Programming," Addison Wesley, New York, 1979.
4. J. A. Bloom. *IEEE Trans. Power Apparatus and Syst. PAS-101*, 797-801 (1982).
5. M. S. Razara, and C. M. Shetty, " Nonlinear Programming : Theory and Application, " John Wiley, New York, 1979.
6. R. Bellman, and S. Dreyfus, "Applied Dynamic Programming, " Princeton University Press, Princeton, 1962.
7. R. T. Jenkins, and D. S. Joy, "WASP-An Electric Utility Optimal Generation Expansion Planning Computer Code", Oak Ridge National Lab, Oak Ridge, 1974.
8. R. T. Jenkins, and T. C. Vorce, *Proc. 2nd WASP Conference*, 35-44 (1977).
9. R. T. Jenkins, and T. C. Vorce, *Proc. 2nd WASP Conference*, 2-6 (1977).
10. Electricitie de France, "La Nouvelle Version du NMI", Working Paper M. 333, EDF, Paris, 1976.

11. Electricitie de France, " Une Representation Simplifies de la Getion Annuelle de Moyens de Production, Etudes Economiques Generales," Working Paper M 334, EDF, Paris, 1976.

12. Electricitie de France, "Lois de Disponbilite du Parc Thermique dans les Modeles de Gestion, Working Paper M. 458, EDF, Paris, 1977.

13. H. Balexriaux, E. Jamoulle, and Linard de Guertechin, Fr., "Simulation de l Exploitation dun Parc de Machines Thermiques de Production Electricite Couple a esd Stations de Pompage, " Revue E (edition S.R.B.E.) Vol. 5, No. 7, pp. 1-24, S.A. EBES, Brussels, 1967.

14. R. R. Booth, *IEEE Trans. Power Apparatus and Systems, PAS-91,* 70-77 (1972).

15. K. Y. Lee, J. L. Ortiz, Y. M. Park, and L. G. Pond, *IEEE Trans. Power Syst., PWRS-1,* 153-159 (1986).

16. EPRI, " Synthetic Electric Utility Systems for Evaluating Advanced Technologies," Report No. EM 285, Section 4, EPRI, Palo Alto, 1977.

17. EPRI, " Probabilistic Simulation of Multiple Energy Storage Devices for Production Cost Calculations, " Report EA 1411, Vol.1 and 2, TSA 78-804, EPRI, Palo Alto, 1978.

18. M. Garkt, E. L'Hermite, and M. Y. Levy, " Methods and Models Used by EDF in Choice of its Investments for its Power Generation System", *Electricitie de France-Paris, France TIMS Congress,* Athens, 1977.

DEVELOPMENT OF EXPERT SYSTEMS AND THEIR LEARNING CAPABILITY FOR POWER SYSTEM APPLICATIONS

CHEN-CHING LIU

SHIH-MING WANG

Department of Electrical Engineering
University of Washington
Seattle, WA 98195

This chapter provides a tutorial on expert system applications to electric power systems and reports our recent results on the development of learning capability. Part I starts with a tutorial on expert systems. Issues such as knowledge acquisition, inference engine, development tools, and maintenance are discussed. Following the tutorial, a brief survey of the state-of-the-art is given based on an extended power system model. An on-line operational expert system, CRAFT (Customer Restoration And Fault Testing), is used to illustrate the expert system components and development facilities, including knowledge base, inference engine, maintenance and explanation facilities.

Besides the components and facilities of expert systems described in Part I. an important intelligent function, learning, is discussed in this chapter. Part II contains an explanation of an inductive learning (decision-tree) technique. The learning algorithm is extended to allow the decision tree to be modified when a misclssification occurs. The extended inductive learning method is applied to power systems. One application deals with determination of the control amount to eliminate voltage violations. The other application uses the learning algorithm to acquire high-level knowledge about security of a power system with respect to contingencies.

CONTROL AND DYNAMIC SYSTEMS, VOL. 44
Copyright © 1991 by Academic Press, Inc.
All rights of reproduction in any form reserved.

239

PART I. EXPERT SYSTEMS IN POWER SYSTEMS

I. EXPERT SYSTEM: A TUTORIAL

A. Definition

Expert System (ES) is a branch of artificial intelligence (AI), which is an area dealing with emulation of human intelligence. Besides expert systems, all the following subjects are related to emulation of human intelligent behavior: computer vision, machine learning, natural language processing, neural networks, speech recognition, etc. An expert system is also called a knowledge-based system; it is a computer system developed to emulate the problem solving behavior of human experts in a specific domain. The term "problem solving behavior" refers to the strategy, experience, rules-of-thumb, and judgements used by human experts in solving a difficult problem.

A knowledge-based system contains two basic elements: the knowledge required to solve a problem and the inference procedures which can be followed to identify the applicable knowledge for a given state in the problem solving process [1]. Expert systems should be applied to problems difficult enough that require significant human expertise for their solution. An expert system should contain high quality knowledge which can be used to perform a non-trivial level of reasoning.

Knowledge can be represented in an expert system in different forms. Rules, or IF-THEN structures, are natural representation for rules-of-thumb or judgements. For example, protective devices in a power system are designed by rules. If the current sensed by an overcurrent relay exceeds a threshold, then the relay sends a tripping signal. As a result, much of the knowledge related to diagnosis of faults in a power system used by human experts can be represented by rules. A rule-based expert system has its knowledge represented in rules. A frame is often used to describe a class of objects such as generator units. A frame for generators can include a number of attributes such as MW capability, MVAR capability, output voltage level, etc. Even though the generating units are sharing the same attributes, they can be assigned different values. A semantic net is particularly suitable for language processing.

B. Benefits of Expert Systems

The benefits of expert systems are 1) automation of decision-making tasks, 2) serving as knowledge tanks, and 3) providing new problem solving techniques. Expert Systems can incorporate the knowledge possessed by human experts to provide a computerized decision-making tool. In a power system environment, an emergency condition can result in a large number of alarms generated by the energy management system (EMS), leading to great confusion and inefficiency in decision making and emergency management. An expert system can help to retrieve data, initiate trivial but necessary functional calls and use its embedded knowledge to advise human dispatchers. In the power industry, many senior engineers are retiring, leading to the loss of experience and knowledge about the specific power system. In a survey from Japan in 1988 [2], out of 88 operators in the 500 kV substations, only 7 are 40 years or older. In this situation, the knowledge base of an expert system can serve as a knowledge tank, which can be used to train less experienced engineers. Finally, logic reasoning and heuristics are useful techniques for problem solving; they are complementary to the conventional algorithmic techniques. A numeric algorithm is usually designed to calculate the optimal values of a certain objective function or to solve equations with unknowns. A rule-based system is intended as a decision-making aid rather than a numerical calculation tool, even though the two techniques are often integrated as a combined problem solving tool.

C. Inference Engine

The basic components of a rule-based expert system are the rule base, data base and inference engine. The rule base contains a number of rules. The database stores the necessary information and data about the problem being solved. There are two basic types of inference mechanisms, i.e., forward chaining and backward chaining. These two mechanisms can be combined to tailor for problems that require more flexible inference methods. The forward chaining method is an algorithm which is used to match data to the conditions (the left hand side) of a rule. If a combination of data elements satisfies the conditions, the data combination is stored in a conflict set. Among all data combinations in the conflict set, the one with the highest priority is executed, i.e., the action is taken or the conclusion is reached. The procedure of selecting the highest priority rule from the

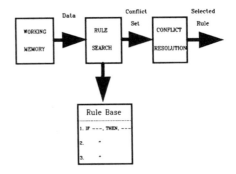

Figure 1 Forward Chaining Mechanism.

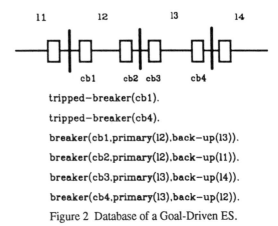

tripped−breaker(cb1).
tripped−breaker(cb4).
breaker(cb1,primary(l2),back−up(l3)).
breaker(cb2,primary(l2),back−up(l1)).
breaker(cb3,primary(l3),back−up(l4)).
breaker(cb4,primary(l3),back−up(l2)).

Figure 2 Database of a Goal-Driven ES.

conflict set is referred to as conflict resolution. Figure 1 illustrates the forward chaining mechanism of a rule-based system.

The forward chaining can be considered as a "data-driven" method. In contrast, the backward chaining method is "goal-driven." The inference engine starts with a given "goal" or conclusion. Suppose an expert system is asked to identify the fault location of a power system. The ES is given the information on the coordination between pairs of protective devices. In Figure 2, it is assumed that the ES receives the information that two circuit breakers, cb1 and cb4, tripped. The ES has in its database the coordination information, e.g., breaker cb1 is the primary protection device for line $l2$ and also the back-up device for a fault on line $l3$. The other pairs can be interpreted in a similar manner. A

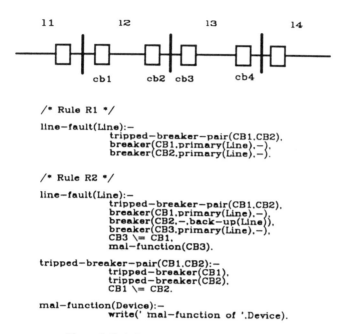

```
/* Rule R1 */

line-fault(Line):-
            tripped-breaker-pair(CB1,CB2),
            breaker(CB1,primary(Line),-),
            breaker(CB2,primary(Line),-).


/* Rule R2 */

line-fault(Line):-
            tripped-breaker-pair(CB1,CB2),
            breaker(CB1,primary(Line),-),
            breaker(CB2,-,back-up(Line)),
            breaker(CB3,primary(Line),-),
            CB3 \= CB1,
            mal-function(CB3).

tripped-breaker-pair(CB1,CB2):-
            tripped-breaker(CB1),
            tripped-breaker(CB2),
            CB1 \= CB2.

mal-function(Device):-
            write(' mal-function of '.Device).
```

Figure 3 Rule Base of a Goal-Driven ES.

small rule base in Figure 3 is developed for this example. The rule labeled, "tripped_breaker_pair (CB1, CB2)," states that if a pair of breakers, CB1 and CB2, (both being variables for breakers), are different breakers and both tripped, then they form a tripped breaker pair. Rule R1 is based on the tripped breaker pair rule. If breakers CB1 and CB2 form a tripped breaker pair and both are the primary protective devices for the line "Line," which is also a variable, then the fault is on the line specified by the value of Line. Rule R2 is more sophisticated. If Rule R1 does not apply, then it is possible that the breaker pair, CB1 and CB2, are primary and back-up devices for the Line, respectively. Breaker CB3 is primary to the Line, but CB3 probably failed to clear the fault and hence the back-up device operated. The "mal-function" rule simply provides an output message for the device ÇB3 that is matched in Rule R2. Figure 4 illustrates the backward-chaining mechanism with the above fault diagnosis example. The inference engine starts with the goal, "?line-fault (Line)." Two tripped breaker pairs, (cb1, cb4) and (cb4, cb1), are retrieved from the database and an attempt is made to match the pairs to Rule R1, but both

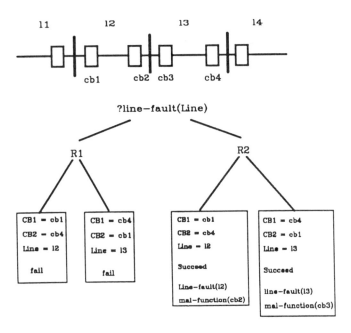

Figure 4 Results of Backward Chaining.

fail. The search procedure backtracks to the goal and attempts to match the pairs to Rule R2. It is found that both (cb1, cb4) and (cb4, cb1) can satisfy Rule R2, but with different results. The fault may be on line *l2* or *l3* and either cb2 or cb3 malfunctions.

D. Building an Expert System

Choosing an appropriate application area is critical for the success of an expert system project. Three criteria are helpful: Need, Suitability, and Feasibility. Building an expert system can be time-consuming. Keeping the knowledge base accurate and updated also requires a substantial amount of effort. Unless the need of an expert system is clearly indicated by the users, it is difficult to justify the cost of the development.

Suitability is related to the nature of the power system problem being solved. ES techniques are suitable for problems that involve decision-making based on strategies, rules, logic, judgements, and heuristics. The number crunching capability of an ES development tool is very limited. As mentioned in Section B, some problems require the combination of algorithmic and rule-based methods.

Feasibility is an evaluation of the likelihood of success in an expert system project. A primary issue here is whether a fairly complete and accurate expert system can be achieved. An ES may have to deal with a large number (or sometimes an infinite number) of possible scenarios in the real-world environment. Testing a few scenarios in the ES prototyping stage does not guarantee the success of the ES in other scenarios. It is always a good idea to narrow down the problem domain by making suitable assumptions initially. The ES capability can be enhanced by gradual addition of knowledge modules to relax some assumptions.

Developing an expert system requires much interaction between domain experts and knowledge engineers. The knowledge engineer acquires knowledge from domain experts. In a power system environment, the knowledge involves power system characteristics/behaviors and the decision making process. Before and during the knowledge acquisition, it is necessary to analyze the suitability of the expert system approach. To facilitate the efficiency of knowledge acquisition, the knowledge engineer should be familiar with the problem domain, even though he/she does not have to be an expert. Other variations are possible; for example, an expert in a problem domain can develop the ES without a knowledge engineer if he/she understands knowledge engineering and has the necessary skills to use the ES development tools.

Knowledge acquisition is an iterative process. Based on our experience, the following method is effective. The knowledge engineers or researchers interview the experts in a problem domain. The discussion is substantiated with specific examples. In the process, the expert(s) explain(s) the decisions and justifications, judgements, and rules, etc. The knowledge engineers/researchers attempt to generalize applicability of the knowledge from specific examples to a larger class of scenarios. This step requires the knowledge engineers or researchers to be fairly knowledgeable in the problem domain. The results of generalization are shown to the experts in the project team. The expert can usually point out the weakness of the generalized knowledge, which often leads to exceptions. Identification of exceptions to generalized rules is a refinement to the acquired knowledge.

Implementation of an expert system is more effective with an ES development software tool. Most of the existing expert systems are rule-based. Some expert systems are designed with the object-oriented data structure. Some of the available ES development tools are:

Real-time shells: G2, R*Time

Non-real-time shells: KES, NEXPERT

Rule-based inference engine: OPS83

Logic based inference engine: PROLOG

Symbolic processing language: LISP

The above tools are ordered from the most sophisticated to the most basic. Real time shells are provided with the necessary interface with the real time data environment. They may provide sophisticated user-interface and convenient graphic facilities for ES development.

ES shells such as KES and NEXPERT are provided with development facilities such as some of the object-oriented programming capabilities. The developer can view the rule-base and edit rules with the embedded ES development facilities. OPS83 is a rule-based inference engine. Due to the use of a fast-pattern matching algorithm for rule/data matching, the C-based tool is very fast. However, the user-interface is very primitive and programming is performed at a level lower than those shells mentioned above. PROLOG is a logic programming tool with a built-in backward-chaining inference engine. LISP is a symbolic processing language with a high level of flexibility but requires an extensive amount of programming effort.

Some considerations in the selection of ES development tools are availability, user-interface, C or LISP in which tool is developed, speed of rule matching algorithm, how close the rule representation is to the natural language, price, portability, memory requirements, development/maintenance effort, knowledge base extendibility, convenience of interface to external procedures, software reliability, etc.

The other issue related to implementation is the rule and data structures. For a given rule/data matching algorithm, a rule design or format may be more efficient than others in terms of the inference speed. For example, the OPS83 inference engine matches the data elements to each condition of a rule from the first condition and then continue to the next condition and so on. Suppose the first condition is a generic one which can be satisfied by many data elements while the last condition is very restrictive, i.e., only a small number of data elements can be matched. The rule chaining algorithm would waste much time matching the first condition only to find that a small number of data elements can satisfy the last rule condition. In this case, it is more advantageous to reorder the conditions, i.e., starting from the most restrictive. There are systematic guidelines for evaluation of the efficiency of a rule-based system design [3].

The performance evaluation of an ES usually involves simulation of a number of scenarios in the problem domain. The developer verifies the reasoning process and the conclusion or suggestion of an ES. It is also important to ensure that the knowledge embedded in the ES is consistent with the expert's knowledge. The user-interface is often crucial for acceptance by the users. It is desirable to involve the users early in the ES development so that the user expectation is fully understood. It is difficult or impossible to simulate all possible scenarios during the testing process. In a practical environment, the number of scenarios can be very large or infinite. Therefore, it is important to maintain the generic nature of the knowledge incorporated in the ES; generic rules can apply to a larger class of scenarios. Modification of the existing knowledge or addition of new knowledge is necessary to keep an ES updated.

In general, there are three stages in the development of an ES: prototype, practical system, and maintenance. Fast prototyping is possible with the existing ES development tools. However, the prototype may still be far from being practical since it is usually time consuming to perform thorough testing of an ES. Maintenance of an expert system refers to the tasks of maintaining the data base, user-interface and rule base. Since the first two components are common to many software systems, only the rule base maintenance is discussed here. Modifying a large rule base may require understanding of the entire rule base. For example, a new rule can be in conflict with the existing rules or rule chains. The relation among rules can be hard to detect. A rule can be derived from a chain of rules in the existing rule base. In general, it is hard to detect the relation without a computer "relation-checking" tool. Typical relations are conflict, redundancy, unnecessary conditions, subsumption, implication, cause-effect, etc. Several relation checking algorithms have been developed, e.g., [4, 5, 6, 7].

For further information on the development of an expert system, see [8].

Table 1 Frequently Investigated Areas[1]

Alarm Reduction & System Diagnosis
Security Assessment
Environments for Operational Aids
Restoration
Remedial & Emergency Controls
Substation Monitoring & Control
Scheduling
Expert System Implementation Issues
Neural Network Applications
Power System Planning
Distribution System Design & Planning
Expert System Development Methodologies

II. STATE-OF-THE-ART

Expert system has been one of the most active research fields over the last decade. According to a recent survey of the Japanese institutions [9], the number of Japanese companies involved in the development of expert systems has increased from 124 in the year 1988 to 151 in 1989 and then to 190 in 1990. The number of expert systems being developed or used also increased from 250 (1988) to 332 (1989) and then to 401 (1990). Out of these systems, the percentage of expert systems actually in use is 34.4 % (1988), 38.0 % (1989) and 47.1 % (1990), respectively. It is noted that the number of expert systems actually in use so far has increased to almost half. These expert systems are developed and used in many domains: banking, insurance, steel, construction, power, chemical, machinery, electronics, manufacturing, etc. As the author of the survey [9] pointed out, "AI is spreading into almost every Japanese industry...."

In the power system area, the knowledge-based system application has also grown substantially during the last decade. A special conference established in 1988, Symposium on Expert Systems Application to Power Systems (ESAPS) [10, 11], is

---

1. Sources of information:

    IEEE Transactions on Power Systems.

    Proceedings of the International Workshop on AI for Industrial Applications, Hitachi City, Japan, 1988.

    Proceedings of the First Symposium on Expert Systems Application to Power Systems, Stockholm-Helsinki, Sweden-Finland, Aug. 1988.

    Proceedings of the Second Symposium on Expert Systems Application to Power Systems, Seattle, U.S.A., Jul. 1989.

    Proceedings of the Power Systems Computation Conference, Graz, Austria, 1987.

dedicated to this field. Based on our survey of the literature, the more popular application areas are listed in Table 1. These areas are ordered according to the number of papers in the literature.

More explanation regarding some of the areas in Table 1 is provided in the remaining of this section.

A. Extended Power System Model

Traditional power system analysis has been based primarily on the steady-state power flow solutions and dynamic simulation of the behavior of machines, excitation systems, controls, etc. Most of the analytical tools developed so far are numerical in nature; typically the existing algorithms deal with solution of the nonlinear network equations or optimization of an objective function subject to equality or inequality constraints. In the operation area, an energy management system may possess software packages such as state estimation, topology processor, contingency screening/evaluation, steady-state security enhancement, external equivalent, short-term load forecast. The steady-state power flow model plays an important role in these methods. Due to the heavy computational burden, dynamic security assessment is usually performed off-line; the computational results may be used during the on-line operation. For power system planning, the power flow, stability analysis and electromagnetic transient simulation packages are often used.

From an on-line operational point of view, the protective devices play an important role. When a fault occurs, such as short circuits coupled with miscoordinated relays, the dispatchers need to understand the design and operation of protective devices in order to perform system diagnosis and restoration. At present, the on-line EMS does not include the models of protective devices. (Dispatcher training simulators do provide such capabilities.) Standard power flow methods assume that the system topology is known: in other words, the EMS can not provide meaningful assistance to power system dispatchers when the system is in a disturbed operating condition.

To develop the ES problem-solving capabilities in the abnormal operating conditions, it is necessary to extend the simple power flow and dynamic models to include the protective devices. It is also necessary to incorporate the alarming scheme of the EMS.

An extended model of the power system is proposed as follows:

$$\frac{d\underline{x}}{dt} = \underline{h}(\underline{x}, \underline{f}, \underline{y}, \underline{d}) \tag{1}$$

$$\underline{f}(N, \underline{x}) = \underline{y}(\underline{g}, \underline{l}) \tag{2}$$

$$N = N(\underline{Y}, \underline{s}) \tag{3}$$

$$\underline{s} = (\underline{s}_1, \underline{s}_2) \tag{4}$$

$$\underline{s} = \underline{s}(\underline{R}, \underline{c}, \underline{\varepsilon}) \tag{5}$$

$$\text{IF } x_i \in \Omega_i \text{ THEN } R_i = \text{TRIP} \tag{6}$$

$$\text{IF } x_i \in \Theta \text{ THEN } s_i = \overline{s}_i \tag{7}$$

where

$\underline{x}$: system state variables (voltages, angles, frequency)
$\underline{h}$: functions of the dynamic models
$\underline{f}$: power flow functions
$\underline{y}$: bus injections
$\underline{d}$: disturbances
N: network topology/parameters
$\underline{g}$: generations
$\underline{l}$: load demands
$\underline{Y}$: line/shunt admittances
$\underline{s}$: breaker/switch status, tap positions, capacitor bank status
$\underline{s}_1$: status of supervised switches/breakers, taps, capacitors
$\underline{s}_2$: status of unsupervised switches/breakers, taps, capacitors

$\overline{s}_i$: complement of status s_i
$\underline{R}$: status of protective relays
Θ: opening or closing condition of automatic switches, e.g. ODL+2 (open dead line, if line deenergized for 2 seconds)
$\underline{c}$: authorized switching actions or tap adjustments (manual, remote)
$\underline{\varepsilon}$: relay/breaker/switch failures
Ω_i: tripping condition of relay R_i

Eq. (1) represents the power system dynamics. The system states, i.e., voltages, angles, and frequency, vary upon the occurrence of disturbances $\underline{d}$ such as a short circuit and their variations depend on the power system operating condition. Eq. (2) denotes the steady-state power flow equations. The power flow functions f are equated to bus injections, including the generations and loads. The network topology $\underline{N}$ depends on the breaker/switch status $\underline{s}$ and the line/shunt parameters $\underline{Y}$. This is shown in Eq. (3). The first three equations are fundamental in the traditional steady-state and dynamic power system analyses.

The breaker/switch status $\underline{s}$ changes if relay tripping signals $\underline{R}$ are received. Some breakers/switches may be supervised by the SCADA (Supervisory Control And Data Acquisition) system, while others may not. Unsupervised breakers/switches, denoted by $\underline{s}_2$, can change status but the energy management system does not receive any message for the status change. The other two relevant parameters in this equation are automatic and/or manual switching actions and relay/breaker failures. The last two "equations," in fact "rules," are the basic design principles of the protective devices. When a system state x_i satisfies an operation condition, e.g., under frequency, impedance thresholds, a breaker tripping signal is generated by the relay. An automatic switch may be used in the subtransmission level of the power system. Rule (7) represents functions such as "ODL+3 (open dead line)," i.e., open if the dead line condition persists for 3 seconds, or "ATDB+2," i.e., a pair of normally open and normally close switches exchange their status if the dead bus condition exists for 2 seconds.

B. Energy Management System Alarm Model

To better reflect the view of a power system dispatcher in a control room, a simplified model of the EMS alarming scheme is given in Figure 5. Only two types of alarms are shown. An alarm can be triggered by violation of an operating constraint; for example, a line overload can generate an alarm. An alarm of the second type is generated by a status change of a breaker or switch. Many other alarming situations exist; for example, fire alarms, break-in alarms, etc. Some alarms are triggered by actions authorized by system dispatchers and hence may not indicate a problem in the power system.

Violation of operating constraints
IF $x_i \notin \Pi_i$, THEN Generate alarm a_i

Change of breaker/switch status
IF $s_j(t)$ not equal to $s_j(t - t_0)$, THEN Generate alarm a_j
(R_i TRIP $\Rightarrow$ Generate message a_k)

Figure 5 EMS Alarming Scheme.

Upon the occurrence of a fault or an abnormal operating condition (disturbance $\underline{d}$), the system is in a dynamic stage and its behavior is governed by Eqs. (1-2). An operating condition such as overcurrent may trigger relays according to Rule (6). Breakers and relays may fail and hence the final breaker/switch status may deviate from the intended functions. Eqs. (3-4) determine a new topology N of the power system, which in turn results in a new operating condition. The events can be cascaded and, therefore, the system may go through different topologies. These cascaded events can lead to numerous EMS alarms, bringing pieces of power system information to the dispatchers in the control room.

C. Model of Operational Constraints

A number of constraints may be considered in the planning and operation of a power system; for example,

Equipment capability,	(C1)
e.g., line current capacity, transformer capacity.	
Abnormal voltages,	(C2)
e.g., bus voltage limits, voltage transients.	
Stability limits,	(C3)
e.g., transfer MW limits over a tie line.	
Communication & computer system availability,	(C4)
e.g., remote terminal unit failures, communication link failures.	
Environmental constraints,	(C5)
e.g., emission limits, water level in a river.	
Generator capability,	(C6)
e.g., ramping rates, boiler and turbine conditions after a blackout.	
Load constraints,	(C7)
e.g., cold load pick-up, load characteristics.	
Field personnel availability,	(C8)
e.g., man-power needed to operate switches and/or breakers.	

Depending on the operating condition of the power system, some of the constraints will be more critical than others. The relevance of these constraints to expert system applications will be discussed later in this chapter.

D. Human Tasks Modeled by Expert Systems

Most of the existing expert system applications to power system planning and operation can be categorized into the following types:

1. Reasoning based on logic, judgements

One of the dispatchers' tasks is to interpret alarms generated by an energy management system due to events such as faults. As seen from Rules (6-7), protective devices are designed by logic; hence it is natural that the dispatchers use similar logic to identify the fault location in the power system [12]. According to the EMS alarm model in Figure 5, the EMS can generate a large number of alarms in an abnormal operating condition. It is impossible for a dispatcher to quickly interpret the alarm messages and determine the root causes of the problems. Hence, it is desirable to prioritize the alarms and filter out the redundant or unnecessary ones before they are presented to the dispatchers [13]. System diagnosis deals with the identification of the root causes. An ES can analyze the alarms $\underline{a}$ (Figure 5), system state $\underline{x}$, relay status $\underline{R}$, breaker/switch status $\underline{s}$, etc., to help determine the status of unsupervised breakers/switches $\underline{s}_2$, fault locations $\underline{d}$, and malfunctioning devices $\underline{e}$. Diagnosis is the task of the simple ES in Figure 3. Examples of existing expert systems for power system diagnosis are [14, 15].

2. Action planning

Remedial actions may be necessary when a constraint, such as the line current (C1) or voltage limits (C2), is violated. A dispatcher would look at the severity of the problem and consider the control options. Low voltages can be raised by adjusting transformer tap positions, generator var outputs, or switching capacitor banks. These controls are represented by $\underline{s}$ and $\underline{Y}$ in Eq. (3). Local controls are usually more effective than the remote ones. Dispatchers may attempt to make step-by-step corrections until the violation is eliminated or they may estimate the amount of adjustment based on their experience [16].

Following a blackout, system dispatchers follow pre-specified restoration procedures to re-energize the power network. Black start units such as combustion turbines provide cranking power, i.e., MW injections $\underline{g}$, to other fossil units. The condition (cold, hot) of the boiler and turbine in a plant determines the amount of MW that can be generated as time goes on. This implies that the constraint (C6) can vary with time. The reactive power absorption capability of generating units, (C6), should also be considered since the power system has little or no load to consume the reactive power produced by long lines or underground cables. The switching actions rely on the availability of field personnel, i.e., (C8). Besides these considerations, the capacity and voltage constraints (C1-C2) should also be satisfied. Other constraints (C3, C5, C7) may also be imposed. Unexpected events, such as loss of equipment, can take place during the restoration process, and hence the dispatcher and the expert system need to adapt to the new situations quickly [17].

3. Design

Designers of a power system considers the design parameters, constraints and criteria in order to develop a feasible plan. In a transmission design problem, one needs to consider generation, distance between generation and load, equipment and construction costs, etc. The detailed design parameters may be the number of parallel circuits, voltage level, var compensation, number of circuit breakers, etc. Any design has to meet requirements on system security, reliability and costs [18]. These requirements include some of the constraints in (C1-C8).

The other example is related to the design of an electrical auxiliary system in a power plant [19]. The design task involves selection of bus configuration, load division, voltage ratings, etc. In selecting the voltage rating, it is necessary to consider the loading, distance from the main power supply, safety, codes and standards, etc.

The above survey of the state-of-the art is by no means complete. The reader is referred to [20, 21] for more detailed descriptions.

III. EXAMPLE EXPERT SYSTEM

A rule-based expert system, CRAFT (Customer Restoration And Fault Testing) [22, 23, 24] will be used to illustrate our results in the development of intelligent systems. Since July 1988, CRAFT has been operating as an on-line tool for power system dispatchers at Puget Sound Power & Light (PSPL) Company, Bellevue, WA. CRAFT is designed for location of faults on subtransmission lines (115 kV) with multiple taps and automatic switches. The automatic switches are designed to quickly sectionalize a line and restore customers when a fault occurs. Alarms will be generated by the alarm processor on the SCADA computer when a fault occurs. These alarms will also be sent to CRAFT, which resides on a separate computer linked to SCADA. CRAFT will be activated when the first alarm arrives. After waiting for a certain period of time, say one minute, for all automatic switches to complete their actions, CRAFT starts to analyze the alarms. In many cases, automatic switches on a faulted line are not supervised. Therefore, the status of these switches has to be determined first. The rule module called AUTOSTATUS [23] will be activated by a set of control rules to determine the status of these unsupervised automatic switches. In doing so, it is assumed that the information on breaker operations is known. As soon as the status of these unsupervised automatic switches is determined, CRAFT starts the reasoning process to locate the fault. The switching actions for restoration of the customer substations can be suggested by CRAFT.

The knowledge acquisition for CRAFT was done by interviewing power dispatchers and EMS engineers. After building the knowledge base, CRAFT was tested extensively for the PSPL system, which involves 370 fault locations and 1177 switch status problems. As soon as the testing and implementation of CRAFT was accomplished, CRAFT began to operate in the on-line control center environment. A critical issue to be addressed for CRAFT is the rule-base maintenance. A relation-checking tool has been developed and integrated into the CRAFT environment. Explanation facilities have been developed for CRAFT in the off-line environment. More details will be provided later.

This section summarizes the CRAFT capabilities; more details can be found in [24]. A power line chosen from the PSPL system, shown in Figure 6, will be used to demonstrate our experience in the ES development related to:

- knowledge base,

- data base,

- inference engine,

- explanation facility, and

- maintenance facility.

In terms of the extended power system model, CRAFT deals with Eqs. (3-5) and Rule (7), which are related to power system topology and breaker/automatic switches. CRAFT relies on the alarming scheme of the EMS (Figure 5) to provide real time information of the power system. CRAFT performs logic reasoning to determine the status of unsupervised automatic switches. This function can be represented by

$$N' = N'(\underline{Y}, \underline{s}_1)$$

Note that the status of supervised switches is available to CRAFT. The line status N' reconstructed by CRAFT may be different from the true status N, since the information CRAFT receives from the alarms may be erroneous.

In Figure 6, there are 4 automatic switches, out of which 546 and 558 are supervised and 525 and 526 are unsupervised. A supervised switch is indicated by a symbol s in the line diagram. Therefore, if a fault occurs, the dispatchers will not know the status of these two unsupervised automatic switches from the data acquisition system. However, the status of these two switches can be determined from the breaker activities.

Assume a fault occurs between switches 1145 and 546. Several alarms will be generated according to the functions and time settings of the switches. In Figure 7, a listing of the alarms is shown. Note that there is no alarm associated with automatic switches 525 and 526.

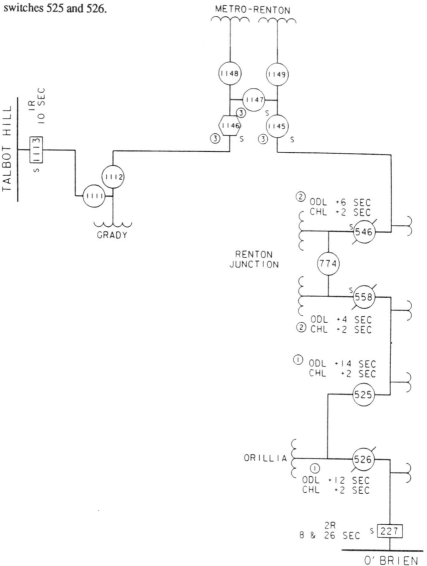

Figure 6 Example Power Line.

Talbot_Obrien1113	open	unauthorized
Talbot_Obrien227	open	unauthorized
Talbot_Obrien558	open	unauthorized
Talbot_Obrien546	open	unauthorized
Talbot_Obrien227	closed	unauthorized
Talbot_Obrien1113	closed	unauthorized
Talbot_Obrien1113	open	unauthorized
Talbot_Obrien558	closed	unauthorized

Figure 7 Alarms Received by On-Line CRAFT.

Since automatic switches 525 and 526 are unsupervised, the status has to be determined by AUTOSTATUS, a rule module for determining the status of unsupervised automatic switches using information on breaker activities. AUTOSTATUS determines that 525 and 526 are both closed. As soon as the status of these two switches is determined, CRAFT starts the reasoning process. In this case, two substations, Metro-Renton and Grady, are out of service after the final configuration is reached. Therefore, test of the line will be considered. By opening switch 1145 and then closing breaker 1113 (both remote-controllable), it can be found that breaker 1113 remains closed. At this point, the fault location is found to be between 1145 and 546 and, furthermore, the two substations are restored at the same time.

A. Knowledge Base

Almost all rules incorporated in CRAFT are general. The generality results in several advantages. First of all, the ES is more portable since the rules do not depend on the specific devices or configuration of a power line or system. Secondly, the expert system is more capable of adapting to changes of the system configuration. As a result, the maintenance of ES is simplified. Finally, generic rules can solve a class of problems and hence the size of the rule base can be reduced.

The generic nature of the rule base will be demonstrated by examples from the AUTOSTATUS module. Recall that AUTOSTATUS has to be executed to determine the status of 525 and 526. Two rules will be given with examples.

Rule 1
The "ODL+CHL" switch is closed if no open switch is located on its sensor side and the breaker on its sensor side does not have more reclosure activities than the normal setting, otherwise, it is locked out.

Rule 2
Two "ODL+CHL" switches have the same status if their sensors sense the same line segment and there is no open switch between these switches.

From the alarm list in Figure 7, it can be seen that breaker 227 recloses successfully at first reclosure (8 seconds after the breaker opened). Since there is no other device between 227 and switch 526, this line section must be energized. Furthermore, since 227 attempted only one reclosure even though a second reclosure (26 seconds) is possible if the first one should fail. Hence, conditions of Rule 1 are satisfied for 526 and, consequently, 526 is judged to be closed. On the other hand, 558 is a supervised automatic switch and its final status is known to be closed from the alarms. Therefore, Rule 2 can be used to conclude that the status of 525 is closed. As soon as the status of both automatic switches is known, CRAFT starts the reasoning to locate the fault. Five rules shown below are used in the reasoning.

Rule 3
For any unsuccessful breaker reclosures, the fault must lie between the breaker and the nearest open device **at the time the reclosure is attempted.**

Rule 4
If any customer service is interrupted and the fault location is not known, then testing of the line is considered.

Rule 5
As many customers as possible should be brought on line by any successful test.

Rule 6
When energizing a line which may be faulted, it is preferable to close a breaker into the line rather than switching in load breaks or section switches which, in general, are more likely to be damaged by closure into a fault.

Rule 7
A line fault must lie between open breakers or switches.

From the final status of all devices, breaker 1113 recloses (at 10 seconds) unsuccessfully and remains open. Also, switch 546 is opened at 6 seconds. Based on Rule 3, the fault must lie between 1113 and 546, i.e., section 1113-1111, 1112-1146 or 1145-546. Also, two customers, Grady and Metro-Renton, are out of service. According to rule 4, testing will be considered to locate the fault and restore customers. There are several alternatives for testing this line. Based on Rule 5, 1145 will be opened because both customers will be restored if the test succeeds. Based on Rule 6, CRAFT suggests opening 1145 and then closing 1113.

In this example, the testing will succeed since the fault is located between 1145 and 546. As soon as the test action is complete, the fault is found by Rule 7 since only 1145 and 546 are open at this point.

B. Data Base

A data base management system, UWRIM, (University of Washington Relational Information Management system), is chosen. Five tables, "NAMES", "LINES", "SUBS", "DEVICES" and "DISPLAYS" are used. All the identification numbers of relevant PSPL lines are stored in the table, "NAMES". The table "LINES" contains the geographical regions to which the lines belong and also names of the two sides of each line. "SUBS" is a table of substation names, lines and switches connected to the substation. "DEVICES" is a table of device information including the adjacent devices and the functions and time settings of the switches. "DISPLAYS" stores the line diagrams for display.

C. Inference Engine

Since CRAFT is intended for on-line operation, the speed of rule-processing is important. The production system language, OPS83, is selected for its efficiency and ease of interface with the FORTRAN language. The forward chaining mechanism is used in OPS83. That is, given a set of WMEs (working memory elements) and a set of rules, the OPS83 inference engine matches these WMEs to the left hand sides of the rules. All matched rules are placed in the conflict set. Note that more than one instantiation of the same rule can exist in the conflict set. OPS83 inference engine allows the user to specify the priority for the choice of the rule to be fired. Referring to Rule 7 shown in Section III.A, an OPS83 version of the rule is shown in Figure 8.

```
rule lf1
-- IF the task is to locate a single fault
-- AND device 1 is the furthest open device up the line
-- AND device 2 is the furthest open device down the line
-- AND the possible line faults have not been identified
-- THEN obtain as a list of possible faults the sections of
-- line between these 2 open devices
  {
   &task (task solving=y; id=test_hypothesis; status=locate_fault;);
   &hypo (hypothesis subject=&task.subject; id=single_fault; status=test;);
   &fault (line_fault subject=&task.subject; status=radial; start=0; end=0;);
   &device1 (device subject=&task.subject; status=open;);
   &device2 (device subject=&task.subject; location>&device1.location;
status=open;);
   [int_real(&device2.location-&device1.location)]
   -->
   modify &fault (start=&device1.location; end=&device2.location; );
  };
```

Figure 8 OPS83 Version of Rule 7.

The first three conditions, &task, &hypo and &fault, are used for task control whereas the next two conditions are for matching two open switches in the working memory. In this example, 1113 and 546 will be assigned to the variables &device1 and &device2, respectively. The last condition is used to find a pair of open switches and their distance. Since in some cases there may be more than one pair of open switches at the final configuration, the inference engine needs to compare the distances of the pairs and identify the pair that is furthest apart.

D. Explanation Facility

The explanation facility is intended to provide a summary of the key decisions in the reasoning. The proposed approach is as follows:

1. Identify key rules in the rule base which correspond to higher level decisions/actions.
2. For each key rule, a template in concise English is stored.
3. Given the list of all rules that are fired during the chaining process, the key rules will be extracted together with instantiations that caused the rules to be fired.
4. The instantiation of a key rule is used to fill in the template to generate an explanation of the rule firing.

A different PSPL line shown in Figure 9 will be used to demonstrate the CRAFT explanation facility. The explanation facility provides a summary of the reasoning, which is shown in Figure 10.

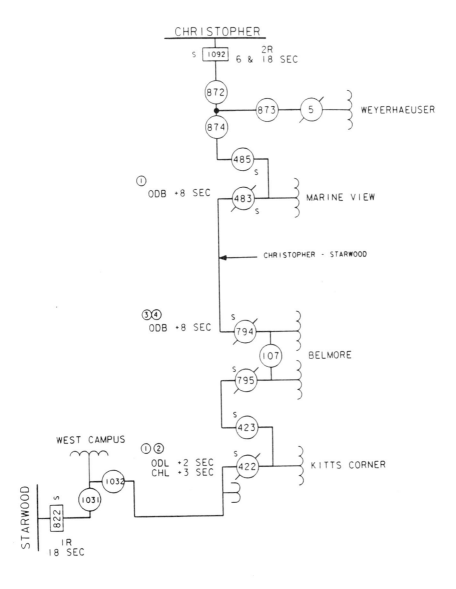

Figure 9 Example PSPL Line for Explanation Facility.

```
1092  872   874   485   483      794 107 795   423   422   1032   1031 822
XX----X--+--X-----X--+--O-------O-+-X-+-X-----X--+--O------X--+--X----XX
          |             |        | |           |              |

     Weyerhaeuser  Marine View     Belmore   Kitts Corner   West_Campus
Line No.:  105                     Cristopher_Starwood
```

Fault appears to lie between 795 - 423

Proposed Solution
Open 795 by supervisory.
Open 423 by supervisory.
Device 422 locked_out. Close locally
Close 483 by supervisory.
Close 794 by supervisory.

why
Fault lies between open switches furthest apart.
Fault lies between 483 - 422.

Bus fault on bus Kitts is not considered.
Fault lies to the left of 423.

Breaker 822 closed successfully on first attempt.
It tripped again 3 seconds later.
Switch 794 was the closest open switch at that time.
Fault must lie to the PCB 822 side of switch 794.

Bus fault on bus Belmore_H is not considered.
Fault lies to the right of 107

Bus fault on bus Belmore_L is not considered.
Fault lies to the right of 795

Figure 10 Output of CRAFT Explanation Facility.

E. Maintenance Facility

As mentioned previously, CRAFT has been on-line operational since July 1988. One of the primary issues to be addressed at present is maintenance of the rule base. A relational checking algorithm [4] has been developed to help maintain consistency of rules as new rules are added to the system. Several relations may exist between a pair of rules: cause-effect, mutual exclusion, redundancy, conflict, subsumption, implication and inclusion of one or more unnecessary condition. Four steps take place in this algorithm: (1) sorting of attribute tests, (2) construction of the RETE nets, (3) relation checking and

(4) grafting the new rule onto the existing LHS (left hand side) and RHS (right hand side) nets [4]. An example with two CRAFT rules shown below demonstrates how the relation checking is performed.

```
rule Rule10
{
  &task (task solving=y; id=test_hypothesis; status=locate_fault; );
  &hype (hypothesis subject=&task.subject; id=single_fault; status=test; );
  &fault (line_fault subject=&task.subject; status=radial; );
  &device (device subject=&task.subject; automatic=yes; location>&fault.start; );
  -->
  make(work subject=&task.subject; solving=y; id=check_timing; );
}

rule Rule11
{
  &task (task solving=y; id=test_hypothesis; status=locate_fault; );
  &hype (hypothesis subject=&task.subject; id=single_fault; status=test; );
  &fault (line_fault subject=&task.subject; status=branch; );
  &device (device subject=&task.subject; automatic=yes; location <= &fault.end; );
  -->
  make(work subject=&task.subject; solving=y; id=check_timing; );
}
```

Suppose a new rule, RuleNew, is to be added into the CRAFT rule base.

```
RuleNew
{
  &task (task solving=y; id=test_hypothesis; status=locate_fault; );
  &hypo (hypothesis subject=&task.subject; id=single_fault; status=test; );
  &fault (line_fault subject=&task.subject; status=radial; );
  ~(device subject=&task.subject; automatic=yes; location > &fault.start; );
  -->
  make(work subject=&task.subject; solving=y; id=check_timing; );
}
```

The LHS and RHS relations between RuleNew and Rule10, Rule11 are shown in Table 2.

Table 2 Relations between Sample Rules

Relation between	LHS	RHS
RuleNew and Rule10	COMPLEMENT	EQUIVALENT
RuleNew and Rule11	DIFFERENT	DIFFERENT

The relation checking algorithm is based on set theory and simple logic. In this case, the last condition of RuleNew is the COMPLEMENT of the corresponding condition of

Rule10. Since the conclusions of the two rules are EQUIVALENT, the condition, &device is unnecessary. Suppose both RuleNew and Rule10 are correct. Since Rule10 and RuleNew contain an unnecessary condition, i.e.,

(device subject=&task.subject; automatic=yes; location > &fault.location;);

This condition of Rule10 can be removed without affecting the validity of the rule. The relation checking facility provides a convenient way to help detect inconsistencies when a new rule is to be added into the rule base or a rule in the rule base is modified.

PART II: APPLICATION OF INDUCTIVE LEARNING

IV. INTRODUCTION

The quality of the knowledge base determines the problem-solving capability of an ES. However, practically, most expert systems will be imperfect since real world situations are complex and may not be predicable. Hence, the ES should be able to learn by itself.

Up to the present, almost all applications of the ES techniques to power system problems are concentrated on the development of knowledge bases for specific problem domains [20, 21]. The application of machine learning methods to power systems has not received much attention. The pattern recognition approach has been applied to power system security assessment in earlier work [25]. In the area of distribution systems, reference [26] reported an application of discrete decision theory to customer classification for direct load control. In a recent paper, an AI framework based on inductive learning was applied to the transient stability assessment and enhancement [27].

The second part of this chapter reports our experience with the application of an inductive learning method, ID3, [28] to power system problems. In this part, Section V introduces the underlying mechanism of ID3 and presents a proposed tree-modification algorithm for enhancing ID3's learning capability. The first application to a voltage control expert system is reported in Section VI. Section VII presents another application to the contingency selection problem in a power system. Section VIII gives the conclusion.

V. INDUCTIVE LEARNING

Inductive learning is the process of acquiring plausible assertions from a given set of examples. The acquired assertions are applied to predict the unknown properties of new examples. Particularly, inductive learning can be applied to the task of classification. In this case, the learning process derives a set of classification rules from given preclassified examples. This section describes the underlying mechanism of ID3 for generating classification rules. A tree-modification algorithm is proposed which modifies the decision tree in case of classification errors.

In ID3, classification rules are represented by a decision tree which is constructed from a **training set**, or preanalyzed objects. Each object in the training set is described in terms of its corresponding **class** and a fixed set of **attributes**. Attributes may be either **discrete** or **continuous**. A discrete attribute has a finite number of unordered values, whereas a continuous one is real-valued. For example, in contingency selection, the attribute *contingency type* is discrete; it is restricted to outages of lines, transformers, and generators. The attribute, *total reactive power demand,* is a continuous one, since its value is given by a real number.

ID3 applies a top-down, divide-and-conquer strategy to construct a decision tree. Starting from the entire training set (the root node), ID3 recursively selects a "good test" to partition a non-leaf node into several subsets until all leaf nodes contain single-class objects.

ID3 uses the index of "gain of information" to evaluate the "merit" of a test; a "good test" is thus the one which yields the maximum gain of information. This criterion enables the algorithm to build a decision tree in such a way that each single-class leaf node requires a minimal number of tests. This can be achieved since a test is selected to best reduce the degree of uncertainty about the class to which a non-leaf node belongs.

A summary of the concept of "information gain" and ID3 is provided in the following.

A. Information Gain Concept and ID3

When a decision tree is used to classify an object, the class of the object is assigned at a leaf node. Thus, a leaf node can be viewed as an information source always emitting a particular type (class) of messages. By this analogy, the root node can be regarded as an information source emitting n types of messages, where n is the number of classes involved in the classification task. Apparently the "degree of uncertainty" about the type of a message emitted by a leaf node is zero. On the other hand, a root node should be associated with a positive degree of uncertainty about the types of emitted messages.

To measure the degree of uncertainty about the outcome of an information source V, a quantity called **entropy** is defined as

$$H(V) = \sum_{i=1}^{n} P_i \log P_i \tag{8}$$

where n is the number of classes of messages, P_i is the relative frequency of messages which belong to the class i, and the base of the log function is 10. Notice that the entropy of a leaf node is 0 since the relative frequency of its class is 1 and those of other classes are 0.

In building a decision tree, a test is applied at each non-leaf node to reduce the uncertainty about the class of an instance. In terms of the entropy concept, the information gain of a test T at node V is defined as

$$I(V;T) = H(V) - \sum_{j=1}^{m} r_j H(W_j) \tag{9}$$

where node W_j is a child of V corresponding to a possible outcome of the test T (m=2 for a binary tree), r_j is the relative frequency of W_j's members at node V. If a test succeeds in partitioning the instances into two leaf nodes (single-class sets), then $H(V_j)$ is 0 for both children nodes. Hence, from Eq. (9), the gain of information is equal to the entropy $H(V)$. In other words, all uncertainty is removed. On the other hand, if in the worst case the entropy of the children nodes is as high as that of the parent, then $I(V;T)$ will be 0, which corresponds to a useless test. A **deadend** is a node for which no test can reduce uncertainty. A decision tree with many deadends implies that the selected attributes/thresholds are not capable of describing the nature of instances.

A test is a query for the value of a specific attribute of an instance. If a test queries the value of a continuous attribute A, the test takes the form

$$[A <= t \, ?]$$ (Test Type 1)

where t is a threshold. Usually, a continuous attribute would be compared with a sequence of thresholds to determine the best threshold in terms of the gain of information. On the other hand, if attribute A is discrete, the test checks against all possible values of A:

$$[A = a_i \, ?], i = 1, ..., r$$ (Test Type 2)

where r is the number of possible values. Note that a test of type 1 partitions a set into two subsets; one satisfies the test and the other fails. A type-2 test results in r partitions, depending on the number of discrete values a variable can take.

The major tasks for building a decision tree are as follows.

1. Search for the optimal test for each non-leaf node. This involves the computation of information gains for all possible tests. If no test yields any gain of information, the non-leaf node is designated as a deadend.

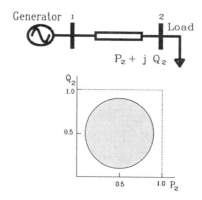

Figure 11 A Two-Bus Power System and Its Security Region.

2. Determine the type (leaf or non-leaf) of new nodes. A leaf node contains members in a single class, while a non-leaf node consists of multiple-class members and hence needs to be further expanded.

To illustrate the application of inductive learning, a simple example of learning the security region[2] of a two-bus power system is given below. The power system shown in Figure 11(a) consists of a load bus whose load demand is served by a generator through a transmission line. In this example, the security region is characterized by the real power (P_2) and reactive power demand (Q2) at the load bus. Assume that a valid security region of this system is inside a circle as shown in Figure 11(b). The circle centers at (0.5, 0.5) with 0.4 as its radius. This example applies the learning method to produce rules to approximate the security region.

A training set with 20 instances is generated via a random number generator with a uniform distribution from zero to one. Each instance represents an operating point in the two-bus system and is characterized by two attributes, P2 (in per unit) and Q2 (in per unit). Two consecutive numbers drawn from the random number generator represent their values. If a point is inside the circle, it is designated as class 1 (secure); otherwise, its class

---

2. The security region of a power system is a set of operating conditions under which no operating constraint is violated. Usually, operating constraints include limits on bus voltages, real power and reactive power generations, and line flows. For a general treatment of a power system's security region, see [29].

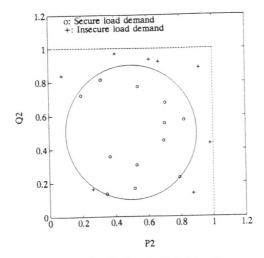

Figure 12 Distribution of 20 Training Instances.

is 0 (insecure). The distribution of the training set is shown in Figure 12. The distribution is sparse because of the small number of training instances generated for illustration here.

Since the selected attributes are continuous, the test space for each non-leaf node includes only type-1 tests. To generate the candidate tests for each non-leaf node, an increment is calculated by dividing an attribute's range in a non-leaf node into ten steps. Then, a sequence of ordered thresholds is generated for each attribute by sequentially incrementing the attribute until the maximum is reached. The tree-building procedure is then used to compute the information gain for each candidate test to determine the best partition of a non-leaf node.

The resultant decision tree is shown in Figure 13 which includes five leaves and 4 non-leaf nodes. Figure 14(a) displays the security region represented by the decision tree. The learned security region is by no means a good approximation of the true one. This is because the training set does not cover enough representative training instances[3], as shown in Figure 12.

---

3. In this example, an instance (a point in X-Y plane) is fully characterized by the selected attributes. This implies that the attributes are adequate for the learning task. In general, a sufficient number of representative training instances is vital to generate a good approximation of the target concept.

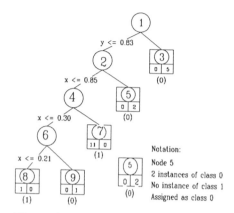

Figure 13 Decision Tree from 20 Instances.

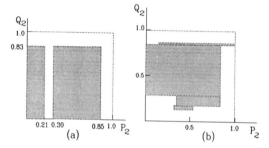

Figure 14 Security Regions from Decision Tree (20 and 50 Instances).

For comparison, 30 more instances are added into the original training set. The distribution of the new training set is shown in Figure 15 with more instances around the circle. The learned region from the new training set is shown in Figure 14(b). Indeed, the approximation is closer to the true security region, although it is still not good enough.

B. Tree-Modification Algorithm

The effectiveness of inductive learning relies heavily on the "representativeness" of a training set. If a training set does not include a particular pattern of instances, the constructed decision tree can misclassify input instances with that pattern. However, it is practically impossible to find a complete set of training instances for some real-world classification tasks. Hence, our goal is to

- First construct a decision tree with a selected set of instances.

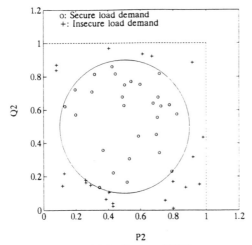

Figure 15 Distribution of 50 Instances.

- Then, modify the tree every time a misclassification occurs. In doing so, previously classified instances must still be remembered.

Clearly, this can be done by rebuilding the decision tree completely with all instances. However, this would be prohibitive from the computational point of view. To avoid the complexity of complete reconstruction, a "tree-modification" algorithm is proposed, which can be used to identify the minimal portion of the decision tree that needs to be modified.

The tree-modification algorithm is presented in Figure 16. The idea of the proposed modification algorithm is to start from the leaf node with the misclassified instance and back-track to its parent nodes, one level at a time, until a node is reached where the test does not have to be changed. Note that the new (misclassified) instance will cause at least one node to change (modify) its test for classification.

More specifically, the misclassified instance is first added to the example set. Then, starting from the node whose child misclassifies the instance, the algorithm checks if the optimal test is changed due to insertion of the new instance. If the test is changed at a node, the check continues to its parent; otherwise, the reconstruction starts from the child node visited previously.

```
                        Tree-Modification Algorithm

Input:      x - an instance with correct class information
            (x denotes a new instance classified by the decision tree.)
            v - the leaf node where x belongs to.
Output: a modified decision tree.

Method:
1. (a) Add x to the training set;
   (b) Add x to the member lists of those nodes
       in the path from root node to v;
2. (a) IF x is not correctly classified THEN
          previous_node <-- v;
          current_node <-- v's parent node;
          Find the new_test with maximum information gain at current_node;
       ELSE
          Add x to training set;
          Return;
       END IF
   (b) WHILE new_test is different from the test stored in current_node DO
          IF current_node is root THEN
             Build the decision tree using the new
             training set;
             Return;
          END IF
          previous_node <-- current_node;
          current_node <-- current_node's parent node;
          Find the new_test with maximum information gain at current_node;
       END WHILE
3. Rebuild the subtree starting from previous_node;
```

Figure 16 Tree-Modification Algorithm.

To illustrate the capability of the tree-modification algorithm, the 30 instances in the first example are used to test the preliminary decision tree in Figure 13. During the testing, the tree-modification algorithm is invoked each time a misclassification occurs. The testing results in five misclassifications which in turn triggers five times of tree modifications. (Apparently, one can collect a number of misclassified examples before the tree-modification is performed.) Figure 17 shows five approximated security regions, each of which representing the resultant decision of each modification. During the testing, the second, third, and fourth modifications rebuild the decision tree from scratch, while the

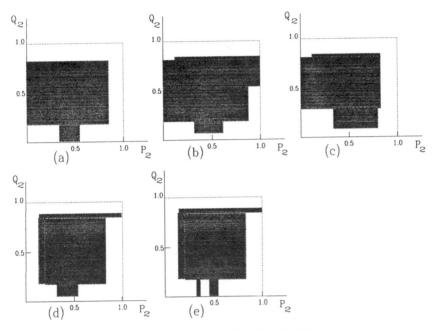

Figure 17 Security Regions During Tree-Modifications.

first and fifth modify only a portion of the decision tree. In reality, if a decision tree is well built, the modification should be minor.

VI. SELF-LEARNING EXPERT SYSTEM

This section reports an application of inductive learning to voltage control of power systems. A voltage control expert system (VCES) has been developed at the University of Washington [30]; the ES applies heuristic rules for adjusting reactive controls, e.g., shunt capacitors, tap changing transformers and generator voltages, etc. An extended version of VCES has been implemented in the on-line environment of the National Control Center of Portugal [31]. In its basic form, VCES uses sensitivity factors to estimate the amount of controls required to correct an abnormal voltage profile. In this study, it is observed that the error resulting from linearization becomes more obvious when the voltages are far away from the nominal operating points, which is an expected result. In view of the weakness, the learning technique is proposed to improve accuracy of the estimation of control amounts. Our approach is indirect. By raising the low voltage limit used by VCES, VCES is forced to select a larger amount of control. As a result, the control amount can be adjusted by varying the low voltage limit used by VCES. The strategy proposed here is to assess the severity of the voltage problems first, and then determine the appropriate low voltage limit for use by VCES. The following results are quite interesting:

- Attributes most relevant to severity assessment of voltage problems have been identified.

- The integrated learning system has the capability of learning from past mistakes, which makes VCES a self-learning expert system.

In the following, the capabilities and limitations of VCES are summarized. Then, its learning environment is presented. Numerical results are given to demonstrate the effectiveness of the proposed learning system.

A. VCES and its Limitations

To ensure proper and efficient power supply to customers, voltages at the load buses of a power system should be maintained within a normal range. This is indicated by the constraint (C2) in Part I. In this study, the normal range for the load voltages is assumed to be from 0.95 p.u. to 1.05 p.u. Should any load voltage fall outside this range, reactive controls via shunt capacitors, transformer tap changers and generator voltages are adjusted to bring the load voltages back to normal. The controls are represented by $\underline{c}$ in Eq. (5). VCES incorporates heuristic rules for the reactive power/voltage control; it is developed to serve as an operator's aid in solving the voltage problems [16, 30].

If a voltage violation is detected, VCES will execute the following tasks to suggest the corrective actions.

1. Classifying the severity (high or low) of the voltage problem.
2. Prioritizing controls for a problem bus based on the type and sensitivity of devices.
3. Determining the amount of control to be applied.
4. Estimating the effect of a control on the voltages of all problem buses.
5. Verifying suggested controls by running the power flow. All problem buses should be within the normal range after the control actions are taken. The power flow is represented by Eq. (2) in the power system model.

Tasks 2, 3, and 4 are based on the linearized sensitivity model, which is given by the equation:

$$\Delta V_i = d_{ij} * \Delta U_i \qquad (10)$$

where ΔV_i is the voltage deviation, ΔU_i denotes the amount of control to be applied. The sensitivity factor d_{ij} is found from the inverse of the Jacobian containing the partial derivatives of reactive power with respect to voltages.

In a normal operating condition, good decoupling exists between the real power/angle and the reactive power/voltage relations. Thus, the sensitivity model provides good behavior prediction when the voltage deviations are not severe. In other words, VCES is capable of determining which control(s) to apply and its (their) amount. For these minor problems, VCES usually solves the problem in one iteration, i.e., the set of controls suggested by VCES passes the verification with the power flow (Task 5). If the verification of Task 5 still shows voltage violations, VCES will be invoked again to select more controls. This second attempt to solve the voltage problem will be referred to as the second "iteration."

In most cases that have been tested so far, VCES solves the problems in one iteration. However, the error introduced by the use of sensitivities (first order approximations) becomes obvious for severe cases. For these cases, the amount of controls selected based on the low voltage limit (0.95 p.u.) is insufficient to correct the voltage problem in one iteration. Therefore, the second or third iterations will be needed.

Based on our experience, the above-mentioned deficiency of VCES can be corrected by manual adjustments of the low voltage limit. By raising the low voltage limit, say, from 0.95 to 0.98, VCES is forced to choose a higher level of reactive control. At first, this may seem to result in "overestimation" of the needed control. However, this is exactly what is needed to correct the problem of "underestimation" that VCES demonstrated. The purpose of "learning" in this study is to help VCES identify the appropriate low voltage limit so that VCES can solve problems in one iteration, which increases the capabilities and efficiency of the expert system. Depending on the severity of the voltage problem, the low voltage limit will be assigned one of the following numbers: 0.95, 0.96, 0.97, 0.98, 0.99 p.u. An inductive learning module added to the original VCES is responsible for the assignment of the low voltage limit.

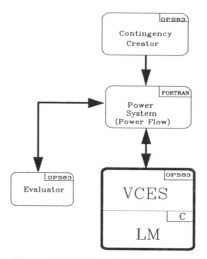

Figure 18 VCES and Its Learning Module.

B. Enhanced VCES and its Training Environment

In Figure 18, the enhanced VCES and its training environment is shown. This study results in a learning module (LM) added to the original VCES. The module LM performs the functions of tree-building, tree-modification and assignment of the low voltage limit. A voltage problem of the power system will be solved by VCES, which utilizes the decision tree built by the LM module. In order to test the capability of the VCES/LM system, a large number of contingencies are created in the module "Contingency Creator." Contingencies with line outages, generator trips, load changes, etc. are simulated by the power flow program. If a scenario results in a voltage problem, VCES is invoked to suggest control devices and their respective amounts. Controls suggested by VCES/LM are evaluated by the module labeled, "Evaluator." This program calls the power flow to check if all voltage violations are removed completely. If the answer is yes, the controls are further evaluated in terms of the total number of controls used, etc. [30]. Also, the scenario will be added to the training set serving as a training instance. In case of a failure to solve the voltage problem in one iteration, VCES will perform the following steps:

1. Find a new low voltage limit for the scenario which enables VCES solve the problem in one iteration.

2. Run the tree-modification procedure to update the decision tree.

The first task can be performed by changing the low voltage limit in the VCES and verifying with the power flow. In the second step, the minimal subtree of the original decision tree will be identified and reconstructed.

C. Inductive Learning Application and Results

1. Attribute Selection and Classification Parameter

In order to effectively discriminate the severity of a voltage problem, relevant attributes have to be chosen to characterize a voltage problem instance. Based on our experience, the following attributes are effective for the classification:

n : total number of problem buses,
av: average voltage deviation,
Q : Var demand of the load bus with the largest voltage deviation.

Obviously, as the number of problem buses increases, the voltage problem is more severe. The same argument applies to the average voltage deviation. The Var demand of the load bus with the largest voltage deviation is chosen as one of the attributes because it is an indication of how much control is needed to correct the problem. If this most severe bus is corrected, violations at other buses with less severe problems may also be eliminated.

As mentioned in Section VI.A, adjustment of the low voltage limit used inside VCES was made to correct the error introduced by the sensitivity model. The adjustment of the low voltage limit will be represented by the deviation from 1.0 p.u. The deviation parameter, denoted by L, is chosen for classification. Thus, for a minor voltage problem, L would be set to 0.05, meaning that a low voltage limit of 0.95 p.u. is to be used by VCES in selecting controls. For a more severe problem, L may be 0.02 which corresponds to a limit of 0.98 p.u. The value of L ranges from 0.05 to 0.01 with a decrement of 0.01 p.u. Therefore, the allowable values for the deviation parameter are 0.05, 0.04, 0.03 0.02, and 0.01, in ascending order of problem severity.

2. Preparation of Training Set

For voltage control, a good training set should be obtained from voltage problems over a wide range of severity. It is impossible, however, to prepare a training set containing all possible instances for the voltage problem since two of the attributes are continuous.

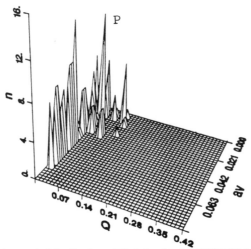

Figure 19 Distribution of Training Set for VCES/LM.

In this study, the test system is the modified IEEE 30-bus system [16]. A total of 96 instances are generated which serve as the training set. Their distribution in the attribute space is shown in Figure 19. The point P represents a total of 2 instances which correspond to an identical value of n and very close values of av and Q. For simplicity, these instances are lumped and represented by one point P.

3. Construction and Modification of Decision Tree

In the tree-building task, the thresholds of continuous attributes are discretized; an increment is selected for each attribute. Thus,

1. Increment of n : 1,
2. Increment of av: 0.00809,
3. Increment of Q : 0.011.

The increments of Q and av are a tenth of the difference between their respective maximum and minimum in the training set. Since n is integer-valued, the increment is 1.

Starting from the whole training set (root node), the tree-building algorithm iteratively finds the optimal test for each non-leaf node until all leaf nodes contain objects of a single class or are deadends. Then, each single-class node is assigned the same value of the deviation limit L. On the other hand, a deadend is assigned the lowest (most stringent)

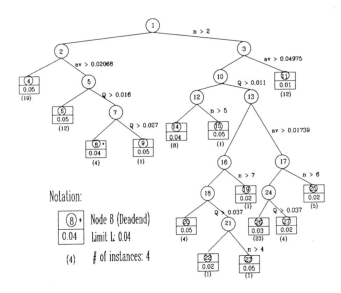

Figure 20 Preliminary Decision Tree for VCES.

value of L among its members. This "worst-case" assignment enables VCES to solve all instances contained in the deadend. The preliminary decision tree is shown in Figure 20, which contains 27 nodes and 14 leaves. Node 8 is the only deadend. Three instances correspond to the L value of 0.04 and the fourth instance has a value of 0.05, hence the worst case value of 0.04 is used.

Due to deficiency of the initial training set, the preliminary decision tree may misclassify the severity levels of some voltage problems. A misclassification invokes the tree-modification algorithm. The modification process is demonstrated using three misclassified instances in Table 3, all associated with Node 9 of Figure 20. Figure 21 shows the step-by-step modification of the subtree with Node 9 as the root. Each misclassified instance necessitates further branching or reconstruction. In Figure 21(a), an extra child node, 28, is created for the misclassified instance, which has a correct L value of 0.03 instead of 0.05.

Table 3 Examples of Misclassified Instances

Case	n	av	Q	Classified limit	Correct limit
1	2	0.025846	0.081158	0.05	0.03
2	2	0.027070	0.327658	0.05	0.04
3	2	0.049540	0.184745	0.05	0.03

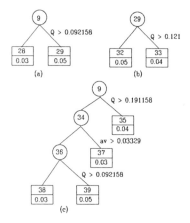

Figure 21 Examples of Tree Modifications.

4. Test Cases

To test the performance of the VCES/LM system, a base case is selected first, then 1089 contingency cases are created by removing 1 or 2 lines, lowering generator voltages, increasing reactive power demands of load buses, tripping generators or combining the above contingencies. They are summarized in Table 4.

Table 4 Summary of Test Scenarios

Scenario	Total
Load increased	323
Load increased with generator(s) tripped	20
Single line and generator(s) tripped	40
Single line tripped with load increased	529
Double lines and generator(s) tripped	6
Double lines tripped with load increased	56
Generator voltages reduced and generator(s) tripped	7
Generator voltages reduced with load increased	107
Generator(s) tripped	1

All the 1089 cases with voltage violations are created by the Scenario Creator. Their distribution in the attribute space is displayed in Figure 22. The interpretation of the diagram is identical to that of Figure 19. It can be seen that the range of reactive power demands Q is much wider than in Figure 19.

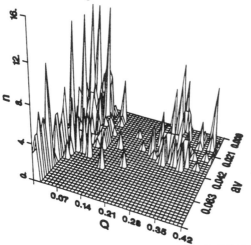

Figure 22 Distribution of Test Set for VCES/LM.

Table 5 Misclassified Instances in Task 2

Instancen	av	Q	Classified limit	Correct limit
1	2 0.025846	0.081158	0.05	0.03
2	5 0.016669	0.315000	0.05	0.04
3	2 0.027070	0.327658	0.05	0.04
4	6 0.029309	0.010000	0.05	0.04
5	2 0.049540	0.184745	0.05	0.03
6	1 0.039666	0.083621	0.05	0.04
7	1 0.034766	0.254998	0.04	0.03
8	1 0.039709	0.083638	0.04	0.03
9	6 0.036119	0.010000	0.04	*
10	7 0.016226	0.312046	0.05	*
11	3 0.040199	0.279980	0.02	*

5. Test Results

Two simulation tasks are performed to show the effectiveness of the learning method:

Task 1: Simulation without tree-modification algorithm.

In this task, 1052 cases are assigned appropriate deviation limits which enable VCES to solve these problems in one iteration. The total number of misclassified cases is 37, which corresponds to a misclassification rate of 3.4 %. Eleven of the 37 misclassified instances are listed in Table 5.

Task 2: Simulation with tree-modification algorithm.

The tree-modification algorithm is incorporated into LM to take into account the misclassified instances shown in Task 1. Again, the 1089 test cases are simulated in this task. As a result, Table 5 shows eleven misclassified instances in the simulation, which is reduced from the previous number of 37. Note that the three instances marked with * in Table 5 are not assigned any deviation limits since VCES can not solve these three cases in one iteration even with the tightest limit, 0.01. The eight remaining cases initiates 8 executions of the tree-modification algorithms. The modified portions of the decision tree are indicated in Figure 23. The major reconstruction occurs at Node 9. The remaining part of the tree performs well with the additional test instances.

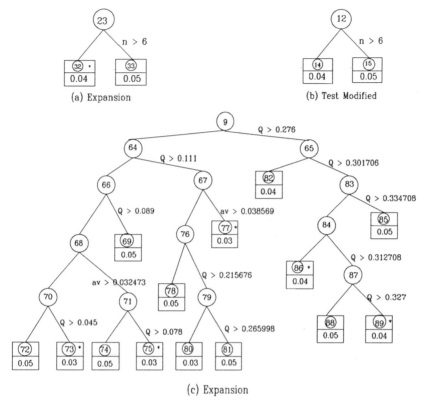

(c) Expansion

Figure 23 Modified Subtrees in Task 2 (Nodes 23, 12, 9).

Due to the limitation of VCES itself, the final decision tree fails in three instances. Yet the result is impressive considering the variety of the contingencies imposed on the 30-bus system.

The tree modification algorithm identifies the minimal portion of the subtree to be rebuilt. Therefore, only the test instances falling within the subtree need to be computed. For the case of Task 2 mentioned above, 115 test instances are used in the reconstruction of the subtree under Node 9. The modification of the subtrees under nodes 12 and 23 requires computation of 2 and 12 instances, respectively.

VII. LEARNING TO SELECT CONTINGENCIES

This section describes the application of decision trees to acquire knowledge for selecting power system contingencies which cause voltage violations. Traditionally, insight into selecting contingencies may be gained by operational planners through numerous simulations. The purpose here is to study whether inductive learning can generate such high-level knowledge automatically from a large number of examples.

The remaining part of this section describes the terms for classification, selected attributes, training and test sets, and numerical results. The IEEE 30-bus system, shown in Figure 24, is also used here .

A. Performance Measures

To measure the performance of a decision tree in selecting contingencies, the outcome of a classification is categorized as follows. If the outcome shows a correct classification of an instance, it is a **hit** [33]. On the other hand, a **miss** means that a contingency causing voltage violations is misclassified as harmless, whereas a **false alarm** indicates the contrary.

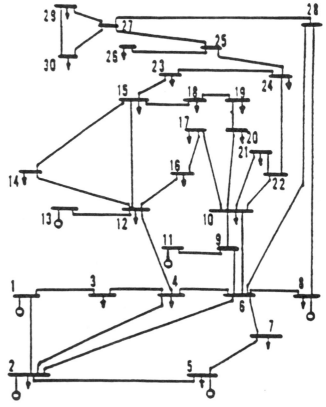

Figure 24 IEEE 30-Bus System.

B. Selected Attributes

The process of contingency selection involves two basic tasks: (1) evaluation of a power system's operating condition and (2) characterization of contingencies. Thus, the selected attributes should be able to represent the operating condition of a power system and to characterize a contingency. In this classification task, the attributes are selected as follows:

1. Contingency type (ctgtype).

   The considered contingencies are: (1) single generator tripping, (2) single line tripping, and (3) single transformer tripping.

2. Location of a contingency (ctgrgn).

   One interesting feature of the IEEE 30-bus system is that the system can be divided into three areas: (1) buses 1 through 13, containing all generations in the system, (2) buses 14 through 24, containing most of the system load, and (3) the rest of the system which is far from the generation area. This attribute has eight possible values, depending on whether each area has a load increase or generator tripping.

3. Reactive power carried by the outaged equipment (carryq).

   If a contingency is tripping of a generator, the value of this attribute is defined as the reactive power generation of the generator. For a line or transformer outage, this value is the reactive power flow on the line or through the transformer.

4. Maximum reactive load deviation (relative to the base case) at a bus (mxqldinc).

5. Total reactive load in the system (sumqld).

Note that the first three attributes give the information about a contingency, whereas the remaining attributes describe the operating condition of a power system.

Table 6 Training Scenarios for Contingency Selection

Scenario	Number of Modifications of Operating Condition
No initial outage in system	15
Line 12-16 out of service	15

Table 7 Test Scenarios for Contingency Selection

Scenario	Number of Modifications of Operating Condition
Line 14-15 out of service	20
Line 16-17 out of service	20
Line 15-18 out of service	20
Line 18-19 out of service	20
Line 10-17 out of service	20

C. Generation of Training and Test Sets

Two scenarios used to generate the training set are listed in Table 6. Each scenario is modified 15 times by increasing every load bus demand by the same percentage (from 0 to 30 %). A number of contingencies are simulated for each of those modified scenarios without any voltage violation in the test system.

A total of 443 instances are generated for the training set. Based on the training set, a preliminary decision tree for contingency selection is constructed. This tree consists of 56 non-leaf nodes and 63 leaves (including one dead-end). Due to the size of the decision tree, only a portion is shown in Figure 25. Interestingly, the decision-making hierarchy

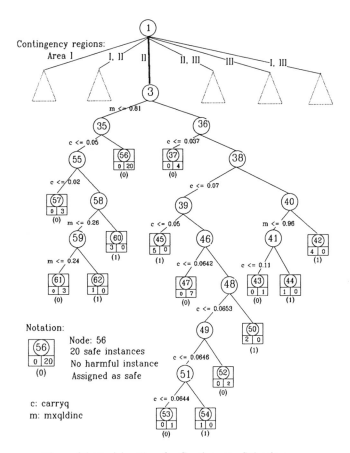

Figure 25 Decision Tree for Contingency Selection.

represented by the constructed decision tree is similar to an operational planner's approach to contingency selection. Given a contingency, the decision tree first determines which area the contingency is located in. Then, other attributes related to the operating condition and the contingency characteristics are evaluated.

To evaluate the performance of the acquired knowledge for contingency selection, two experiments are performed below: (1) test without the tree-modification algorithm and (2) test with the tree-modification algorithm.

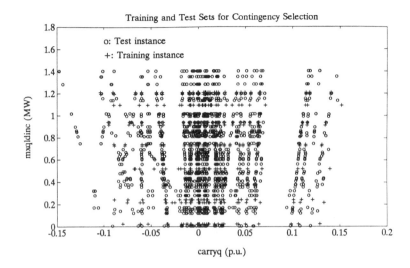

Figure 26 Distribution of Training and Test Sets for Contingency Selection.

The test scenarios are listed in Table 7. They are different from the training set. The test set consists of 1600 instances. For comparison, Figure 26 displays the distribution of the training instances and test instances with respect to two attributes, carryq and mxqldinc.

D. Tests

For the first experiment, the performance measures of contingency selection are shown in Figure 27. With a hit rate as high as 89.4 %, the results are quite encouraging. This indicates that the proposed learning method has the capability to acquire some high-level knowledge for contingency selection. Even though the acquired knowledge is system specific, the learning procedure is generic.

The second experiment is designed to test the decision tree with the tree-modification algorithm. During the testing, the tree-modification procedure modifies the decision tree 74 times, of which nine modifications result in complete rebuilding of the decision tree. The final tree consists of 147 non-leaf nodes and 154 leaf nodes (including two deadends). Also, the attribute ctgrgn is used to partition the root node. The performance measures

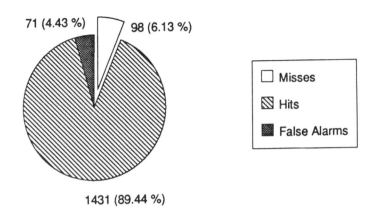

71 (4.43 %)   98 (6.13 %)

□ Misses
▧ Hits
■ False Alarms

1431 (89.44 %)

Figure 27 Performance Measures for Experiment 1 (without Tree Modifications).

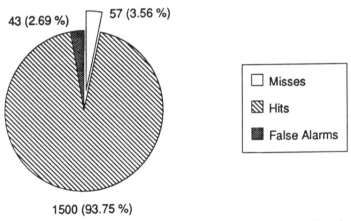

43 (2.69 %)   57 (3.56 %)

□ Misses
▧ Hits
■ False Alarms

1500 (93.75 %)

Figure 28 Performance Measures for Experiment 2 (with Tree Modifications).

for this experiment are shown in Figure 28. The tree-modification algorithm eliminates 41 misses and 28 false alarms, increasing the hit rate by 4.4 %. For a more detailed comparison, a breakdown of the performance measures of these two experiments according to contingency regions is shown in Tables 8 and 9. Notice that with the tree-modification algorithm, the number of misclassifications is reduced in almost every tree (except the subtree for area II).

Table 8 Performance of Subtrees (without Tree-Modification)

Contingency Regions	Hits	Misses	False Alarms
I	634 (39.6 %)	54 (3.4 %)	46 (2.9 %)
II	233 (14.6 %)	15 (0.9 %)	3 (0.2 %)
III	242 (15.1 %)	2 (0.1 %)	2 (0.1 %)
I & II	171 (10.7 %)	19 (1.2 %)	15 (0.9 %)
I & III	74 (4.6 %)	5 (0.3 %)	3 (0.2 %)
II & III	77 (4.8 %)	3 (0.2 %)	2 (0.1 %)

Table 9 Performance of Subtrees (with Tree-Modification)

Contingency Regions	Hits	Misses	False Alarms
I	678 (42.4 %)	28 (1.8 %)	28 (1.8 %)
II	239 (14.9 %)	8 (0.5 %)	4 (0.3 %)
III	243 (15.2 %)	2 (0.1 %)	1 (0.1 %)
I & II	183 (11.4 %)	14 (0.9 %)	8 (0.5 %)
I & III	79 (4.9 %)	2 (0.1 %)	1 0.1 %)
II & III	78 (4.9 %)	3 (0.1 %)	1 (0.1 %)

VIII. CONCLUSION

The first part of this chapter is a discussion of the basic issues on the development of knowledge-based systems for power systems. The criteria of need, suitability and feasibility are proposed for the selection of an ES application area. ES development tools available at present are much improved in terms of their inference capability, environment and user-interface. However, none of the tools are ideal; it is difficult to find a tool that excels in all performance criteria. Maintenance and verification of rule-based systems remain important areas for the practicality of expert systems. During the last decade, most of the ES applications are prototypes developed on stand-alone machines. To fully integrate expert systems into the power system operation and planning environment, it is essential to develop better methods for interface among database, software, and hardware facilities.

Our attempt to apply inductive learning to power system operations has shown some promise of the technique. The learning module that is integrated with VCES suggests a low voltage limit for VCES to determine the appropriate amount of controls for a voltage problem. In the other application, the learning method has shown the capability to acquire high-level knowledge for contingency selection. These results are interesting for two reasons. First, they demonstrate that high-level knowledge can be generated from preanalyzed examples. This is an approach to reducing the computation time of numerical algorithms in the real-time environment. Secondly, in both applications, the proposed tree-modification algorithm is effectiveness in reducing the number of misclassifications. This is important since it is always difficult to prepare a complete training set for a real-world application; a self-learning system is more robust to a changing, uncertain environment.

ACKNOWLEDGEMENTS

The authors would like to acknowledge the support from National Science Foundation through grant ECS 8657671. The CRAFT project was sponsored by Electric Power Research Institute (EPRI) and Puget Sound Power and Light (PSPL) Company. The CRAFT and VCES projects involve a number of colleagues at UW, PSPL, and EPRI. We particularly appreciate the contributions of J. M.-S. Tsai, H. Marathe, L. Wong, M. J. Damborg, and M. Lauby for the work reported in this chapter.

REFERENCES

1. A. Barr and E. A. Feigenbaum (Eds.), *The Handbook of Artificial Intelligence,* Vol. II, William Kaufmann, Inc., 1982, (Chapters VII-IX).

2. S. Ito, I. Hata et al., "Application of Expert System to 500 KV Substation Operation Guide System," *Proc. First Symp. on Expert Systems Application to Power Systems* (ESAPS), pp. 6.14-6.21, 1988 .

3. K. Tomsovic and C. C. Liu, "Bounding the Computation Time of Forward-Chaining Rule-Based Systems," *Proc. IEEE Int'l Symp. on Circuit and Systems,* pp. 421-424, Jun. 1988 .

4. H. Marathe, T. K. Ma, and C. C. Liu, "An Algorithm for Identification of Relations among Rules," *Proc. IEEE Int'l Workshop on Tools for AI*, pp. 360-367, 1989.

5. H. Sasaki, et al., "A Novel Scheme for Validation and Verification of Rule Bases in Expert Systems," *Proc. Second Symp. on ESAPS*, pp. 416-422 Jul. 1989.

6. M. Suwa, A.Scott, and E. Shortliffe, "An Approach to Verifying Completeness and Consistency in a Rule-Based Expert System," *AI Magazine*, pp. 16-21, Fall 1982.

7. T. Nguyen, W. Perkins, T. Laffey, and D. Pecora, "Checking an Expert System Knowledge Base for Consistency and Completeness," *Proc. Int'l Joint Conference on Artificial Intelligence*, pp. 375-378, 1985.

8. F. Hayes-Roth, D. A. Waterman, and D. B. Lenat, *Building Expert Systems*, Addison-Wesley, Reading, Mass., 1983.

9. H. Motoda, "The Current Status of Expert System Development and Related Technologies in Japan," *IEEE EXPERT*, pp. 3-11, Aug. 1990.

10. *Proceedings of the First Symposium on Expert Systems Application to Power Systems*, Stockholem-Helsinki, Aug. 1988.

11. *Proceedings of the Second Symposium on Expert Systems Application to Power Systems*, Seattle, Jul. 1989.

12. E. Cardozo and S. N. Talukdar, "A Distributed Expert System for Fault Diagnosis," *IEEE Trans. on Power Systems*, pp. 641-646, May 1988.

13. R. Bijoch, S. Harris, and T. Volkmann, "Intelligent Alarm Processor at Northern States Power," *Proc. Second Symp. on ESAPS*, pp. 79-83, Jul. 1989.

14. Y. Sekine, H. Okamoto, and T. Shibamoto, "Fault Section Estimation Using Cause-Effect Network," *Proc. Second Symp. on ESAPS*, pp. 276-282, Jul. 1989.

15. P. Fauquembergue, P. Brezillon, and Y. Harmand, "Synthesis of Events in an EHV Substation: an Expert System Approach," *Proc. IFAC Symp. on Power Systems: Modelling and Control Applications*, pp. 16.6.1- 16.6.5, Sep. 1988.

16. C. C. Liu and K. Tomsovic, "An Expert System Assisting Decision-Making of Reactive Power/Voltage Control," *IEEE Trans. on Power Systems*, pp. 195-201, August 1986.

17. Y. Kojima, S. Warashina, et al., " Development of a Guidance Method for Power System Restoration," *IEEE Trans. Power Systems*, pp. 1219-1227, Aug. 1989.

18. F. D. Galiana and D. McGillis, "Design of a Longitudinal EHV Transmission Network: A Knowledge-Based Approach," *Proc. First Symp. on ESAPS*, pp. 3.9-3.16, 1988 .

19. J. J. Jasen and H. B. Puttgen, "ASDEP: An Expert System for Electric Power Plant Design," *IEEE EXPERT*, pp. 56-66, Spring 1987.

20. B. Wollenberg and T. Sakaguchi, "Artificial Intelligence in Power System Operations," *Proceedings of the IEEE*, pp. 1678-1685 Dec. 1987.

21. C. C. Liu and T. S. Dillon, "State-of-the-Art," *Expert System Applications in Power Systems*, T. S. Dillon and M. A. Laughton (Eds.), Prentice Hall (U.K.), pp. 383-408, 1990.

22. K. Tomsovic, C. C. Liu, et al., "An Expert System as a Dispatchers' Aid for the Isolation of Line Section Faults," *IEEE Trans. Power Delivery*, pp. 736-743, Jul. 1987.

23. H. Marathe, C. C. Liu, et al., " An On-Line Operational Expert System with Data Validation Capabilities," *Proc. Power Industry Computer Application Conference*, pp. 56-63, May 1989.

24. C. C. Liu and M. Damborg, "CRAFT: On-Line Expert System for Customer Restoration and Fault Testing," EPRI EL-6680, Final Report, Vol. 1-5, March 1990.

25. C. K. Pang, F. S. Prabhakara, A. H. El-Abiad, and A. J. Koivo, "Security Evaluation in Power Systems Using Pattern Recognition," *IEEE Trans. on Power Apparatus and Systems*, Vol. PAS-93, May-June 1974.

26. J. R. B. Cockett and J. D. Birdwell, "Inductive Inference of Model Structure Using Hypothesis Feedback," *Proceeding of the IEEE International Symposium on Circuits and Systems*, pp. 1891-94 June 1988.

27. L. Wehenkel, Th. Van Cutsem, and M. Ribbens-Pavella, "An Artificial Intelligence Framework for on-line Transient Stability Assessment of Power Systems," IEEE PES Summer Meeting, Paper 88 SM 699-1, July 1988.

28. J. R. Quinlan, "Induction of Decision Trees," *Machine Learning*, pp. 81-106, 1986.

29. F. F. Wu and S. Kumagai, "Steady State Security Regions of Power Systems," *IEEE Trans. Circuits and Systems*, Vol. CAS-29, No. 11, pp. 703-711, Nov. 1982.

30. C. C. Liu, H. Marathe, and K. Tomsovic, "A Voltage Control Expert System and its Performance Evaluation," *Expert System Applications in Power Systems*, T. S. Dillon and M. A. Laughton (Eds.), Prentice Hall (U.K.), pp. 105-152, 1990.

31. L. Barruncho, J. P. S. Paiva, and C. C. Liu, "Voltage/Var Control Optimization and Knowledge-Oriented Approach: An Application for the Present Run-time Environments, " *Proc. Second Symp. on ESAPS*, Seattle, July, 1989.

32. C. C. Liu, "Knowledge-Based Systems in Power Systems: Applications and Development Methods," *Trans. IEE of Japan*, pp. 241-250, Apr. 1990.

33. S. M. Wang, C. C. Liu, and R. Fischl, "Qualitative Assessment of Voltage Problems due to Contingencies in a Power System," *Proc. Fifth IEEE Int'l Symp. on Intelligent Control*, Philadelphia, Sep. 1990.

Advances in Fast Power Flow Algorithms

Daniel J. Tylavsky, Peter E. Crouch
Leslie F. Jarriel and Hua Chen

Center for Systems Science and Engineering
and Department of Electrical Engineering
Arizona State University
Tempe, Arizona 85287

Abstract

The object of this paper is to survey the recent literature on fast power flow algorithms in order to determine the direction of research which is most likely to lead to new, possibly faster and/or more robust power flow algorithms. The analysis here identifies the central idea running through these algorithms. The orientation provided by this review is used to propose some new algorithms. One of the many candidate algorithms proposed is tested on high R/X ratio systems and shown to be robust and to have low execution times when compared with the BX or XB algorithms. It is expected that there exist more algorithms or variations of current algorithms which have similar properties.

CONTROL AND DYNAMIC SYSTEMS, VOL. 44

295

1   Toward a Central Theory Governing Fast Power Flow Algorithms.

1.0   Motivation.

In years that followed the almost universal acceptance of the fast decoupled power flow (FDPF) algorithm by Stott and Alsaç [3], there was a feeling in many quarters that improvements to the FDPF were unlikely. Research into such algorithms was often discouraged. This attitude began to change somewhat with the paper published by van Amerongen [2] in which the distinction between, what is now known as, the BX and XB algorithms was first drawn. (An indicator of the climate of the time, this important paper was rejected by reviewers but approved for publication because of the recognition of its importance by elements in the review heirarchy.) The performance of the BX algorithm in [2] showed broad based superiority over the XB algorithm. This new BX algorithm received wide scale testing and was becoming accepted widely as "the" fast power flow algorithm. The nascent view of this algorithm as a panacea ceased when a particularly nasty system was shown to converge well for the XB and not the BX algorithm. (This nasty system is the 1655 bus system used in Section 3.) What to do? The solution in the industry was to begin putting both the XB and BX algorithms in power flow packages. The most effective algorithm could then be picked to suit the application. This was a prophesy come true; for years it had been predicted that power flow packages would evolve into having multiple algorithms per package. (Further, using the team schemes proposed by S. Talukdar at Carnegie Mellon, such packages could "easily" be modified for fast and robust performance on parallel processors.) Of course, there were still problems; how do you know which algorithm to use when? If the BX did perform better on most system, why did it and were there perhaps other variants which might perform equally well if not better? This type of questioning became broad based throughout the industry. This lead EPRI to fund a project to examine the current state of art in power flow technology and to chart a course for the directions most fertile for new algorithms. The work reported hear is a synopsis of some of the results obtained in this project.

1.1   Introduction.

One objective of the portion of the work presented here is to bring together, in a unified frame work, the recent work that has been conducted in the area of Fast Decoupled (FD) Power Flow (PF) algorithms. It is hoped that this frame work can then be used to shed light on new directions in which to turn in the design of faster and more robust power flow algorithms. While it has not been possible to include all previous work in this research, it is hoped that the works which are seminal in some (related) sense have been included. The results of this effort are described in sections 1 and 2 in this paper. A summary of our results is included in Section 3.

The approach taken in Section 1 is to quickly establish a base notation, then, one by one, review some of the major works in the area of fast power flow algorithms. Throughout this work, new ideas will be flagged, and where appropriate, new algorithms will be proposed. Also, what appear to be, the most fertile grounds for research will be indicated.

1.2   Application of Newton and Quasi Newton Methods to the Power Flow Equations.

The Newton algorithm solves nonlinear equations of the form

$$f(x) = 0, \text{ or } f_k(x_1, \dots, x_n) = 0, \ 1 \leq k \leq n.$$

This is achieved by the iteration scheme:

$$x^{k+1} = x^k - Df(x^k)^{-1} f(x^k)$$

where D is the derivative operator. Quasi - Newton algorithms, on the other hand, deal with modifications to this iteration schemes in the following general way:

298   DANIEL J. TYLAVSKY *ET AL.*

$$x^{k+1} = x^k - (A^k)^{-1} f(x^k)$$

where A^k are approximations to the Jacobian matrices $Df(x^k)$. These algorithms are usually implemented via the update rules,

$$A^k \Delta x^k = -f(x^k), \quad x^{k+1} = x^k + \Delta x^k.$$

Using standard notation we write the power flow equations for the power injections $S_k = P_k + jQ_k$, into a network with n buses in the form

$$P_k(\theta, V) = \sum_m V_k V_m (g_{km} \cos \theta_{km} + b_{km} \sin \theta_{km}) = P_k^d$$

$$Q_k(\theta, V) = \sum_m V_k V_m (g_{km} \sin \theta_{km} - b_{km} \cos \theta_{km}) = Q_k^d$$

where $Y = G + jB$ is the network admittance matrix, with entries $G_{km} = g_{km}$ and $B_{km} = b_{km}$, the bus Voltages have magnitude V_k and phase θ_k ($\theta_{km} = \theta_k - \theta_k$) and $P_k^d + jQ_k^d = S_k^d$ are the desired values of the power flows. The real and reactive power mismatches are given by,

$$\Delta P_k(\theta, V) = P_k^d - P_k(\theta, V)$$

$$\Delta Q_k(\theta, V) = Q_k^d - Q_k(\theta, V)$$

Newton and Quasi Newton methods are used to solve for zero real and reactive power mismatches. The Newton update rule for the power flow equations can be expressed in the form,

$$\begin{bmatrix} \Delta P(\theta, V) \\ \Delta Q(\theta, V) \end{bmatrix} = \begin{bmatrix} J_{P\theta} & J_{PV} \\ J_{Q\theta} & J_{QV} \end{bmatrix} \begin{bmatrix} \Delta \theta \\ \Delta V/V \end{bmatrix}$$

where the sub matrices J_{xx} are obtained from the Jacobian matrix of the mismatch vector. This matrix can be decomposed into three distinct components,

$$J(\theta, V) = J_0(S, \theta, V) + J_s(B, \theta, V) + J_a(G, \theta, V)$$

where

$$J_0(S,\theta, V) = \begin{bmatrix} [-Q] & [P] \\ [P] & [Q] \end{bmatrix}$$

and, for example, [P] denotes the diagonal matrix with entries on the diagonal equal to the entries of the vector P. The remaining components are given by

$$J_a(G, \theta, V) = \begin{bmatrix} [V] & 0 \\ 0 & [V] \end{bmatrix} \begin{bmatrix} [g_{km}s_{km}] & [g_{km}c_{km}] \\ [-g_{km}c_{km}] & [g_{km}s_{km}] \end{bmatrix} \begin{bmatrix} [V] & 0 \\ 0 & [V] \end{bmatrix}$$

$$J_s(B, \theta, V) = \begin{bmatrix} [V] & 0 \\ 0 & [V] \end{bmatrix} \begin{bmatrix} [-b_{km}c_{km}] & [b_{km}s_{km}] \\ [-b_{km}s_{km}] & [-b_{km}c_{km}] \end{bmatrix} \begin{bmatrix} [V] & 0 \\ 0 & [V] \end{bmatrix}$$

where

$$c_{km} = \cos\theta_{km}, \quad s_{km} = \sin\theta_{km}.$$

We have now established an important fact that seems to have been overlooked in the literature. The Newton update matrix $J(\theta,V)$ for the power flow equations defined by the mismatch vector $[\Delta P(\theta,V), \Delta Q(\theta, V)]$, can be decomposed into three components; $J_0(S, \theta, V)$, a symmetric matrix depending on the power injections S ; $J_s(B, \theta\ V)$, a symmetric matrix depending on the imaginary part B, of the admittance matrix ; and $J_a(G, \theta, V)$, a skew symmetric matrix depending on the real part G, of the admittance matrix.

We are also able to establish another result, which does not seem to appear in the literature either and depends upon applying the Newton algorithm to the following modified mismatch vector:

$$\Delta P'_k(\theta, V) = P^d_k/V_k - P_k(\theta, V)/V_k$$

$$\Delta Q'_k(\theta, V) = Q^d_k/V_k - Q_k(\theta, V)/V_k$$

The update rule for the Newton algorithm using the modified mismatch vector $[\Delta P'(\theta,V), \Delta Q'(\theta, V)]$, may be written as:

$$\begin{bmatrix} \Delta P(\theta, V) \\ \Delta Q(\theta,V) \end{bmatrix} = \begin{bmatrix} J'_{P\theta} & J'_{PV} \\ J'_{Q\theta} & J'_{QV} \end{bmatrix} \begin{bmatrix} \Delta\theta \\ \Delta V/V \end{bmatrix}$$

where the matrix $J'(\theta, V)$ decomposes as follows:

$$J'(\theta, V) = J_0(S^d, \theta, V) + J_s(B, \theta, V) + J_a(G, \theta, V)$$

The only difference between this expression and the expression for $J(\theta, V)$, is the replacement of the injected power vector S by the desired injected power vector S^d. That is, in the Jacobian, terms such as the calculated real and reactive power at a bus, P^{calc} and Q^{calc} are replaced with P^d and Q^d respectively.

This result supports the treatment of the component $J_0(S, \theta, V)$ of the Jacobian matrix as a constant, modulo the scaling by the bus voltage magnitudes, and demonstrates that adding the term $J_0(S^d, \theta, V)$ into algorithms may be beneficial. It might be expected that including S^d vis-a-vis S terms provides some stability in the J_0 matrix when initial bus voltage profile estimates may yield values of S which are far from the solution value. This modification was tested and the results are presented in Table 1.2.1.

Table 1.2.1 Number of Iterations Versus X Factor for a Newton Algorithm using Fixed or Variable $J_0(S)$

	$J_0(S^d)$			$J_0(S)$		
X FACTOR	14	30	57	14	30	57
1.0	5	5	6	4	4	4
0.5	5	5	5	4	4	4
0.25	5	5	6	4	4	4
0.2	5	5	6	4	4	4
0.166	5	5	7	4	4	4
0.147	5	6	7	4	4	4
0.125	5	6	8	4	4	4
0.111	6	7	9	4	4	5
0.1	6	7	10	4	4	5
0.083	7	9	15	4	4	5
0.071	8	12	28	4	5	6
0.063	10	17	NC(>30)	5	5	NC(>30)
0.056	14	NC(>30)		5	NC(>30)	
0.05	NC(>30)			6		
X FACTOR	14	30	57	14	30	57
0.5	4	5	5	4	4	4
1.0	5	5	6	4	4	4
1.5	5	6	6	4	4	4
2.0	6	7	7	4	4	4
2.5	7	8	8	4	4	4
3.0	9	11	9	4	4	4
4.0	18	NC(>30)	18	5	7	5
5.0	NC(>30)		NC(>30)	NC(>30)	NC(>30)	NC(>30)

These tests involved multiplying each line reactance by the X FACTOR shown in the first column of Table 1.2.1 and observing the number of iterations to convergence for the IEEE 14, 30, and 57 bus systems. These results show that the traditional method of using S vis-a-vis S^d is preferred. (Although, some success using S^d was obtained in marginally convergent cases encountered with low impedance line models. However, better methods are available for such problems.)

1.3   Critical Review of Monticelli et al.[1], Amerongen [2], and Stott and Alsaç [3] on the BX and XB Algorithms.

In the work by Monticelli et al. [1], a justification of the BX and XB algorithms was given, which is fundamental to the understanding of the two algorithms. We briefly give the main points.

The starting point for the discussion is the Newton update rule expressed in polar coordinates as,

$$\begin{bmatrix} \Delta P(\theta, V) \\ \Delta Q(\theta, V) \end{bmatrix} = \begin{bmatrix} D_\theta\, P(\theta,V) & D_V\, P(\theta,V) \\ D_\theta\, Q(\theta, V) & D_V\, Q(\theta, V) \end{bmatrix} \begin{bmatrix} \Delta\theta \\ \Delta V \end{bmatrix}$$

1.3.1

Up to second order terms in $\Delta\theta$ and ΔV this set of equations may be re-expressed in two distinct ways, as,

$$\begin{bmatrix} \Delta Q(\theta,V) \\ \Delta P(\theta,V + (D_V Q)^{-1}\Delta Q(\theta,V)) \end{bmatrix}$$

$$= \begin{bmatrix} D_V Q & D_\theta Q \\ 0 & D_\theta P - D_V P\,(D_V Q)^{-1} D_\theta Q \end{bmatrix} \begin{bmatrix} \Delta V \\ \Delta\theta \end{bmatrix}$$

1.3.2(N1)

and

$$
\begin{bmatrix} \Delta P(\theta, V) \\ \Delta Q(\theta + (D_\theta P)^{-1} \Delta P(\theta, V), V) \end{bmatrix}
$$

$$
= \begin{bmatrix} D_\theta P & D_V P \\ 0 & D_V Q - D_\theta Q \, (D_\theta P)^{-1} D_V P \end{bmatrix} \begin{bmatrix} \Delta \theta \\ \Delta V \end{bmatrix}
$$

1.3.3(N2)

In these expressions all Jacobian matrices, $D_V Q$ for example, are evaluated at (θ, V). These fundamental observations yield the required decoupling from which the BX and XB algorithm are derived. From experience one constructs Quasi-Newton algorithms from these exact Newton type algorithms by making the following approximations in the above update rules

$$
D_V Q \cong R_1(V), \; D_\theta Q \cong T_1(V), \; D_\theta P - D_V P \, (D_V Q)^{-1} D_\theta Q \cong S_1(V)
$$

1.3.4

and

$$
D_\theta P \cong R_2(V), \; D_V P \cong T_2(V), \; D_V Q - D_\theta Q \, (D_\theta P)^{-1} D_V P \cong S_2(V)
$$

1.3.5

Combining these rules and approximations we obtain two, three stage algorithms, with update rules given by,

$$
\Delta V' = R_1(V)^{-1} \Delta Q(\theta, V), \; \Delta \theta = S_1(V)^{-1} \Delta P(\theta, V + \Delta V')
$$

$$
\Delta V = \Delta V' - R_1(V)^{-1} T_1(V) \Delta \theta.
$$

1.3.6(QN1)

and

$$
\Delta \theta' = R_2(V)^{-1} \Delta P(\theta, V), \; \Delta V = S_2(V)^{-1} \Delta Q(\theta + \Delta \theta', V)
$$

$$
\Delta \theta = \Delta \theta' - R_2(V)^{-1} T_2(V) \Delta V.
$$

1.3.7(QN2)

The **XB algorithm** may be expressed in terms of the update rules,

$$
[1/V_k] \Delta Q(\theta_k, V_k) = B'' \Delta V_k
$$

$$
[1/V_k] \Delta P(\theta_k, V_k + \Delta V_k) = [1/x]' \Delta \theta_k
$$

1.3.8

and the **BX algorithm** expressed in terms of the update rules

$$[1/V_k] \Delta P(\theta_k, V_k) = B' \Delta\theta_k$$

$$[1/V_k] \Delta Q(\theta_k + \Delta\theta_k, V_k) = [1/x]'' \Delta V_k \qquad 1.3.9$$

where the B' and B" matrices are defined as follows:

XB Method

$$[1/x]'_{i\,j} = B'_{ij} = \frac{-1}{X_{ij}} \qquad \forall\ i \neq j$$

$$[1/x]'_{i\,i} = B'_{ii} = \sum_{j \neq i} \frac{-1}{X_{ij}} \qquad 1.3.10$$

$$B'' = [B] \qquad 1.3.11$$

Note that the B' diagonal elements include no effect from any shunt elements which includes lines charging, capacitive and reactive shunts, and transformer losses due to off-nominal tap ratio's. The B" matrix is exactly the B portion of the Y matrix. (Note that B" does not include doubled shunts as suggested in Appendix 4 of [9]. The use of doubled shunts makes very little difference in the IEEE systems, however, a 1655 bus system used in the testing would not converge with the doubled shunts.)

BX Method

$$B'_{ij} = B_{ij} \quad \forall\ i \neq j \ \text{(excluding effect o fo ff-nominal taps)}$$

$$B'_{ii} = \sum_{j \neq i} B_{ij} \quad \text{(excludes all shunt effects)} \qquad 1.3.12$$

$$[1/x]''_{i\,j} = B''_{ij} = \frac{1}{X_{ij}} a_{ij}$$

$$[1/x]''_{ii} = B''_{ii} = B_{shunt} + \sum_{j \neq i} \frac{1}{X_{ij}} a'_{ij} + \frac{S_{ij}}{2}$$

1.3.13

where:

a_{ij} = the inverse of the tap ratio (1.0 for transmission lines).

a' = a^2 if i is the tap side.

= 1.0 if i is the impedance side.

S_{ij} = the charging susceptance of the branch.

Monticelli et al. [1] give two arguments supporting the transition from the algorithms QN1 (Quasi-Newton 1) and QN2 to the XB and BX algorithms respectively. The most important observation is the fact that for radial networks (including the effect of shunts) or networks with branches having the same r/x ratios, we have for algorithm QN1

$$\left[D_\theta P - D_V P (D_V Q)^{-1} D_\theta Q \right]^{V=1}_{\theta=0} \cong [1/x] \cong \left[S_1(V) \right]^{V=1}_{\theta=0}$$

$$\left[D_V Q \right]^{V=1}_{\theta=0} \cong B'' \cong \left[R_1(V) \right]^{V=1}_{\theta=0}$$

1.3.14

and for algorithm QN2

$$\left[D_V Q - D_\theta Q (D_\theta P)^{-1} D_V P \right]^{V=1}_{\theta=0} \cong [1/x] \cong \left[S_2(V) \right]^{V=1}_{\theta=0}$$

$$\left[D_\theta P \right]^{V=1}_{\theta=0} \cong B' \cong \left[R_2(V) \right]^{V=1}_{\theta=0}$$

1.3.15

These facts show that under the stated circumstances the XB and BX algorithms are closely approximated by the first two stages of the Quasi - Newton algorithms QN1 and QN2 respectively. Monticelli et al.[1] then justifies the omission of the third stage in each of the algorithms QN1 and QN2 by combining two consecutive updates of the respective Quasi-Newton Algorithms. This can be summarized in the case of the algorithm QN1, by setting $U = V + R_1(V)^{-1} DQ(\theta, V)$, in which case we obtain the modified update rule,

$$\Delta\theta = S_1 (V)^{-1} \Delta P(\theta, U) , \ \Delta U = R_1 (V)^{-1} \Delta Q(\theta + \Delta\theta, U) \qquad 1.3.16(MQN1)$$

In the case of the algorithm QN2, by setting $\eta = \theta + R_2(V)^{-1}DP(\theta,V)$, we obtain the modified update rule:

$$\Delta V = S_2 (V)^{-1} \Delta Q(\eta, V) , \ \Delta\eta = R_2 (V)^{-1} \Delta P (\eta, V + \Delta V). \qquad 1.3.17(MQN2)$$

The update rules MQN1 and MQN2 do not coincide with the XB and BX algorithms respectively, as claimed in Monticelli et al. [1]. In fact it seems that the update rule MQN1 (respectively MQN2) is as closely related to the BX (respectively XB) algorithm as it is to the XB (respectively BX). The approximations made by ignoring the third stages of the algorithms QN1 and QN2 still seem to be real approximations.

It is also true that Monticelli et al. [1] provide no analytical argument why the approximations 1.3.14 and 1.3.15 should be made in the case where the network is not radial or the branches do not have constant r/x ratios.

1.4 Results of Theoretical Analysis of BX and XB Algorithms as per Wu [4].

In this section we use two elementary convergence theorems to obtain qualitative information about the convergence.of the generalized BX and XB algorithms. The work by Wu [4], is one of the few papers to consider the convergence question in this manner, but in that work the added insight obtained from the observations by Monticelli et al. [1], was of course absent. In particular the important restatements of the Newton algorithms in N1(1.3.2) and N2(1.3.3) were not employed. So a re-examination of that work is now justified. The work of Wu[4] also paid much attention to detailed estimates, which were subsequently shown to be less effective than anticipated in practice by Rao et al. [5]. It is well known that the estimates obtained in such analyses are often overly conservative. The main point of our work is to point to the main issues that govern convergence.

We start by considering two generalized XB and BX algorithms, which we formulate in the following way; The generalized XB algorithm is written in the form,

$$(\theta^{k+1}, V^{k+1}) = \phi^1 (\theta^k, V^k)$$

where,

$$\phi^1 (\theta, V) = \begin{bmatrix} \phi_V^1 (\theta, V) \\ \phi_\theta^1 (\theta, V) \end{bmatrix} = \begin{bmatrix} V + R_1 (V)^{-1} \Delta Q(\theta, V) \\ \theta + S_1 (V)^{-1} \Delta P(\theta, \phi_V^1 (\theta, V)) \end{bmatrix}$$

and the generalized BX algorithm is written in the form,

$$(\theta^{k+1}, V^{k+1}) = \phi^2 (\theta^k, V^k)$$

where,

$$\phi^2 (\theta, V) = \begin{bmatrix} \phi_\theta^2 (\theta, V) \\ \phi_V^2 (\theta, V) \end{bmatrix} = \begin{bmatrix} \theta + R_2 (V)^{-1} \Delta P(\theta, V) \\ V + S_2 (V)^{-1} \Delta Q(\phi_\theta^2 (\theta, V), V) \end{bmatrix}$$

The values of the matrices $R_k(V)$ and $S_k(V)$, k = 1,2, at V= 1 will be actually determined as in the assumptions 1.3.14 and 1.3.15. The traditional dependence on V is pre multiplication by the matrix [V], but for this analysis it is not necessary to be specific.

Two results are important to understand the convergence of an algorithm $x^{k+1} = \phi(x^k)$ for x in an n dimensional vector space.

In the first case we assume that f has a fixed point α, $\phi(\alpha) = \alpha$. Let $B_\rho(a) = \{x; \|x -a\| < \rho \}$ where $\|x\|$ is the infinite norm of $x = (x_1, ..., x_n)$, $\|x\| = \max.\{ x_k; 1 \le k \le n \}$. If $\|D\phi(x)\| \le \lambda/n$, where $\lambda < 1$, for all x belonging to $B_\rho(\alpha)$, then for any initial state x_0 in $B_\rho(\alpha)$ the algorithm $x^{k+1} = \phi(x^k)$ converges to a.

In the second case we assume that $\|D\phi(x)\| < c < 1$ for x belonging to $B_\rho(x_0)$ and that $\|\phi(x_0) - x_0 \| < (1 - c) \rho$. The algorithm $x^{k+1} = \phi(x^k)$ then converges to a unique fixed point α, from the initial point x_0. This is the result that is used in Wu[4].

Note that the first result guarantees convergence of the algorithm in a region surrounding the fixed point a, while the second result guarantees convergence in a region around the initial state x_0. For quantitative results the second result is more useful because one does not have to know the fixed point before hand. However the first result is interesting for qualitative results because it shows that the algorithm will converge for some initial state if $\|D\phi(\alpha)\| < \lambda/n$, where $\lambda \leq 1$.

Clearly to apply these results one has to calculate the Jacobian matrix $D\phi$. We claim that we may decompose these Jacobian matrices as follows,

$$D\phi^k(q,V) = A^k(\theta,V) + B^k(\theta,V), \quad k=1,2.$$

where $B^k(\theta^\#,V^\#) = 0$ and $\alpha = (\theta^\#,V^\#)$ is the fixed point of ϕ. Thus convergence of the algorithm at all, is dependent on $\|A^k(\theta^\#,V^\#)\|$, while the convergence of the algorithm from a specific initial state $x_0 = (\theta_0,V_0)$ is determined by both $\|A^k(\theta_0,V_0)\|$ and $\|B^k(\theta_0,V_0)\|$. The results below show that the generalized XB and BX algorithms described above, make $\|A^k(\theta_0,V_0)\|$ very small, so that convergence of the algorithm is determined almost totally by the size of $\|B^k(\theta_0,V_0)\|$ and $\|\phi(x_0) - x_0\|$.

We may write

$$D\phi^1 = \begin{bmatrix} D\phi^1_V \\ D\phi^1_\theta \end{bmatrix} = \begin{bmatrix} D_V\,\phi^1_V & D_\theta\,\phi^1_V \\ D_V\,\phi^1_\theta & D_\theta\,\phi^1_\theta \end{bmatrix} = \begin{bmatrix} A^1_{11} & A^1_{12} \\ A^1_{21} & A^2_{22} \end{bmatrix} + \begin{bmatrix} B^1_{11} & B^1_{12} \\ B^1_{21} & B^1_{22} \end{bmatrix}$$

and,

$$D\phi^2 = \begin{bmatrix} D\phi^2_\theta \\ D\phi^2_V \end{bmatrix} = \begin{bmatrix} D_\theta\,\phi^2_\theta & D_V\,\phi^2_\theta \\ D_\theta\,\phi^2_V & D_V\,\phi^2_V \end{bmatrix} = \begin{bmatrix} A^2_{11} & A^2_{12} \\ A^2_{21} & A^2_{22} \end{bmatrix} + \begin{bmatrix} B^2_{11} & B^2_{12} \\ B^2_{21} & B^2_{22} \end{bmatrix}$$

We now tabulate the coefficient matrices of the matrix A for both algorithms.

$$A^1_{11} = I - R_1 (V)^{-1} D_V Q(\theta, V)$$

$$A^1_{12} = - R_1 (V)^{-1} D_\theta Q(\theta, V)$$

$$A^1_{21} = - S_1 (V)^{-1} D_V P(\theta, V) (I - R_1 (V)^{-1} D_V (\theta, V))$$

$$A^1_{22} = I - S_1(V)^{-1} (D_\theta P(\theta, V) - D_V P(\theta, V) R_1 (V)^{-1} D_\theta Q(\theta, V))$$

and

$$A^2_{11} = I - R_2 (V)^{-1} D_\theta P(\theta, V)$$

$$A^2_{12} = - R_2 (V)^{-1} D_V P(\theta, V)$$

$$A^2_{21} = - S_2 (V)^{-1} D_\theta Q(\theta, V) (I - R_2 (V)^{-1} D_\theta P(\theta, V))$$

$$A^2_{22} = I - S_2 (V)^{-1} (D_V Q(\theta, V) - D_\theta Q(\theta, V) R_2 (V)^{-1} D_V P(\theta, V))$$

It is clear from the assumptions 1.3.4 and 1.3.5 that all but the matrices A^k_{12} are small in norm, and can be made to have zero norm at the flat start, $V = 1$ and $\theta = 0$, as long as the we have equality in the expression 1.3.4 and 1.3.5. The effect of the terms A^k_{12} may be made equally as small if the third stage of the Quasi- Newton algorithms QN1 and QN2 is included. The necessary modification to the mapping ϕ^1, describing the generalized XB algorithm is given by,

$$\phi^1_V = V + R_1 (V)^{-1} \Delta Q - R_1 (V)^{-1} T_1(V) S_1 (V)^{-1} \Delta P(\theta, V + R_1 (V)^{-1} \Delta Q)$$

$$\phi^1_\theta = \theta + S_1 (V)^{-1} \Delta P(\theta, V + R_1 (V)^{-1} \Delta Q)$$

Similarly the necessary modification to the mapping ϕ^2, describing the generalized BX algorithm is given by,

$$\phi^2_\theta = \theta + R_2 (V)^{-1} \Delta P - R_2 (V)^{-1} T_2 (V) S_2 (V)^{-1} \Delta Q(\theta + R_2 (V)^{-1} \Delta P, V)$$

$$\phi^2_V = V + S_2 (V)^{-1} \Delta Q(\theta + R_2 (V)^{-1} \Delta P, V)$$

This change effects the submatrices of A^1 and A^2 as follows:

$$A^1_{11} = I - R_1 (V)^{-1} D_V Q + R_1 (V)^{-1} T_1 (V) S_1 (V)^{-1}(D_V P(I - R_1 (V)^{-1} D_V Q))$$

$$A^1_{12} = - R_1 (V)^{-1} D_\theta Q + R_1 (V)^{-1} T_1 (V) S_1 (V)^{-1} (D_\theta P - D_V P R_1 (V)^{-1} D_\theta Q)$$

$$A^2_{11} = I - R_2 (V)^{-1} D_\theta P + R_2 (V)^{-1} T_2 (V) S_2 (V)^{-1} (D_\theta Q(I - R_2 (V)^{-1} D_\theta P))$$

$$A^2_{12} = -R_2 (V)^{-1} D_V P + R_2 (V)^{-1} T_2 (V) S_2 (V)^{-1} (D_V Q - D_\theta Q R_2 (V)^{-1} D_V P)$$

From these expressions and the assumptions 1.3.4 and 1.3.5 it is clear that these matrices have very small norms (if not zero), at the flat start. It follows that the inclusion of the third stage of the Quasi-Newton algorithms QN1(1.3.6) and QN2(1.3.7) diminishes the effect of the matrices A^k at the flat start. However since they also contribute many other terms to the expressions for the Jacobian of ϕ, it is unlikely to be beneficial , as is found in practice.

1.5 Further justification for BX and XB algorithms as in Carpentier [6] and Rajicic and Bose [7].

In this section we wish to give some motivation for the assumptions 1.3.14 and 1.3.15, in the case of non radial networks. The work is motivated by that of Carpentier [6], who already arrived at an approximation to the Newton algorithm N2(1.3.3), not through deductions but through assumptions. He did this by arguing as follows; the Newton update rule is obtained from the following assumptions,

$$P^d - P(\theta + \Delta\theta, V + \Delta V) \cong 0 \qquad\qquad 1.5.1$$

$$Q^d - Q(\theta + \Delta\theta, V + \Delta V) \cong 0 \qquad\qquad 1.5.2$$

Carpentier suggested that these assumptions should be replaced by the following,

$$P^d - P(\theta + \Delta\theta, V) \cong 0 \qquad\qquad 1.5.3$$

and

$$P(\theta + \Delta\theta, V + \Delta V) - P(\theta , V) \cong 0 \qquad\qquad 1.5.4$$

$$Q^d - Q(\theta + \Delta\theta, V + \Delta V) \cong 0.$$

This gives the update rules,

$$\Delta P(\theta, V) = D_\theta\, P(\theta, V)\Delta\theta \qquad\qquad\qquad\qquad\qquad \text{C1}$$

$$D_\theta\, P(\theta, V)\Delta\theta + D_V\, P(\theta, V)\Delta V = 0$$

$$\Delta Q(\theta, V) = D_\theta\, Q\,(\theta,\ V)\Delta\theta + D_V\, Q(\theta, V)\Delta V$$

$$\Delta Q(\theta, V) = (D_V\, Q - D_\theta\, Q\, (D_\theta\, P)^{-1}\, D_V\, P)\Delta V \qquad\qquad \text{C2}$$

$$\text{1.5.5}$$

The update rules C1 and C2 yield an algorithm similar to the Newton algorithm N2 and the BX algorithm. However Carpentier [6] did not notice that the assumption 1.3.15 could be made, in some instances without significant error, and so considered further assumptions which enable the equations C2(1.5.5) to be replaced by equations that retain the sparseness of the original network equations. In fact he assumed that not only is P held constant during the ΔV update in C2, but that for each bus the real power flows into each branch are also held constant.

To examine this condition and its consequences, we set,

$$P_{k\,m}(\theta, V) = V_k\, V_m\, (g_{k\,m}\, c_{k\,m} + b_{k\,m}\, s_{k\,m}) - V_k^2\, g_{k\,m}\,,\ k \ne m$$

$$Q_{k\,m}(\theta, V) = V_k\, V_m\, (g_{k\,m} s_{k\,m} - b_{k\,m}\, c_{k\,m}) + V_k^2\, b_{k\,m}\,,\ k \ne m.$$

Clearly we have,

$$P_k\,(\theta, V) = \sum_{m \ne k} P_{k\,m}\,(\theta, V)\,, \quad Q_k\,(\theta, V) = \sum_{k \ne m} Q_{k\,m}\,(\theta, V).$$

Assumptions 1.5.4 are now replaced by the assumptions,

$$P_{k\,m}\,(\theta + \Delta\theta, V + \Delta V) - P_{k\,m}\,(\theta,\ V) \cong 0$$

$$Q_k^d - Q_k\,(\theta + \Delta\theta,\ V + \Delta\theta) \cong 0$$

$$\text{1.5.6}$$

Thus during the ΔV update we have,

$$P_{k\,m}\,(\theta, V) = P'_{k\,m} = \text{const.}$$

These constraints allow us to view V_k as independent variables and θ_{km} as dependent variables with $\theta_{k\,m} = \theta_{k\,m}\,(P'_{k\,m},\ V_k, V_m)$. For the current purposes, for each fixed k, we ignore the other representation of

$$\theta_{km} = -\theta_{mk} = -\theta_{m\,k}\,(P'_{m\,k},\ V_k, V_m).$$

Now $Q_{k\,m}(V_k, V_m) = Q_{k\,m}(\theta_m(P'_{k\,m}, V_k, V_m), V_k, V_m)$ is a function of V_k and V_m only. To obtain the modified update rule for ΔV we must compute the following derivatives,

$$\frac{\partial Q_k}{\partial V_m} = \sum_{i \neq k} \frac{\partial Q_{k\,i}}{\partial V_m} = \frac{\partial Q_{k\,m}}{\partial V_m}, \quad \frac{\partial Q_k}{\partial V_k} = \sum_{m \neq k} \frac{\partial Q_{k\,m}}{\partial V_k}$$

We can in fact show that,

$$\frac{\partial Q_{k\,m}}{\partial V_m} = \frac{V_k}{x_{k\,m}} \frac{1}{c_{k\,m} + \dfrac{r_{k\,m}}{x_{k\,m}} s_{k\,m}}, \quad \frac{\partial Q_{k\,m}}{\partial V_k} = \frac{-1}{x_{k\,m}} \frac{2 V_k c_{k\,m} - V_m}{c_{k\,m} - \dfrac{r_{k\,m}}{x_{k\,m}} s_{k\,m}}$$

1.5.7

These formulas are a re-formulation of formulas found in the paper by Carpentier [6], but the notation used in that work makes it difficult to understand their significance. In particular it is clear from these expressions that the entries of the Jacobian matrix $D_V Q(\theta,V)$ computed in this way are equal to multiples of $1/x_{k\,m}$, and that the matrix retains the sparseness of the original admittance matrix. Indeed computed at the flat start the Jacobian matrix $D_V Q(\theta,V)$, calculated as above, coincides exactly with the matrix $[1/x]$. Thus we have analytically justified the use of the matrix $[1/x]$ in the BX algorithm under the assumptions 1.5.6.

We now make a similar analysis to the one above, but for the case of the XB algorithm. The natural assumption to make, which mirrors the assumption (1.5.6), is that the $\Delta\theta$ update is performed under the following assumptions,

$$Q_{k\,m}(\theta + \Delta\theta, V + \Delta V) - Q_{k\,m}(\theta,V) \cong 0$$

$$P_k^d - P_k(\theta + \Delta\theta, V + \Delta V) \cong 0$$

Thus during the $\Delta\theta$ update we have,

$$Q_{k\,m}(\theta, V) = Q'_{k\,m} = \text{const.}$$

These constraints allow us to view $\theta_{k\,m}$ as independent variables and V_k and V_m as dependent variables with $V_m = V_m(Q'_{k\,m}, Q'_{m\,k}, \theta_m)$ and $V_k = V_k(Q'_{k\,m}, Q'_{m\,k}, \theta_{k\,m})$. For the current purposes, for each fixed k, we ignore the other representations of $V_m = V_m(Q'_{im}, Q'_{mi}, \theta_{im})$ for all $i \neq k$. Now

$$P_{km}(\theta_{km}) = P_{km}(\theta_{km}, V_k(Q_{km}, Q_{mk}, \theta_{km}), V_m(Q_{km}, Q_{mk}, \theta_{km}))$$

is a function of θ_{km} only. To obtain the update rule for $\Delta\theta$ we must compute the following derivatives,

$$\frac{\partial P_k}{\partial \theta_m} = \sum_{i \neq k} \frac{\partial P_{ki}}{\partial \theta_m} = \frac{\partial P_{km}}{\partial \theta_m}, \quad \frac{\partial P_k}{\partial \theta_k} = \sum_{m \neq k} \frac{\partial P_{km}}{\partial \theta_k}$$

Note that in this case we also have,

$$\frac{\partial P_{km}}{\partial \theta_k} = - \frac{\partial P_{km}}{\partial \theta_m},$$

since $\theta_{km} = \theta_k - \theta_m$. We can in fact show that:

$$\frac{\partial P_{km}}{\partial \theta_m} = \frac{V_k V_m}{x_{km}} \frac{2 V_k V_m c_{km} - (V_k^2 + V_m^2)}{2 V_k V_m - c_{km}(V_k^2 + V_m^2) \frac{r_{km}}{x_{km}} s_{km}(V_m^2 - V_k^2)}$$

$$1.5.8$$

This formula again shows that the entries in the Jacobian matrix $D_\theta P(\theta, V)$ computed in this way are equal to multiples of $1/x_{km}$ and that the matrix retains the sparsity of the admittance matrix. Moreover it is again apparent that under the flat start conditions, the matrix coincides with the matrix $[1/x]$. This again provides a partial analytical justification for the use of the matrix $[1/x]$ in the XB algorithm. Once again it would be interesting to examine approximations to the Jacobian matrix as computed above when the flat start condition is not assumed.

It is also interesting to tie in this result with that of the paper by Rajicic and Bose [7], where a modification of the XB algorithm was considered. The entries described in [7] of the $S_1(V)$ matrix, (recall the generalized XB algorithm), were formed from a combination of the elements b_{km}, g_{km} and $1/x_{km}$ in the form

$$B'_{km} = -.3/x_{km} - .7b_{km} - .4b_{km}$$

$$B'_{kk} = - \sum_m B'_{km}$$

1.5.9

This motivates the need for further investigations to examine the possibility of making better approximations to the entries of the Jacobian matrix, as computed above, rather than just taking the flat start conditions.

1.6 Summary of Decoupled Formulations with Polar Variables.

It is important to note that in all of the works discussed thus far the polar form of the bus voltage variables and rectangular form of the line impedances have been used. The different results obtained in these works have been based on slight variations of the same formulation or on the identical variations which are made unrecognizable by use of a different derivation (based on different assumptions). It is clear in this analysis that there are many variations which have yet to be tested and further work is needed in this area. We will show (see Section 1.9) that one simple variation of the above work can lead to significantly enhanced performance. In the next section we shall look at decoupled methods based on rectangular form of the bus voltage variables and the polar form for line impedance representation.

1.7 The Super Decoupled Algorithm by Haley and Ayres [8].

In this section we briefly review the main points of the super decoupled algorithm described by Haley and Ayres [8], since it inspires a promising line of new algorithms. In their paper, the analysis incorporated the distributed slack bus. We shall leave out this added complexity to make the principle of the algorithm more clear. Their approach uses the Cartesian form of the power flow equations, rather than the polar form. We therefore write the complex power injection S_k into bus k in the form,

$$S_k^* = \sum_m v_k^* Y_{k\,m} v_m$$

where v_k is the complex voltage at bus k and * represents complex conjugate. The Newton update rule for the complex mismatches $\Delta S_k(v) = S_k^d - S_k(v)$ is now given by,

$$\Delta S_k^* = \Delta v_k^* / v_k^* \, S_k^* + \sum_m v_k^* Y_{k\,m} \Delta v_m$$

As in the polar representation we write,

$$v_k = V_k \, e^{j\theta_k}$$

and also set,

$$Y_{k\,k} = g_{k\,k} + j\,b_{k\,k} = \sqrt{g_{k\,k}^2 + b_{k\,k}^2}\; e^{j\,\eta_k}, \quad \varphi_k = \tfrac{1}{2}\tan^{-1}(g_{kk}/b_{k\,k}).$$

It follows that,

$$e^{j\,2\,\varphi_k}\, Y_{k\,k} = j\,\sqrt{g_{k\,k}^2 + b_{k\,k}^2}$$

Letting,

$$\Delta\hat{v}_k = e^{-j\,(\theta_k + \varphi_k)}\, \Delta v_k$$

we obtain the following modified update rule,

$$e^{j\,\varphi_k}\frac{\Delta S_k^*}{V_k} = \frac{S_k^*}{V_k^2}\,\Delta\hat{v}_k^* + \sum_m Y_{k\,m}\, e^{-j\,(\theta_k - \theta_m)}\, e^{-j\,(\varphi_k + \varphi_m)}\, \Delta\hat{v}_m$$

Define new admittance matrices by setting,

$$Y'_{k\,m} = Y_{k\,m}\, e^{j\,(\varphi_k + \varphi_m)} = g'_{k\,m} + j\,b'_{k\,m} = \frac{1}{r'_{k\,m} + j\,x'_{k\,m}}$$

and,

$$Y''_{k\,m} = \frac{1}{x'_{k\,m}}$$

Note that,

$$Y'_{k\,k} = Y_{k\,k}\, e^{j\,2\,\varphi_k}$$

is pure imaginary, so that in the process of obtaining $Y''_{k\,m}$ from $Y'_{k\,m}$ no approximation is made in obtaining the diagonal elements, by neglecting the resistive components $r'_{k\,m}$. The update rule is decoupled by first taking the real and imaginary parts of the equation and approximating the matrix with coefficients,

$$Y_{k\,m} e^{-j\,(\theta_k - \theta_m)}\, e^{-j\,(\varphi_k + \varphi_m)}$$

by the matrix with coefficients $-j\,Y''_{k\,m}$. The resulting update rule takes the form,

$$\text{Re.}\left(\frac{e^{j\,\varphi_k}}{V_k}\, \Delta S_k^* \right) \cong \text{Re.}\left(\frac{S_k^*\, \Delta \hat{v}_k^*}{V_k^2} \right) + \sum_m Y''_{k\,m}\, \text{Im.}(\Delta \hat{v}_m)$$

and

$$\text{Im.}\left(\frac{e^{j\,\varphi_k}}{V_k}\, \Delta S_k^* \right) \cong \text{Im.}\left(\frac{S_k^*\, \Delta \hat{v}_k^*}{V_k^2} \right) - \sum_m Y''_{k\,m}\, \text{Re.}(\Delta \hat{v}_m)$$

To finish the decoupling procedure, further approximations are made in the form

$$\text{Re.}\left(\frac{S_k^*}{V_k^2}\, \Delta \hat{v}_k^* \right) \cong [Y_\theta]_{k\,k}\, \text{Im.}(\Delta \hat{v}_k)$$

and

$$\text{Im.}\left(\frac{S_k^*\, \Delta \hat{v}_k^*}{V_k^2} \right) \cong [Y_V]_{kk}\, \text{Re.}(\Delta \hat{v}_k)$$

where Y_q and Y_V are real constant diagonal matrices. The final update equations can then be written in the following form

$$\text{Re.}\left(\frac{e^{j\,\varphi_k}\, \Delta S_k^*}{V_k} \right) = \sum_m [Y'' + Y_\theta]_{k\,m}\, \text{Im.}(\Delta \hat{v}_m)$$

$$\text{Im.}\left(\frac{e^{j\,\varphi_k}\Delta S_k^*}{V_k} \right) = \sum_m [-Y'' + Y_V]_{k\,m}\, \text{Re.}(\Delta \hat{v}_m)$$

The first set of approximations are justified by the fact that the contributions of the phase terms $\theta_{k\,m} = \theta_k - \theta_m$ is usually small and the fact that the contribution of the term

$$e^{j\,(\varphi_k + \varphi_m)}$$

was calculated to ensure that the diagonal terms in Y' are pure imaginary. Since the row sum of the off diagonal terms yields the diagonal terms, it is a good approximation to assume that they too are pure imaginary. The remaining decoupling assumptions were also justified, but in a more ad-hoc manner.

The work by Haley and Ayres [8] was accompanied by an interesting comments by W. Hubbi. This motivated the authors to reconsider their work in the more general setting where the update equations are subject to diagonal rotation matrices both on the left and the right. Their original work assumes that these rotation matrices, with diagonal entries given by

$$e^{j\,\varphi_k}$$

are the same. In particular, rotations which make the diagonal entries of the admittance matrix purely real is also given some consideration. These and related modifications to the super decoupled algorithm would seem to warrant further investigation.

1.8   Block Decoupling and Rotations.

In this section we consider the possibilities of combining both the observations of Monticelli et al. and the use of rotations in developing further algorithms which may prove helpful in the case of networks containing branches with high r/x ratios. We begin our discussion by considering the non-decoupled Quasi-Newton algorithm for solving the power flow equations in which the Jacobian matrix is approximated by its value at the flat start, and the mismatches are normalized by the magnitude the bus voltage. We write the corresponding update equations in the form,

$$\begin{bmatrix} [1/V]\Delta P(\theta,V) \\ [1/V]\Delta Q(\theta,V) \end{bmatrix} = \begin{bmatrix} -B & G \\ -G & -B \end{bmatrix} \begin{bmatrix} \Delta\theta \\ \Delta V \end{bmatrix}$$

The diagonal terms of the Jacobian involving $P^{calc.}$ and $Q^{calc.}$ are ignored. Testing of this method showed only marginally good results and usually took more iterations to converge than the decoupled method. However, the following variation of this method, as reported by Amerongen [1] and Nagendre Rao et al. [9], shows very robust qualities:

$$\begin{bmatrix} \Delta P/V \\ \Delta Q/V \end{bmatrix} = \begin{bmatrix} B' & G \\ -G^T & B'' \end{bmatrix} \begin{bmatrix} \Delta\theta \\ \Delta V \end{bmatrix}$$

In this formulation, the B' is that of the BX method and the B" is that of the XB method. *(In subsequent analysis the "prime" and "double prime" superscripts are not included; however it is to be understood that the submatrices used are those obtained from the B' and B" matrices.)* This algorithm, hereafter called the BGGB algorithm, provided excellent robustness for systems with one or more high r/x ratio branches as will be shown in the test results in section 2.3. This comparison follows the comparison with the full Newton, XB and BX algorithms in the paper by Amerongen [2], and it was found that its performance usually lay somewhere between that of the full Newton and the XB and BX algorithms. The paper by Rao et al. [5] argued that the reason for the improvement over the XB and BX algorithms lay in the fact that the algorithm took the non-decoupled nature of the equations into account. Clearly the analysis by Monticelli et al.[1] discounts this argument to some extent. However it is true that the XB and BX algorithms are still based on solving the linear update equations by "pivoting" on the -B blocks. For networks where there are branches with high r/x ratios it is still to be expected that this will not result in convergence performance which competes with the non-decoupled Quasi-Newton algorithm above. Making this fact more quantitative is one of the main contributions of the paper by Rao et al.[5].

However no solution to the problem of solving the power flow equations is given in the case where their criterion fails.

If we were given a network where the x/r ratios were <u>uniformly</u> low and less than one, then the update equations above would be re-written in the form,

$$\begin{bmatrix} [1/V]\Delta P(\theta,V) \\ [1/V]\Delta Q(\theta,V) \end{bmatrix} = \begin{bmatrix} G & -B \\ -B & -G \end{bmatrix} \begin{bmatrix} \Delta V \\ \Delta\theta \end{bmatrix}$$

Clearly now one could construct fast decoupled algorithms by "pivoting" on the G and -G blocks and repeating the analysis of Monticelli et al. [1]. This would yield "GR" and "RG" algorithms.

We conclude from the above arguments that it is likely to be advantageous to construct algorithms in which the update rules include "pivoting" on G blocks and/or inversion of non-decoupled blocks, but only for a selected set of buses where the r/x ratios are known to be high.

There is clearly no unique way of determining which branches do have "high" r/x ratios, but the sensitivity properties of such algorithms to variations of this sort will need to be investigated simultaneously. Assuming that we can devise some way of selecting branches with "high" r/x ratios, we divide the set of buses into two groups: group one buses having "low" r/x ratios and group two buses having "high" r/x ratios. The non decoupled Quasi-Newton update equations can now be written in the form,

$$\begin{bmatrix} [1/V^1]\Delta P^1 \\ [1/V^2]\Delta P^2 \\ [1/V^1]\Delta Q^1 \\ [1/V^2]\Delta Q^2 \end{bmatrix} = \begin{bmatrix} -B_{11} & -B_{12} & G_{11} & G_{12} \\ -B_{21} & -B_{22} & G_{21} & G_{22} \\ -G_{11} & -G_{12} & -B_{11} & -B_{12} \\ -G_{21} & -G_{22} & -B_{21} & -B_{22} \end{bmatrix} \begin{bmatrix} \Delta\theta^1 \\ \Delta\theta^2 \\ \Delta V^1 \\ \Delta V^2 \end{bmatrix}$$

Here V^k and θ^k represent the vectors of bus voltage magnitudes and phase angles in the group k, k = 1 or 2. ΔP^k and $\Delta\Theta^k$ denote similar assignments.

The blocks in the matrix are dimensioned accordingly. Our first step in deriving new algorithms is to rewrite this equation in the form,

$$\begin{bmatrix} [1/V^1]\Delta P^1 \\ [1/V^1]\Delta Q^1 \\ [1/V^2]\Delta P^2 \\ [1/V^2]\Delta Q^2 \end{bmatrix} = \begin{bmatrix} -B_{11} & G_{11} & G_{12} & -B_{12} \\ -G_{11} & -B_{11} & -B_{12} & -G_{12} \\ -B_{21} & G_{21} & G_{22} & -B_{22} \\ -G_{21} & -B_{21} & -B_{22} & -G_{22} \end{bmatrix} \begin{bmatrix} \Delta\theta^1 \\ \Delta V^1 \\ \Delta V^2 \\ \Delta\theta^2 \end{bmatrix}$$

We may view the block matrix above as obtained from the previous one through pre- multiplication by the rotation matrix S_1 and post multiplication by the rotation matrix S_2 where,

$$S_1 = \begin{bmatrix} I & 0 & 0 & 0 \\ 0 & 0 & I & 0 \\ 0 & I & 0 & 0 \\ 0 & 0 & 0 & I \end{bmatrix}, \quad S_2 = \begin{bmatrix} I & 0 & 0 & 0 \\ 0 & 0 & 0 & I \\ 0 & I & 0 & 0 \\ 0 & 0 & I & 0 \end{bmatrix}$$

Here I represents an identity matrix of the correct dimension. These rotation matrices are very different from those considered by Haley and Ayres [8], even when one accounts for the fact that their's is expressed in Cartesian coordinates and ours are expressed in real polar coordinates.

We now consider various methods to decouple the update equations so far obtained. The most straight forward method is to simply ignore the off diagonal blocks. This gives rise to the following update rules:

New Algorithm

$$\begin{bmatrix} [1/V^1]\Delta P^1 \\ [1/V^1]\Delta Q^1 \end{bmatrix} = \begin{bmatrix} -B_{11} & G_{11} \\ -G_{11} & -B_{11} \end{bmatrix} \begin{bmatrix} \Delta\theta^1 \\ \Delta V^1 \end{bmatrix}$$

$$\begin{bmatrix} [1/V^2]\Delta P^2 \\ [1/V^2]\Delta Q^2 \end{bmatrix} = \begin{bmatrix} G_{22} & -B_{22} \\ -B_{22} & -G_{22} \end{bmatrix} \begin{bmatrix} \Delta V^2 \\ \Delta\theta^2 \end{bmatrix}$$

One would now treat the first set of update rules as in the BX or XB algorithms, where of course, the set of buses is now in group one; and one would treat the second set of update equations as they stand, since in typical situations, the number of group two buses will be relatively small.

The decoupling mechanism described above does not take into account any of the coupling between the two groups of buses, which in any particular problem may well be significant. The exact decoupling technique described in Monticelli et al. [1] may be written as the following algorithm. The first step in this process is to rewrite the update equations in the form,

$$
\begin{bmatrix} [1/V^1]\Delta P^1 \\ [1/V^1]\Delta Q^1 \\ [1/V^2]K^2 \\ [1/V^2]\Lambda^2 \end{bmatrix} = \begin{bmatrix} -B_{11} & G_{11} & G_{12} & -B_{12} \\ -G_{11} & -B_{11} & -B_{12} & -G_{12} \\ 0 & 0 & \Theta_{11} & \Theta_{12} \\ 0 & 0 & \Theta_{21} & \Theta_{22} \end{bmatrix} \begin{bmatrix} \Delta\theta^1 \\ \Delta V^1 \\ \Delta V^2 \\ \Delta\theta^2 \end{bmatrix}
$$

where

$$
\begin{bmatrix} K^2 \\ \Lambda^2 \end{bmatrix} = \begin{bmatrix} \Delta P^2 \\ \Delta Q^2 \end{bmatrix} - \begin{bmatrix} -B_{21} & G_{21} \\ -G_{21} & -B_{21} \end{bmatrix} \begin{bmatrix} -B_{11} & G_{11} \\ -G_{11} & -B_{11} \end{bmatrix}^{-1} \begin{bmatrix} \Delta P^1 \\ \Delta Q^1 \end{bmatrix}
$$

and

$$
\begin{bmatrix} \Theta_{11} & \Theta_{12} \\ \Theta_{21} & \Theta_{22} \end{bmatrix} = \begin{bmatrix} G_{22} & -B_{22} \\ -B_{22} & -G_{22} \end{bmatrix} - \begin{bmatrix} -B_{21} & G_{21} \\ -G_{21} & -B_{21} \end{bmatrix} \begin{bmatrix} -B_{11} & G_{11} \\ -G_{11} & -B_{11} \end{bmatrix}^{-1} \begin{bmatrix} G_{12} & -B_{12} \\ -B_{12} & -G_{12} \end{bmatrix}
$$

As in Monticelli et al.[1] we may implement this update rule as a three stage algorithm in the following way:

New Algorithm

$$
\begin{bmatrix} [1/V^1]\Delta P^1(\theta^1, V^1, V^2, \theta^2) \\ [1/V^1]\Delta Q^1(\theta^1, V^1, V^2, \theta^2) \end{bmatrix} = \begin{bmatrix} -B_{11} & G_{11} \\ -G_{11} & -B_{11} \end{bmatrix} \begin{bmatrix} \Delta\hat{\theta} \\ \Delta\hat{V} \end{bmatrix}
$$

$$\begin{bmatrix} [1/V^2]\Delta P^2(\theta^1 + \Delta\hat{\theta}, V^1 + \Delta\hat{V}, V^2, \theta^2) \\ [1/V^2]\Delta Q^2(\theta^1 + \Delta\hat{\theta}, V^1 + \Delta\hat{V}, V^2, \theta^2) \end{bmatrix} = \begin{bmatrix} \Theta_{11} & \Theta_{12} \\ \Theta_{21} & \Theta_{22} \end{bmatrix} \begin{bmatrix} \Delta V^2 \\ \Delta\theta^2 \end{bmatrix}$$

$$\begin{bmatrix} \Delta\theta^1 \\ \Delta V^1 \end{bmatrix} = \begin{bmatrix} \Delta\hat{\theta} \\ \Delta\hat{V} \end{bmatrix} - \begin{bmatrix} -B_{11} & G_{11} \\ -G_{11} & -B_{11} \end{bmatrix}^{-1} \begin{bmatrix} G_{12} & -B_{12} \\ -B_{12} & -G_{12} \end{bmatrix} \begin{bmatrix} \Delta V^2 \\ \Delta\theta^2 \end{bmatrix}$$

In the implementation of such an update procedure one would of course apply a BX or XB type of algorithm to effect the first stage of the process. The second stage would be implemented as it is, because of the assumed small number of type two buses. The third stage of the algorithm may well prove unnecessary, as in the third stage of the Quasi-Newton algorithms discussed above. The second stage of the algorithm requires the inversion of a matrix in the evaluation of the matrix Q. The matrix Q may be evaluated using the partition matrix inversion formula as follows,

$$\begin{bmatrix} X & U \\ V & Y \end{bmatrix}^{-1} = \begin{bmatrix} R^{-1} & -R^{-1}UY^{-1} \\ -Y^{-1}VR^{-1} & Y^{-1} + Y^{-1}VR^{-1}UY^{-1} \end{bmatrix},$$

$$R = X - UY^{-1}V$$

It would be interesting to obtain some network meaningful interpretation of the resulting matrix expression for Q; which is similar to the interpretation of the matrix $-B - GB^{-1}G$ as $[1/x]$ in the case of radial networks. However this seems to be a hard problem in general.

1.9   Critically Coupled XB (CCXB) and BX (CCBX) Algorithms.

Several new algorithms have been proposed in the previous section. In this section we develop an additional pair of algorithms based on the BX and XB algorithms. Consider first the following theorem and proof.

DANIEL J. TYLAVSKY *ET AL.*

<u>Theorem</u>

Consider the system of equations

$$\begin{bmatrix} A & B & C \\ D & E & F \\ G & H & I \end{bmatrix} \begin{bmatrix} x \\ y \\ z \end{bmatrix} = \begin{bmatrix} e \\ f \\ g \end{bmatrix}$$

(1.9.1)

and the related set of equations

$$\begin{bmatrix} A & B & 0 & C \\ D & E & 0 & F \\ 0 & 0 & R_{11} & R_{12} \\ 0 & 0 & R_{21} & R_{22} \end{bmatrix} \begin{bmatrix} x' \\ y' \\ y'' \\ z' \end{bmatrix} = \begin{bmatrix} e \\ f \\ f - D A^{-1} e \\ g - G A^{-1} e \end{bmatrix}$$

(1.9.2)

where

$$\begin{bmatrix} R_{11} & R_{12} \\ R_{21} & R_{22} \end{bmatrix} = \begin{bmatrix} E & F \\ H & I \end{bmatrix} - \begin{bmatrix} D \\ G \end{bmatrix} A^{-1} [B \ \ C]$$

(1.9.3)

Assume that A through I are matrices of appropriate dimensions, and e,f and g are vectors with corresponding dimensions. Additionally assume that A, E and R are invertible and hence square matrices. Then both sets of equations have unique solutions satisfying:

   $x=x'$, $y=y'=y''$ and $z=z'$.

<u>Proof</u>

   We first consider the equation,

$$\begin{bmatrix} A & 0 & B & C \\ D & E & 0 & F \\ D & 0 & E & F \\ G & 0 & H & I \end{bmatrix} \begin{bmatrix} x \\ y \\ y' \\ z \end{bmatrix} = \begin{bmatrix} e \\ f \\ f \\ g \end{bmatrix}$$

If E is invertible then the second and third block equations force $y = y'$, and hence give identical equations. It follows easily that this set of equations simply repeats the middle set of equations in equation (1.9.1) of the theorem. If, in addition, A is invertible then the first two block equations can be multiplied by the matrix,

$$\begin{bmatrix} D & 0 \\ G & 0 \end{bmatrix} \begin{bmatrix} A & 0 \\ D & E \end{bmatrix}^{-1}$$

and then subtracted from the last two block equations, to yield the system of equations:

$$\begin{bmatrix} A & 0 & B & C \\ D & E & 0 & F \\ 0 & 0 & R_{11} & R_{12} \\ 0 & 0 & R_{21} & R_{22} \end{bmatrix} \begin{bmatrix} x \\ y \\ y' \\ z \end{bmatrix} = \begin{bmatrix} e \\ f \\ f - DA^{-1}e \\ g - GA^{-1}e \end{bmatrix}$$

where the matrix R coincides with that in equation (1.9.3). Notice that if the matrix R is also invertible then the solution of this set of equations is unique. The only difference between this set of equations and the set of equations (1.9.2), appearing in the statement of the theorem, is the placement of the matrix B. However once we note that $y = y'$ it is easy to see that the two sets of equations have identical solutions. Q.E.D.

Next apply equation (1.9.1) to the BGGB update rule by viewing the 4 X 4 block matrix appearing in the rule as the following 3 X 3 block matrix:

$-B_{11}$	$-B_{12}$	G_{12}	G_{11}
$-B_{21}$	$-B_{22}$	G_{22}	G_{21}
$-G_{21}$	$-G_{22}$	$-B_{22}$	$-B_{21}$
$-G_{11}$	$-G_{12}$	$-B_{12}$	$-B_{11}$

Note that in this case equation (1.9.3) implies that $R = -(B + G B^{-1}G)$ and so equation (1.9.2) yields the following update rule:

$$\begin{bmatrix} -B_{11} & -B_{12} & G_{12} & 0 & G_{11} \\ -B_{21} & -B_{22} & G_{22} & 0 & G_{21} \\ -G_{21} & -G_{22} & -B_{22} & 0 & -B_{21} \\ 0 & 0 & 0 & R_{11} & R_{12} \\ 0 & 0 & 0 & R_{21} & R_{22} \end{bmatrix} \begin{bmatrix} \Delta\theta^1 \\ \Delta\theta^2 \\ \Delta V^2 \\ \Delta V^2 \\ \Delta V^1 \end{bmatrix} = \begin{bmatrix} \Delta P^1 \\ \Delta P^2 \\ \Delta Q^2 \\ \Delta Q^{\#2} \\ \Delta Q^{\#1} \end{bmatrix}$$

(1.9.4)

where,

$$\begin{bmatrix} \Delta Q^{\#2} \\ \Delta Q^{\#1} \end{bmatrix} = \begin{bmatrix} \Delta Q^2 \\ \Delta Q^1 \end{bmatrix} - \begin{bmatrix} G_{21} & G_{22} \\ G_{11} & G_{12} \end{bmatrix} \begin{bmatrix} B_{11} & B_{12} \\ B_{21} & B_{22} \end{bmatrix}^{-1} \begin{bmatrix} \Delta P^1 \\ \Delta P^2 \end{bmatrix}$$

(1.9.5)

If we set,

$$\hat{B} = \begin{bmatrix} -B_{11} & -B_{12} & G_{12} \\ -B_{21} & -B_{22} & G_{22} \\ -G_{21} & -G_{22} & -B_{22} \end{bmatrix}$$

and,

$$\hat{\theta} = \begin{bmatrix} \theta^1, \theta^2, V^2 \end{bmatrix}^T, \quad \hat{P} = \begin{bmatrix} P^1, P^2, Q^2 \end{bmatrix}^T$$

this leads directly to the Critically Coupled BX algorithm (CCBX):

$$\hat{B}' \Delta \hat{\theta} = [1/V] \Delta \hat{P}$$

$$[1/X]'' \Delta V = [1/V] \Delta Q(\hat{\theta} + \Delta \hat{\theta}, V)$$

(1.9.6)

The relation between this algorithm and that described by the update rule (1.9.4) involves ignoring the third step of the algorithm defined by (1.9.4) and approximation of the the matrix R by the matrix [1/X], just as in the justification of the BX algorithm detailed above. Also the ΔV update in the rule (1.9.6) differs from that indicated by the relation (1.9.5), which suggests that the correction term $\Delta \theta$, derived as in the usual BX algorithm should be used. The notation Z' and Z" applied to the matrices above are applied exactly as in the BX algorithm.

In a similar manner the corresponding critically coupled XB algorithm (CCXB) can easily be derived. Letting

$$\tilde{B} = \begin{bmatrix} -B_{11} & -B_{12} & -G_{12} \\ -B_{21} & -B_{22} & -G_{22} \\ G_{21} & G_{22} & -B_{22} \end{bmatrix}$$

and

$$\tilde{V} = \begin{bmatrix} V^1, V^2, \theta^2 \end{bmatrix}^T, \quad \tilde{Q} = \begin{bmatrix} Q^1, Q^2, P^2 \end{bmatrix}^T$$

then the algorithm can be written in the form:

$$\tilde{B}'' \, \Delta \tilde{V} = [1/V] \, \Delta \tilde{Q}$$

$$[1/X]' \Delta \theta = [1/V] \, \Delta P(\theta, \, \tilde{V} + \Delta \tilde{V})$$

Finally in this section we make some observations on two limiting cases in these two newly proposed algorithms. In the case where the set of group I buses is vacuous, we may set,

$$\hat{B} = \tilde{B} = \begin{bmatrix} -B & G \\ -G & -B \end{bmatrix}$$

and

$$\hat{V} = \tilde{V} = [\theta, \, V]^T, \quad \tilde{Q} = \hat{P} = [P, Q]^T$$

It then follows that the first stage of both the critically coupled XB and BX algorithms become the same as the BGGB algorithm. It can be shown that the effect of the second stage in each of these algorithms, does not have an appreciable effect on the robustness of the algorithms and they behave as if they were simply BGGB algorithms. Of course this extreme is not typical of the envisaged applications, in which the number of group II buses is relatively small. In the other extreme, where there are no group II buses, the critically coupled XB and BX algorithms reduce exactly to the XB and BX algorithms respectively.

The advantage of the proposed algorithms is that they adapt themselves to the situation dictated by the choice of "high" r/x ratio lines, resulting in an algorithm which has the qualities of both the BX(XB) and BGGB algorithms. Of course the disadvantage of the algorithms is that they require the solution of a larger number of equations at each iteration. This trade off is investigated experimentally in the next section.

1.10   Summary.

The algorithms thus far discussed by Monticelli et al.[1], Amerongen [2], Stott and Alsaç [3], Wu [4], Carpentier [6], Rajicic and Bose [7], Haley and Ayres [8], and those proposed in sections 1.8 and 1.9 all fit into the category of Quasi-Newton Methods in which the approximated Jacobian is fixed throughout the solution process. The advantage of such schemes is that they allow short-execution-time solutions provided the problems fit the assumptions used by each method. When such assumptions are not satisfied, these algorithms may require a large number of iterations or fail to converge. In this situation, often a full Newton-Raphson (with variable "exact" Jacobian) approach is used to obtain a solution. If both methods fail then it is often assumed by the user that no solution exists; this may be the case and is often all but proven to be the case. This conclusion is of course not supported by the observation that some cases may not be solvable by either. The area between the two extremes of the fixed and variable Jacobian (e.g., methods involving infrequent (approximate) Jacobian updates) appears to be fertile ground for research.

The primary objective of this discussion has been to identify new algorithms which show potential a high-speed power-flow equation solvers. Many new Quasi-Newton algorithms have been proposed. Provided in this work has been a unifying framework for describing these algorithms. The next step in identifying potentially useful algorithms is to test some of these candidate algorithms. These tests are reported in the next section.

2.0   Fast Algorithm Test Results.

The goal of this research was to propose new algorithms with a large potential for being able to solve the power flow problem. Included here are the results of testing some of these algorithms together with a comparison with similar test results for the BGGB, full Newton Raphson, BX, and XB algorithms.

2.1   Test Systems.

The IEEE 14, 30, 57, and 118 bus systems are used for testing of these methods. Testing their robustness requires that r/x ratios be varied by both multiplying the line X or R values by X factors and R factors respectively. In certain testing only the R or X values of one line with the highest r/x ratio is scaled. In some cases 10 lines are used. Duplication of this work requires that these line be identified as in tables 2.1.1 and 2.1.2.

Table 2.1.1. Details of the buses connected to the lines in IEEE networks with highest r/x ratio, in standard IEEE format.

14			30		
FROM BUS	TO BUS	R/X RATIO	FROM BUS	TO BUS	R/X RATIO
12	13	1.105	14	15	1.107
6	13	0.507	25	26	0.669

57			118		
FROM BUS	TO BUS	R/X RATIO	FROM BUS	TO BUS	R/X RATIO
32	33	1.089	23	24	0.640
56	41	1.007	83	84	0.473
53	54	0.809	-	-	-

Table 2.1.2. Details of the buses connected to the 10 lines scaled in testing.

14			30		
FROM BUS	TO BUS	R/X RATIO	FROM BUS	TO BUS	R/X RATIO
4	5	0.317	4	12	0.0
4	7	0.0	12	13	0.0
4	9	0.0	12	14	0.481
5	6	0.0	12	15	0.508
6	11	0.478	12	16	0.476
6	12	0.480	14	15	1.107
6	13	0.508	16	17	0.429
7	8	0.0	15	18	0.491
7	9	0.0	18	19	0.495
9	10	0.376	19	20	0.5

57			118		
FROM BUS	TO BUS	R/X RATIO	FROM BUS	TO BUS	R/X RATIO
12	13	0.307	43	44	0.248
12	16	0.221	34	43	0.246
12	17	0.222	44	45	0.249
13	14	0.304	45	46	0.295
13	15	0.310	46	47	0.299
13	49	0.0	46	48	0.318
14	15	0.313	47	49	0.306
14	46	0.0	42	49	0.221
15	45	0.0	42	49	0.221
18	19	0.673	45	49	0.368

2.2   Performance of Carpentier's Method.

Carpentier's method involves the P-θ iteration given by $\Delta P = B' \Delta \theta$ where B' is the same as used in the BX algorithm. The Q-V iteration is governed by $\Delta Q = L' \Delta V$, where L' is defined by (1.5.7).

Table 2.2.1 and 2.2.3 give the results of employing the Carpentier BX method when the line with the highest r/x ratio is scaled. Table 2.2.2 and 2.2.4 report the results when 10 lines have been scaled uniformly. These results indicate that Carpentier's method generally requires more iterations or ceases to converge for smaller R factors or larger X factors than the BX algorithm as shown by the results in the next section. Unfortunately we have not been able to get the Carpentier-like algorithm analogous to the XB method to converge.

Table 2.2.1.   Iteration counts for the Carpentier BX method with the r-factor scaled in the line with highest r/x ratio.

R-SCALE FACTOR	NUMBER OF	ITERATIONS	IN THE	SYSTEM OF
	IEEE14	IEEE30	IEEE57	IEEE118
0.5	6	5	7	5
1.0	6	5	7	5
1.5	6	5	7	5
2.0	6	6	7	5
2.5	7	7	7	5
3.0	7	8	7	5
4.0	9	9	8	5
5.0	10	10	9	6
10.0	12	12	nc	12
20.0	13	14		nc

Table 2.2.2. Iteration counts for the Carpentier BX method with the r-factor uniformly scaled in 10 lines.

R-SCALE FACTOR	NUMBER OF	ITERATIONS	IN THE	SYSTEM OF
	IEEE14	IEEE30	IEEE57	IEEE118
0.5	5	5	7	5
1.0	6	5	7	5
1.5	6	6	7	5
2.0	6	6	7	5
2.5	6	6	7	6
3.0	7	7	7	6
4.0	7	8	7	7
5.0	8	10	8	9
10.0	13	nc	13	nc
20.0	nc		nc	

Table 2.2.3. Iteration counts for the Carpentier BX method with the x-factor scaled in the line with highest r/x ratio.

X-SCALE FACTOR	NUMBER OF	ITERATIONS	IN THE	SYSTEM OF
	IEEE14	IEEE30	IEEE57	IEEE118
1.0	6	5	7	5
0.5	6	6	7	5
0.25	10	10	12	5
0.2	12	12	15	5
0.166	14	14	18	5
0.125	18	19	23	6
0.111	20	21	27	6
0.1	22	23	30	6
0.083	26	27	nc	7
0.071	29	nc		7
0.063	nc			8
0.056				8
0.05				8
0.04				10
0.03				11
0.01				18
0.007				23

Table 2.2.4. Iteration counts for the Carpentier BX method with the x-factor uniformly scaled in 10 lines.

X-SCALE	NUMBER OF	ITERATIONS	IN THE	SYSTEM OF
FACTOR	IEEE14	IEEE30	IEEE57	IEEE118
1.0	6	5	7	5
0.5	8	6	8	5
0.25	9	8	10	6
0.2	10	9	10	6
0.166	11	10	11	6
0.125	14	12	12	7
0.111	15	13	12	8
0.1	17	13	13	9
0.083	20	17	16	10
0.071	nc	nc	17	11
0.063	27		20	12
0.056	nc		23	14
0.05			30	16
0.04			nc	23
0.03				nc
0.01				
0.007				

2.3   Performance of the CCBX and CCXB Algorithms.

In this section we present a selection of our numerical test results giving comparisons between the performance of the Newton, BGGB, BX and XB algorithms and the performance of our critically coupled BX and XB algorithms. Five distinct tests are reported, in which the algorithm's performance is tested in a variety of ways. Most of the tests concern the standard 14, 30, 57 and 118 bus IEEE. networks, and unless otherwise stated the algorithms were deemed to converge if all real and reactive power mismatches were smaller than 0.01 MW/Mvar. For the "two stage" BX, XB and the critically coupled BX and XB algorithms all test results that we report here employ the successive method of iteration, as opposed to the classical method (for the "one stage" Newton and BGGB algorithms there is of course no distinction). Thus for the two stage algorithms one iteration includes one iteration of both stages. The iterations were performed without the checking of VAR limits. The algorithm was deemed to have not converged if the number of iterations exceeded 30 (denoted by "nc" in the following tables). In some cases overflow occurred and the algorithm is said to have diverged (denoted "div" in the following tables). Unless otherwise stated the group II buses employed in the critically coupled BX and XB algorithms are connected to lines whose r/x ratio exceeds or is equal to unity. Thus for the purposes of most of our tests we took the somewhat arbitrary decision that a high r/x value is one which exceeds unity. However this value did achieve good robustness while limiting the number of group II buses. We include one test for comparison, described below, where we vary this critical r/x value.

The first set of tests reported in tables 2.3.1 and 2.3.2, involves scaling (upwards) the resistive component of certain lines in the IEEE networks, for all six algorithms. Table 2.3.1 reports the results when, the line with the highest r/x ratio is scaled as per Table 2.1.1. Table 2.3.2 reports the results when 10 lines have been scaled uniformly. For each network the ten lines selected are detailed in table 2.1.2. We note that not all lines in these groups of 10 have non zero resistances, and only the lines in the 30 bus network contain the line with the highest r/x ratio.

In the following tables r (x)- scale, is the scale factor applied to the resistance (reactance) of the designated line(s), Max r/x is the maximum r/x value found in any line in the network and # II's is the number of group II buses in the critically coupled XB and BX algorithms.

The second set of tests involves scaling (downwards) the reactive component of certain lines in the IEEE networks. The results are reported in tables 2.3.3 and 2.3.4, for all six algorithms. Our comments above concerning the structure of the tests reported Table 2.3.1 and Table 2.3.2, also apply to the tests reported in Table 2.3.3 and Table 2.3.4 respectively.

Table 2.3.1. Iteration counts for IEEE networks with the resistance of the highest r/x ratio line(s) scaled(tolerance of 0.01 MW/Mvar; successive iteration scheme).

R-SCALE	MAX R/X	# II'S	NEWTON	BGGB	XB	BX	CCXB	CCBX
IEEE	14 bus							
0.5	0.6	0	3	5	5	5	4	5
1.0	1.1	2	3	5	5	5	4	5
1.5	1.7	2	3	5	5	5	4	5
2.0	2.2	2	3	5	5	6	4	5
2.5	2.8	2	3	5	5	7	4	5
3.0	3.3	2	3	5	5	8	4	5
4.0	4.4	2	3	5	5	9	4	5
5.0	5.5	2	3	5	5	10	4	5
10.0	11.1	2	3	5	7	12	4	5
20.0	22.1	2	3	5	9	13	4	5
IEEE 30 bus								
0.5	0.7	0	3	5	4	5	4	5
1.0	1.1	2	3	5	4	5	4	5
1.5	1.7	2	3	5	4	5	4	5
2.0	2.2	2	3	5	4	6	4	5
2.5	2.8	2	3	5	5	7	4	5
3.0	3.3	2	3	5	4	8	4	5
4.0	4.4	2	3	5	4	9	4	5
5.0	5.5	2	3	5	5	10	4	5
10.0	11.1	2	3	5	8	12	4	5
20.0	22.1	2	3	5	10	13	4	5
IEEE 57 bus								
0.5	1.0	2	3	5	5	5	5	5
1.0	1.1	4	3	5	5	5	5	5
1.5	1.6	4	3	5	5	5	5	5
2.0	2.2	4	3	5	5	5	5	5
2.5	2.7	4	3	5	5	5	5	5
3.0	3.3	4	3	5	5	7	5	5
4.0	4.4	4	3	5	6	7	5	5
5.0	5.4	4	3	5	6	8	5	5
10.0	10.9	4	3	5	8	12	5	5
20.0	21.8	4	3	5	20	div	5	5
IEEE 118 bus								
0.5	0.5	0	3	6	5	5	5	5
1.0	0.6	0	3	6	5	5	5	5
1.5	1.0	0	3	6	7	5	6	5
2.0	1.3	2	3	6	9	5	5	5
2.5	1.6	2	3	6	11	5	5	5
3.0	1.9	2	3	6	14	5	6	5
4.0	2.6	2	3	6	21	5	7	5
5.0	3.2	2	3	6	29	6	9	5
10.0	6.4	2	3	6	nc	9	15	5
20.0	12.8	2	3	6		11	21	5

Table 2.3.2. Iteration counts for IEEE networks with the resistance of 10 lines uniformly scaled(tolerance of 0.01 MW/Mvar; successive iteration scheme).

R-SCALE	MAX R/X	# II'S	NEWTON	BGGB	XB	BX	CCXB	CCBX
IEEE	14 bus							
0.5	1.1	2	3	5	4	5	4	4
1.0	1.1	2	3	5	5	6	4	5
1.5	1.1	2	3	5	5	6	4	5
2.0	1.1	3	3	5	5	6	4	6
2.5	1.3	4	3	5	6	6	5	5
3.0	1.5	6	3	5	6	7	5	5
4.0	2.0	8	3	5	7	7	5	4
5.0	2.5	8	3	5	9	8	5	4
10.0	5.1	8	3	5	16	11	6	5
20.0	10.2	8	3	5	nc	nc	7	8
IEEE 30 Bus								
0.5	0.7	0	3	5	4	5	4	5
1.0	1.1	2	3	5	4	5	4	5
1.5	1.7	2	3	5	5	5	4	5
2.0	2.2	3	3	5	5	6	4	5
2.5	2.8	8	3	5	5	6	4	4
3.0	3.3	8	3	5	5	7	4	5
4.0	4.4	8	3	5	6	8	5	5
5.0	5.5	8	3	5	8	9	6	5
10.0	11.1	8	3	5	15	17	8	6
20.0	22.1	8	3	5	nc	div	10	8
IEEE 57 Bus								
0.5	1.1	4	3	5	5	5	5	5
1.0	1.1	4	3	5	5	5	5	5
1.5	1.1	6	3	5	5	5	5	5
2.0	1.3	6	3	5	6	5	6	5
2.5	1.7	6	3	5	8	5	8	5
3.0	2.0	6	3	5	9	5	9	5
4.0	2.7	10	3	5	11	6	7	5
5.0	3.4	12	3	6	14	8	8	5
10.0	6.7	12	3	8	nc	14	14	7
20.0	13.5	12	4	14		21	nc	14
IEEE 118 Bus								
0.5	0.6	0	3	6	5	5	5	5
1.0	0.6	0	3	6	5	5	5	5
1.5	0.6	0	3	6	6	5	5	5
2.0	0.7	0	3	6	7	5	7	5
2.5	0.9	0	3	6	9	5	9	5
3.0	1.1	2	3	6	11	6	10	6
4.0	1.5	5	3	7	17	7	12	7
5.0	1.8	9	3	8	25	8	10	7
10.0	3.7	9	nc	div	nc	nc	div	div

Table 2.3.3. Iteration counts for IEEE networks with the x-factor scaled in the line with highest r/x ratio (tolerance of 0.01 MW/Mvar; successive iteration scheme).

X-SCALE	MAX R/X	# II'S	NEWTON	BGGB	XB	BX	CCXB	CCBX
IEEE	14 bus							
1.0	1.1	2	3	5	5	5	4	5
0.5	2.2	2	3	5	5	6	4	5
0.2	5.5	2	3	5	9	12	4	5
0.1	11.1	2	3	5	16	22	4	5
0.05	22.1	2	3	5	nc	nc	4	5
0.04	27.6	2	3	5			4	5
0.03	36.8	2	3	5			4	5
0.01	110.5	2	3	5			4	5
0.007	157.9	2	3	5			4	5
IEEE 30 Bus								
1.0	1.1	2	3	5	4	5	4	5
0.5	2.2	2	3	5	5	6	4	5
0.2	5.5	2	3	5	9	12	4	5
0.1	11.1	2	3	5	16	22	4	5
0.05	22.1	2	3	5	30	nc	4	5
0.04	27.7	2	3	5	nc		4	5
0.03	36.9	2	3	5			4	5
0.01	110.7	2	3	5			4	5
0.007	158.1	2	3	5			4	5
IEEE 57 Bus								
1.0	1.1	4	3	5	5	5	5	5
0.5	2.2	4	3	5	5	5	5	5
0.2	5.4	4	3	5	6	5	5	5
0.1	10.9	4	3	5	6	8	5	5
0.05	21.8	4	3	5	6	8	5	5
0.04	27.2	4	3	5	6	10	5	5
0.03	36.3	4	3	5	7	14	5	5
0.01	108.9	4	3	5	10	div	5	5
0.007	155.6	4	3	5	12		5	5
IEEE 118 Bus								
1.0	0.6	0	3	6	5	5	5	5
0.5	1.3	2	3	6	7	5	5	5
0.2	3.2	2	3	6	16	5	5	5
0.1	6.4	2	3	6	nc	6	5	5
0.05	12.8	2	3	6		9	5	5
0.04	16.0	2	3	6		11	5	5
0.03	21.3	2	3	6		14	5	5
0.01	64.0	2	3	6		nc	5	5
0.007	91.5	2	3	6			5	5

Table 2.3.4. Iteration counts for IEEE networks with the x-factor uniformly scaled in 10 lines (tolerance of 0.01 MW/Mvar; successive iteration scheme).

X-SCALE	MAX R/X	# II'S	NEWTON	BGGB	XB	BX	CCXB	CCBX
IEEE	14 bus							
1.0	1.1	2	3	5	5	5	4	5
0.5	1.1	3	3	5	5	7	4	6
0.2	2.5	8	3	5	8	10	4	5
0.1	5.1	8	3	5	14	16	5	5
0.05	10.2	8	3	4	nc	30	5	5
0.04	12.7	8	3	4		nc	6	5
0.03	16.9	8	3	5			6	5
0.01	50.8	8	9	6			11	8
0.007	72.5	8	4	10			18	10
0.004	126.9	8	nc	nc			nc	nc
IEEE 30 Bus								
1.0	1.1	2	3	5	4	5	4	5
0.5	2.2	3	3	5	5	6	4	6
0.2	5.5	8	3	5	6	9	4	6
0.1	11.1	8	3	5	12	16	6	7
0.05	22.1	8	3	5	26	28	7	8
0.04	27.7	8	3	4	nc	nc	8	8
0.03	36.9	8	3	4			8	8
0.01	110.7	8	3	4			9	9
0.007	158.1	8	4	4			9	9
0.004	276.7	8	nc	nc			nc	nc
IEEE 57 Bus								
1.0	1.1	4	3	5	5	5	5	5
0.5	1.3	6	3	5	6	6	6	5
0.2	3.4	12	3	5	12	8	5	6
0.1	6.7	12	4	5	23	13	6	8
0.05	13.5	12	4	5	nc	22	8	12
0.04	16.8	12	4	6		26	8	14
0.03	22.4	12	5	5		nc	10	18
0.01	67.3	12	6	20			nc	nc
0.007	96.1	12	div	nc				
0.004	168.2	12						
IEEE 118 Bus								
1.0	0.6	0	3	6	5	5	5	5
0.5	0.7	0	3	6	7	5	6	5
0.2	1.8	9	3	6	17	6	8	5
0.1	3.7	9	3	6	nc	8	13	5
0.05	7.4	9	3	7		15	22	7
0.04	9.2	9	4	10		18	27	9
0.03	12.3	9	4	20		29	nc	20
0.01	36.8	9	div	div		div		div

The test results presented in tables 2.3.1, 2.3.2, 2.3.3 and 2.3.4 indicate that the critically coupled BX and XB algorithms perform significantly better than the corresponding BX and XB algorithms. Indeed the critically coupled algorithms seem to behave some what (but not uniformly) better than the BGGB algorithm. We note that the difference in the performance of the BX and XB algorithms is generally reflected in the performance of the corresponding CCBX and CCXB algorithm. Moreover this performance is generally obtained at the expense of including relatively few "extra" group II buses.

In the next set of tests the r/x ratios of all lines are scaled in a random fashion, to contrast the previous tests where the r/x ratios of particular line(s) are uniformly scaled. To be specific, the reactance of each line in the network was multiplied by a randomly distributed scale factor lying between 0 and 1. The tests were performed on 20 networks modified in this fashion and the results averaged over the 20 networks. The test results for the CCBX algorithm run on the IEEE 57 and 118 bus networks are displayed in table 2.3.5. In these tests the buses connected to lines whose r/x ratio exceeds a critical value, are included in group II, and the results tabulated for a range of critical r/x values.

Table 2.3.5. Average iteration counts for the CCBX algorithm with a random distribution of x scale factors applied to all lines of the IEEE 57 and 118 bus networks(tolerance of 0.01 MW/Mvar; successive iteration scheme).

# OF BUSES	CRITICAL R/X RATIO	AVERAGE # OF GROUP II BUSES	AVERAGE # ITERATIONS FOR CONVERGED CASES	# OF CASES NOT CONVERGED
57	0.0	57	5	1
	0.5	54	7	1
	1.0	43	8	2
	2.0	27	9	2
	3.0	20	11	2
	4.0	16	12	3
	5.0	13	12	3
	6.0	11	14	3
	7.0	8	14	2
	8.0	7	16	2
	9.0	7	16	2
	10.0	6	17	2
	20.0	3	21	4
	50.0	1	21	11
	100.0	1	21	13
	1000.0	0	21	15
118	0.0	118	6	4
	0.5	92	5	4
	1.0	63	6	5
	2.0	37	7	6
	3.0	26	8	6
	4.0	20	9	6
	5.0	17	10	6
	6.0	15	11	6
	7.0	12	12	6
	8.0	11	12	6
	9.0	10	13	6
	10.0	9	14	6
	20.0	5	20	9
	50.0	2	18	17
	100.0	1	18	17
	1000.0	0	28	19

In evaluating the results of the tests in table 2.3.5 it is interesting to first consider the extreme critical r/x values of 0 and 1000. For a critical r/x value of 0 all buses are group II buses, and the CCBX algorithm becomes the BGGB algorithm with an extra stage, as noted in section 2. For a critical value of 1000 none of the buses are classified as group II buses, so as noted in section 2, the CCBX algorithm is just the BX algorithm. We note that in this case the BX algorithm converges in only 5 cases for the 57 bus network and in only 1 case in the 118 bus case. As the critical r/x ratio is decreased from 1000 to 0 the CCBX algorithm adapts itself from the BX algorithm to essentially the BGGB algorithm, with a corresponding improvement in its performance.

Unfortunately the tests in table 2.3.5 do not immediately indicate what constitutes a "high" r/x ratio line, since the performance of the CCBX algorithm seems to vary smoothly from one extreme critical r/x value to another. At first sight the test results indicate that to obtain a set of group II buses in the region of 5 -10, the critical "high" r/x ratio should be set between 7 and 10, which is much higher than the "high" r/x ratios between 0.5 and 3, quoted in the literature or the value 1 used in the other tests reported here. However it should be noted that the test cases reported in table 2.3.5 are very extreme, since the BX algorithm (corresponding to a critical r/x value of 1000) converged in very few of the cases. Lower "high" r/x values would be expected if the reactance values were randomly scaled between the values of 0.1 and 1, for example, rather than between 0 and 1.

In the next set of tests the BX, XB, CCBX and CCXB algorithms were run on a fairly difficult case, in which the resistance of 10 lines was uniformly scaled by a factor of 5. The ten lines were selected as before and are detailed in table 2.1.2. In this case however the measure of convergence was varied from 1 to 0.001 MW/Mvar in the power mismatches. The results are tabulated in table 2.3.6. The iterations required for convergence of the CCBX and CCXB algorithms vary much less than the corresponding BX and XB algorithms. This again indicates increased robustness of the new algorithms.

Table 2.3.6. Iteration counts for several tolerances with 10 lines scaled by an r scale factor of 5, tolerance in MW/Mvar (successive iteration scheme).

IEEE 14 Bus				IEEE 30 Bus			
TOL.	BX	XB	CCBX	CCXB	BX	XB	CCBX	CCXB
1.0	4	4	2	3	4	3	2	3
0.1	6	6	3	4	6	5	4	4
0.01	8	9	4	5	9	8	5	6
0.001	10	11	5	6	11	10	7	7

IEEE 57 BUS				IEEE 118 BUS			
TOL.	BX	XB	CCBX	CCXB	BX	XB	CCBX	CCXB
1.0	4	8	3	4	5	11	4	5
0.1	5	11	4	6	7	18	5	8
0.01	8	14	5	8	8	25	7	10
0.001	11	18	8	10	11	nc	9	13

For the final set of tests we introduced a more realistic 1655 bus network. This network contains 5 branches with r/x ratio's greater than 2.0, 11 with r/x ratio greater than 1.0 and 98 with r/x ratio greater than 0.5. In table 2.3.7 we detail the performance of the algorithm on this network as before, by scaling the x factors in the 5 lines with with highest r/x ratio. We draw attention to the impressive robustness properties of the CCXB algorithm, in which only 21 (or 1.5%) of the buses are classed as group II buses, having a "high" r/x ratio of over 1.

Table 2.3.8 gives the CPU time, excluding input-output activity, required to solve the power flow equations for the 118 and 1655 bus networks, using the BX, XB and the critically coupled algorithms. The algorithms were implemented using standard sparsity coding techniques. The results indicate that in general the increased number of calculations per iteration, required by the critically coupled algorithms to solve the equations, is more than compensated for by the decreased number of iterations needed for convergence.

Table 2.3.7. Iteration counts for a 1655 bus network with the x-factor uniformly scaled in 5 lines (tolerance of 0.01 MW/Mvar; successive iteration scheme).

X-SCALE	MAX R/X	# II'S	NEWTON	BGGB	XB	BX	CCXB	CCBX
1.0	11.9	21	4	14	11	div	10	10
0.5	23.9	21	4	14	19		10	10
0.2	59.7	21	4	14	nc		10	12
0.1	119.3	21	4	14			10	22
0.05	238.7	21	4	14			10	nc
0.04	298.3	21	4	14			10	
0.03	397.8	21	4	14			10	
0.01	1193.3	21	4	14			10	
0.007	1704.8	21	4	14			10	
0.004	2983.3	21	4	14			10	

Table 2.3.8. A comparison of CPU times, in seconds, for the BX, XB, CCBX and CCXB algorithms, for the IEEE 118 bus network and a 1655 bus network with x values scaled on ten lines for the 118 bus case and five lines in the 1655 bus case (tolerance of 0.01 MW/Mvar; successive iteration scheme; run on a VAX 11/785).

# OF BUSES	X-SCALE FACTOR	BX TIME (SECS.)	CCBX TIME (SECS.)	XB TIME (SECS.)	CCXB TIME (SECS.)
118	1.0	0.18	0.23	0.17	0.22
	0.5	0.18	0.19	0.20	0.22
	0.2	0.20	0.20	0.37	0.28
	0.166	0.18	0.21	0.47	0.30
	0.147	0.20	0.19	0.51	0.31
	0.1	0.22	0.20	nc	0.40
	0.063	0.31	0.21		0.51
	0.04	0.41	0.27		0.68
	0.03	0.60	0.51		nc
1655	1.0	div	6.51	6.16	6.41
	0.5		6.44	7.78	6.56
	0.2		6.96	9.96	6.47
	0.166		7.50	nc	6.30
	0.147		7.95		6.60
	0.1		9.82		6.58
	0.063		nc		6.26
	0.04				6.33
	0.03				6.36

3.0   Summary.

A critical review of the work by Monticelli et al. [1] shows that both of the two-step algorithms analogous to the BX and XB algorithms are indeed approximations to three step algorithms. This conflicts with the conclusion in [1] that these two-step algorithms are exactly equivalent to the three step algorithms derived in his paper. This implies that the effect of the third step in any related algorithms must be carefully considered before such a step is neglected. Indeed, it can be shown by a revised analysis similar to that of Wu [4] using the observations of Monticelli et al. [1], that this third step does play a role in the convergence of the decoupled like algorithm. The important observations from this analysis is that the criteria for selection of an appropriate approximate Jacobian for initiation of convergence are not necessarily the same criteria to use for selection of an approximate Jacobian for completion of convergence.

It is further shown that the Carpentier [6] generated the same approximate Jacobian matrix as did Monticelli [1] but apparently did did not recognize sparsity implications of the results (under the appropriate assumptions of either a radial network of constant r/x ratio network). In order to maintain sparsity, he used the additional assumptions that V updates were based upon assuming that the real power flow over each line is held constant which also guarantees (as is classically assumed) that the real power at each bus is held constant. His notation makes it difficult to understand the significance of his work. When his work is appropriately restated two points become clear. First, to a first order approximation, the approximate Jacobian matrix contained in [6] is identical to that of Monticelli [1]. When a higher order approximation is used, the Jacobian matrix of [6] is reminiscent of the heuristically derived matrices proposed by Rajicic and Bose [7].

The super decoupled algorithm of Haley and Ayres [8] is based on applying diagonal matrix rotations to the rectangular form based power flow Jacobian. This approach is very different in spirit from the rotations suggested by Monticelli or Carpentier [1,6] but bear further examination.

We believe, that through this analysis of the literature, we have identified the common thread which runs through the recent algorithms which have shown superior performance. This analysis has allowed us to propose the use of block decoupling and rotations as described in Section 1.8 which may lead to improved decoupled performance. Test results show that the proposed critically coupled BX (CCBX) and XB (CCXB) algorithms display **improved robustness** characteristics when compared to the BX and XB algorithms on high R/X ratio systems. It has also been demonstrated that the extra complexity of the critically coupled algorithms does not unduly affect the speed at which the algorithms perform, compared with the BX and XB algorithms. The results indicate that the algorithms are worthy of consideration as alternatives to the BX and XB algorithms. **Although further testing is called for, the authors believe that this work demonstrates further substantial improvement on fast power flow algorithms is likely.** In this regard the ideas surrounding the algorithms of Carpentier [2] and Haley and Ayres [5] also warrant further investigation.

4.0   References.

1.   A. Monticelli, A. Garcia and O.R. Saavedra, "Fast Decoupled Load Flow: Hypothesis, Derivations and Testing," I.E.E.E. P.E.S. Winter Meeting, New York, 1989.

2.   R. A. M. van Amerongen, "A General Purpose Version of the Fast Decoupled Loadflow," I.E.E.E. P.E.S. Summer Meeting, Portland, 1988.

3.   B. Stott and O. Alsaç, "Fast Decoupled Load Flow," I.E.E.E. Trans. on Power Apparatus and Systems, Vol. PAS.-93, pp.859-869, 1974.

4.   F.F. Wu, "Theoretical Study of the Convergence of the Fast Decoupled Load Flow," I.E.E.E. Trans. on Power Apparatus and Systems, Vol. PAS.- 96, pp.268-275, 1977.

5.   P.S. Nagendre Rao, K. S. Prakasa Rao and J. Nanda, "An Empirical Criterion for the Convergence of the Fast Decoupled Load Flow Method," I.E.E.E. Trans. on Power Apparatus and Systems, Vol. PAS.-103, pp. 974-981, 1984.

6.   J.L. Carpentier, ""CRIC", A New Active Reactive Decoupling Process in Load Flows, Optimal Power Flows and System Control," I.F.A.C. Symposium on Power Systems and Power Plant Control, Beijing, 1986.

7.   D. Rajicic and A. Bose, "A Modification to the Fast Decoupled Power Flow for Networks with High R/X Ratios," Proc. P.I.C.A. conference, pp. 360-363, 1987.

8.   P. H. Haley and M. Ayres, "Super Decoupled Loadflow with Distributed Slack Bus," I.E.E.E. Trans. on Power Apparatus and Systems, Vol. PAS -104, pp. 104-113, 1985.

9.   S. Deckmann, A. Pizzolante, A. Monticelli, B. Stott, and O. Alsac, "Numerical Testing on Power System Load Flow Equivalents', IEEE Transactions on Power Apparatus and Systems, vol. PAS-99, pp. 2292-2300, 1980.

POWER SYSTEMS STATE ESTIMATION BASED ON LEAST ABSOLUTE VALUE (LAV)

G.S.Christensen   S.A.Soliman and   M.Y.Mohamed

Elect. Eng.Dept   Elect. Power & Machines Dept.
University of Alberta   Ain Shams University
Edmonton,Alberta,CANADA   Cairo , EGYPT

The operation and control of power systems have undergone many changes during the last two decades. The increasing size demand for reliable electric energy by consumers and the size increasing of power systems have added significantly to the complexity of operating power systems.

To improve the assessment of power systems, complete measurements have to be brought into the control room. However, if conventional individual meter display are used to exhibit meter readings, the control room operators will not possibly be able to monitor all readings. Also, it is economically unfeasible to measure all desired quantities. Consequently, it is necessary to measure some of the desired quantities and calculate the remaining quantities based on measurements and system configuration.

Since no measuring instrument is perfect, it is probable that the metering process will add small errors to the measurements. Furthermore, the transmission of these measurements through microwave links or telephone lines are subjected to additional contamination by various noise sources. As a result,

CONTROL AND DYNAMIC SYSTEMS, VOL. 44

there is a high probability that the system status predicted based on the measurements and calculations (which are also based on the measurements) does not truly represent the system.

State estimation is the process of assigning a value to unknown system state variables and filtering out erroneous measurements before they enter into the calculating process. As a result, the system conditions presented to the control room operators are guaranteed to be corrected even if some bad measurements exist.

A commonly used and familiar criterion in state estimation is that of minimizing the sum of the squares of the difference between the estimated and true (measured) value of a function . The ideas of least-squares estimation have been known and used since the middle part of the eighteenth century. In power systems, the state estimates variables are the magnitude of the voltage and relative phase angles at the system nodes. The input to the state estimator is the active power and reactive power, the state estimator is designed to give the best estimate of the voltages and phase angles minimizing the effect of the bad measurements.

Another valuable technique of state estimation is based on minimizing the absolute value of the difference between the measured and calculated quantities, and it is called least absolute value [LAV]. A few authors have developed a least absolute value based on linear programming in the estimation process. These estimators have not been widely accepted because they require excessive memory space and computing time.

A new method for least-absolute value state estimator developed in [33] which depends on minimizing the absolute value of the difference between the estimated variables and the measurements.

Also, it has the property that it detects and rejects the bad measurements in the process of obtaining the solution without any prior knowledge of the presence or locations of the bad measurements.

Thus one of the main objectives of this chapter is to show how the least absolute value estimators are a valuable alternative to least squares estimators, especially in regard to the robustness in the presence of bad measurements.

I. PARAMETER ESTIMATION PROBLEM

1.1. Introduction

The parameter estimation problem can be stated as :

Given the system measurement equation :

$$z = H \theta + v \qquad\qquad (\ 1.1 \ )$$

where z is an m×1 vector of system measurements (known)

θ is an n×1 vector of parameters to be estimated (unknown)

H is an m×n matrix which describes the mathematical relationship between the measurements and the system parameters vector (known)

v is an m×1 vector of measurement errors (unknown) to be minimized.

Determine the parameters vector θ, which minimizes the errors vector v.

If the number of measurements (m) equals to the number of unknown parameters (n), then an estimation of θ can be obtained as in (1.2).

$$\theta = H^{-1} z \qquad\qquad (\ 1.2 \ )$$

For this type of estimation the estimated parameters vector exactly fits the measurements set, i.e.,

$$z - H \theta = v = 0 \qquad\qquad (\ 1.3 \ )$$

Estimates obtained in this manner are of poor quality and not useful since this estimation process assume that the error vector v only contains zeros. Thus this type of estimation does not account for or filter out the measurement errors.

1.2. Least Error Squares (LS)

In most cases the number of measurements (m) exceeds the number of system parameters (n) where (m > n). Thus measurement errors can be filtered out in the estimation process and a good quality can be obtained.

Thus for a good quality estimate the least error squares method is used. In this method the objective is to minimize the sum of the squares of the residuals (the difference between the measured and calculated value). This can be mathematically expressed as :

$$J_2 (\theta) = \sum_{i=1}^{m} v_i^2 \qquad\qquad (1.4)$$

$$\text{or} \quad J_2 (\theta) = \sum_{i=1}^{m} (z_i - H_i \theta)^2 \qquad (1.5)$$

which can be written in vector form as :

$$J_2 (\theta) = [z - H \theta]^T [z - H \theta] \qquad (1.6)$$

Equation (1.6) can be written as :

$$J_2(\theta) = z^T z - \theta^T H^T z - z^T H \theta + \theta^T H^T H \theta \qquad (1.7)$$

For the cost function (1.7) to be a minimum, we put the gradient of J (θ) exactly equals zero to obtain :

$$\nabla J_2 (\theta) = - H^T z - H^T z + 2 H^T H \theta = 0 \qquad (1.8)$$

or $H^T H \theta = H^T z$

which gives

$$\theta^* = [H^T H]^{-1} H^T z \qquad (1.9)$$

where θ^* represents the least error squares estimates of θ.   To obtain θ^*, H H^T must be invertible. This is always true when H is of full rank.

1.3. Weighted Linear Least Squares Estimation

In least error squares if all measurements are treated equally,   then the less accurate measurements will affect the calculating process as much as the more accurate measurements. As a result, the final set of data obtained from the least squares estimation process will still contain large errors due to influence of bad measurements.

By introducing a weighting matrix to distinguish the more accurate measurements from the less accurate measurements, the calculating process can then force the results to coincide with the more accurate measurements. A sensible way of choosing the weights is to make them inversely proportional to the variances of the measurements. This approach means that a larger weighting is placed on these measurements with smaller variance (more accurate) and a smaller weighting on these measurements with larger variance (less accurate).

Thus the cost function that is to be minimized in this case is given as :

$$J_2 (\theta) = \sum_{i=1}^{m} \frac{(z_i - H_i \theta)^2}{\sigma_i^2}$$

$$J_2 (\theta) = \sum_{i=1}^{m} w_i (z_i - H_i \theta)^2 \qquad (1.10)$$

where σ_i is the standard deviation of the
            ith measuring device
            σ_i^2 is the variance of the ith measurements
            w_i is the weight assigned to the ith
            measurements

Equation (1.10) can be written in the form of :

$$J_2 (\theta) = [Z - H \theta]^T W [Z - H \theta] \qquad (1.11)$$

where   W is an m×m diagonal weighting matrix
          (inverse of the covariance matrix).

Similarly as shown in least error squares derivation, it can be shown that the weighted least squares estimation is given by :

$$\theta^* = [\ H^T \ W \ H \ ]^{-1} \ H^T \ W \ Z \qquad (\ 1.12 \ )$$

1.4. The Constrained Least Squares Estimation

In this section we discuss the solution of linear parameter estimation problem having linear constraints. To solve the problem we adjoin the equality constraints to the cost function via Lagrange's multiplier [10,30].

The constrained linear least squares problem can be stated as: Minimize :

$$J_2 (\ \theta \ ) = [\ z \ - \ H \ \theta \ ]^T \ [\ z \ - \ H \ \theta \ ] \quad (\ 1.13 \ )$$

subject to satisfying the linear constraints :

$$C \ \theta = d \qquad\qquad (\ 1.14 \ )$$

where   C   is an $\ell \times n$ matrix which represents the relation between θ and d (where $\ell < n$)
         d   is an $\ell \times 1$ vector which represent the constraints measurements.

We can form an augmented cost function by adjoining to the cost function of equation (1.13) the equality constrained of equation (1.14) via Lagrange's multiplier λ to obtain :

$$J_2 (\theta) = \frac{1}{2} [z - H \theta]^T [z - H \theta] +$$

$$+ \lambda^T [C \theta - d] \qquad\qquad (1.15)$$

The cost function of equation (1.15) can be written as :

$$J_2(\theta) = \frac{1}{2} [z^T z - z^T H \theta - \theta^T H^T z + \theta^T H^T H \theta] +$$

$$+ \lambda^T C \theta - \lambda^T d \qquad\qquad (1.16)$$

the cost function of equation (1.16) is a minimum when :

$$\frac{\partial J_2}{\partial \theta} = \frac{1}{2} [- H^T z - H^T z + 2 H^T H \theta] + C^T \lambda = 0$$

or $- H^T z + H^T H \theta + C^T \lambda = 0$ \qquad (1.17)

which gives :

$$\theta = [H^T H]^{-1} [H^T z - C^T \lambda] \qquad (1.18)$$

The Lagrange multiplier λ is to be determined such that the equality constraints of equation (1.14) are satisfied. Premultiplying equation (1.18) by C we obtain :

$$C \theta = C [H^T H]^{-1} [H^T z - C^T \lambda] = d \qquad (1.19.a)$$

$$\text{or} \quad d = C [H^T H]^{-1} H^T z - C [H^T H]^{-1} C^T \lambda \qquad (1.19.b)$$

which yields :

$$\lambda = \left[C [H^T H]^{-1} C^T \right]^{-1} \left[C [H^T H]^{-1} H^T z - d \right] \qquad (1.19.c)$$

Thus the parameter θ can be obtained by substituting equation (1.19.c) into (1.18) to obtain :

$$\theta = [H^T H]^{-1} \left[H^T z - C^T \left[C [H^T H]^{-1} C^T \right]^{-1} \times \right.$$

$$\left. \times \left[C [H^T H]^{-1} H^T z - d \right] \right] \qquad (1.20)$$

It can be noticed that if $C = 0$, there are no constraints, then θ can be written as :

$$\theta = [H^T H]^{-1} H^T z \qquad (1.21)$$

Which is the optimal estimate for the unconstrained least squares estimation.

1.5. Nonlinear Least Error Squares Estimation

In the previous sections we discussed the linear least squares estimation problem where there is a direct linear relationship between the measuring value and the estimation parameters so

that the solution is obtained directly without any
iteration.   If the relationship between the
measurements and the estimate parameters is
nonlinear, we need to linearize the cost function
by using the first order Taylor series expansion to
solve this problem [37].   In this section we discuss
how to solve the nonlinear parameter estimation
problem using the least error squares algorithm
explained earlier in the previous sections.

The nonlinear least squares problem is to
estimate the parameter θ which minimizes :

$$J_2(\theta) = \sum_{i=1}^{m} \frac{(z - f(\theta))^2}{\sigma_i^2} \qquad (1.22)$$

The gradient of $J_2(\theta)$ is given by:

$$\nabla J_2(\theta) = \begin{bmatrix} \dfrac{\partial J_2(\theta)}{\partial \theta_2} \\[2ex] \dfrac{\partial J_2(\theta)}{\partial \theta_2} \\[2ex] \dfrac{\partial J_2(\theta)}{\partial \theta_m} \end{bmatrix}$$

$$\nabla J_2(\theta) = 2 * \begin{bmatrix} \dfrac{\partial f_1}{\partial \theta_1} & \dfrac{\partial f_1}{\partial \theta_2} & - \\[2ex] \dfrac{\partial f_2}{\partial \theta_1} & \dfrac{\partial f_2}{\partial \theta_2} & - \\[2ex] \dfrac{\partial f_m}{\partial \theta_1} & \dfrac{\partial f_m}{\partial \theta_2} & - \end{bmatrix} \begin{bmatrix} \dfrac{1}{\sigma_1^2} & & \\[2ex] & \dfrac{1}{\sigma_2^2} & \\[2ex] & & \dfrac{1}{\sigma_m^2} \end{bmatrix} \begin{bmatrix} z_1 - f_1(\theta) \\[2ex] z_2 - f_2(\theta) \\[2ex] z_m - f_m(\theta) \end{bmatrix}$$

$$(1.23)$$

Equation (1.23) can be written in compact form as

$$\nabla J_2(\theta) = -2 H^T W \Delta z \qquad (1.24)$$

where

$$H = \begin{bmatrix} \dfrac{\partial f_1}{\partial \theta_1} & \dfrac{\partial f_1}{\partial \theta_2} & \dfrac{\partial f_1}{\partial \theta_3} & - - - \\[2ex] \dfrac{\partial f_2}{\partial \theta_1} & \dfrac{\partial f_2}{\partial \theta_2} & \dfrac{\partial f_2}{\partial \theta_3} & - - - \\[2ex] \dfrac{\partial f_m}{\partial \theta_1} & \dfrac{\partial f_m}{\partial \theta_2} & \dfrac{\partial f_m}{\partial \theta_3} & - - - \end{bmatrix}$$

$$= \text{Jacobian of } f(\theta)$$

$$W = \begin{bmatrix} \dfrac{1}{\sigma_1^2} & & \\[2ex] & \dfrac{1}{\sigma_2^2} & \\[2ex] & & \dfrac{1}{\sigma_m^2} \end{bmatrix} = \text{Weighting matrix}$$

$$
\Delta z = \begin{bmatrix} z_1 - f_1(\theta) \\ z_2 - f_2(\theta) \\ \\ z_m - f_m(\theta) \end{bmatrix}
$$

To make $\nabla J(\theta)$ equal zero , we apply Newton Raphson method.   Thus we obtain :

$$
\Delta\theta = \left[\frac{\partial \nabla J_2(\theta)}{\partial \theta} \right]^{-1} \left[-\nabla J_2(\theta) \right] \qquad (1.25)
$$

The Jacobian matrix of $\nabla J_2(\theta)$ is calculated by treating [H] as a constant matrix.   Thus :

$$
\frac{\partial \nabla J_2(\theta)}{\partial \theta} = -2 H^T W [-H]
$$

Then $\Delta\theta = \dfrac{1}{2} [H^T W H]^{-1} [2 H^T W \Delta z]$ (1.26.a)

or $\Delta\theta = [H^T W H]^{-1} [H^T W \Delta z]$ (1.26.b)

The procedures of the algorithm to solve the nonlinear state estimation problem are as follows :
1. Assume an initial guess for θ
2. Calculate Δz by using the initial guess
3. Calculate H matrix at this guess
4. Solve for $\Delta\theta$ by using equation (1.26.b)
5. If $\Delta\theta$ satisfies a specified terminating criterion, terminate the iteration, otherwise go to step
   6.

6. Update the value of θ as :

$$\theta_{new} = \theta_{old} + \Delta\theta \; ;$$

and go to step 2.

1.6. Properties of Least Error Squares Estimates

Least error squares estimation posses a number of interesting properties. The least squares are best estimates (maximum likelihood) when the measurement errors obey Gaussian or normal distribution and the weighting matrix equals the inverse of the covariance matrix. Also, in the case where the measurement error distribution does not obey Gaussian but the number of measurements greatly exceeds the number of unknown parameters, the method of least squares yields very good estimates.

Another valuable feature of least error squares estimation is that least squares estimates can be calculated, and equation of estimation can be easily implemented on the computer.

There are many estimation problems for which the error distribution is not Gaussian and the number of measurements does not greatly exceed the number of unknown parameters. In these cases, least error squares estimation are adversely affected by bad data. This problem has been recognized and addressed by several authors who have proposed different ways of refining the least error squares method to make estimation less affected by presence of bad data. In the next section we discuss an alternative technique which may be the best for the power system state estimation. This technique is based on least absolute value approximation.

1.7. Least Abolute VAlue State Estimation (LAV)

As mentioned earlier the least error squares estimation technique is based on minimizing the squares of the difference between the measured and the calculated values. In contrast the least absolute value estimation is based on minimizing the sum of the absolute values of the residuals. The basic difference in calculating the actual approximation using the least absolute value formulation and least error squares formulation is that with least absolute value, a best approximation is determined by interpolating a minimum subset of the available measurements while with the least error squares the best approximation is derived from the mean of the available measurements.

Least absolute value approximation techniques have been used for a long time. However, it is only recently that algorithms to calculate least absolute value approximation have been become available.

The main purpose of this section is to give some insight into the least absolute value approximation theory. The first section gives a historical perspective on the development of least absolute value approximation in which we discuss which is better least squares or least absolute value approximations. Then we discuss some of the algorithms that are available to obtain LAV state estimation. After that we introduce, based on LAV, a new algorithm to obtain the best state estimation. Finally we compare between least squares method and the new LAV technique through illustrative examples.

1.7.1. Historical Perspective

Kotiuga [22] in his paper goes through the historical development of least squares and its alternatives.

The development of least absolute value method as well as the least squares, can be tracked back to the middle of the eighteenth century. The development of least absolute value and least squares methods was a result of trying to find the best method to summarize the information obtained from a number of measurements. The pioneers in regression analysis, Bascovish (1757), Laplace (1781), Gauss (1809), and Glaisher (1872) proposed criteria for determining the best fitting straight line through three or more points.

In 1793 Laplace presented a procedure for finding the best set of measurements based on minimizing the sum of the absolute deviations namely least absolute value.

Gauss in 1809 showed that the method of least squares follows as a consequence of the Gaussian law of error (normal distribution). Glaisher (1871) later showed that for a Laplacian (double exponential) distribution the least absolute value estimation gives the most probable true value. In the early 1800's the work regression analysis

focussed on conditions under which least squares regression and least absolute value regression give the best estimates. Laplace in 1812 showed that for a large sample size the method of least squares is superior. Houber in 1830 evaluated the least squares and least absolute value estimators and showed that for a Gaussian distribution, the least squares estimator gives the best results but he also noted that one advantage of least absolute value estimator is that unbiased estimates can be obtained for any symmetrical distribution. However, Laurent in 1875 questioned the exactness of the Gaussian distribution and on the basis of actual studies of measurements concluded that the method of least squares should not be used when one has only small number of observations. Jefferys in 1939 showed that the equivalence between the following three statements:
i- the Gaussian distribution is correct
ii- the mean value is the best value
iii- the method of least squares gives the best estimates.

He also pointed out that for other symmetrical distribution, the method of least squares should not be used and stated that there is much to be said for the use of least absolute value when the distribution law is unknown because it is less affected by large residuals than the least squares.

The above debates took place before the advent of computers and fast efficient methods of calculating least squares and least absolute value estimates, and hence the discussions were primarily restricted to relatively small problems. Larger problems were primarily solved using analytical methods. In general, only least squares regression was used since there were no efficient techniques for obtaining least absolute value estimates. In the early 1950's the emphasis turned to computational conservations for solving large problems. However by this time the method of least squares was well established as the method for doing regression analysis.

The popularity of least squares continued to grow even though it was known that it does not lead to the best available estimates of unknown parameters when the law of error (distribution) is other than Gaussian. But if the number of independent observations available is much larger than the number of parameters to be estimated, the method of least squares can usually be counted on to yield nearly best estimates.

In summary, the results of research to date indicate that least absolute technique give better approximation when the error in the measurements set have unknown distribution and also when the sample size is small.

1.7.2. Least Absolute Value Algorithms

There are many algorithms available that can be used to get a best least absolute value state estimation. In this section, we discuss the most efficient use of these algorithms [29].

1.7.2.1. Least Absolute Value Based on Linear Programming

In this method we have an $m \times 1$ vector of measurements z and we try to solve the linear parameter estimation problem using the linear programming to evaluate an $n \times 1$ vector of unknown parameters θ [29].

The cost function in the case of least absolute value is given by :

$$J_1 (\theta) = \sum_{i=1}^{m} | z_i - \sum_{j=1}^{n} H_{ij} \theta_i | \qquad (1.27)$$

$$= \sum_{i=1}^{m} | r_i |$$

where r_i is the residual (error) and is given by :

$$r_i = z_i - \sum_{j=1}^{n} H_{ij} \theta_i \, , \, i = 1 \, , \, ..., \, m \qquad (1.28)$$

The formulation of linear programming problem can be carried out as follows :

Minimize the cost function of

$$J_1 (\theta) = \sum_{i=1}^{m} r_i \qquad\qquad (1.29)$$

subject to satisfying

$$r_i \geq 0 \quad, \quad i = 1, \ldots, \ldots m \qquad\qquad (1.30a)$$

$$r_i + \left| z_i - \sum_{j=1}^{n} H_{ij}\, \theta_i \right| \geq 0 \quad, \quad i = 1, \ldots, m \qquad (1.30b)$$

Equation (1.30) can be written as :

$$r_i \geq - \left| z_i - \sum_{j=1}^{n} H_{ij}\, \theta_i \right| \quad, \quad i = 1, \ldots, m \qquad (1.31)$$

which can be written as :

$$r_i \geq \mp \left(z_i - \sum_{j=1}^{n} H_{ij}\, \theta_i \right) \quad, \quad i = 1, \ldots, m \qquad (1.32)$$

$$\text{or} \quad r_i \geq z_i - \sum_{j=1}^{n} H_{ij}\, \theta_i \quad, \quad i = 1, \ldots, m \qquad (1.33)$$

and $r_i \geq \sum\limits_{j=1}^{n} H_{ij} \, \theta_i - z_j$, $i = 1, \ldots, m$ (1.34)

Thus the linear programming problem is to minimize equation (1.29) subject to satisfying the constraints given by equation (1.33) and (1.34).

From equations (1.33) and (1.34), if one of the two constraints is negative the other will be positive, and r_i must be positive in order to obey the linear programming requirements.

The following example demonstrates the formulation of a simple least absolute value estimation problem as a linear programming problem.

Example 1

Formulate the given least absolute value estimation problem as a linear programming.

Fit the given set of measurements $\left\{ (0 , 2), (0.1 , -2) , (2 , -1) , (2.2 , 1) \right\}$ .with a straight line in the form of $y = a x + b$.

Let $z = \begin{bmatrix} 2 \\ -2 \\ -1 \\ 1 \end{bmatrix}$, $H = \begin{bmatrix} 0.0 & 1 \\ 0.1 & 1 \\ 2.0 & 1 \\ 2.2 & 1 \end{bmatrix}$ and , $\theta = \begin{bmatrix} a \\ b \end{bmatrix}$

The cost function

$$J = \sum_{i=1}^{4} \left| z_i - \sum_{j=1}^{2} H_{ij} \theta_i \right| \qquad\qquad (1.35)$$

The least absolute value estimation problem is to calculate θ such that J in equation (1.35) is a minimum. The linear programming problem formulation is as follows :

define $r_i = z_i - \sum_{j=1}^{2} H_{ij} \theta_i$,   i= 1 , 2 , 3 , 4

minimize $J = \sum_{i=1}^{4} r_i$

subject to : $r_1 + b \geq 2$

$r_2 + 0.1\, a + b \geq -2$

$r_3 + 2\, a + b \geq -1$

$r_4 + 2.2\, a + b \geq 1$

and : $r_1 - b \geq -2$

$r_2 - 0.1\, a - b \geq 2$

$r_3 - 2\, a - b \geq 1$

$r_4 - 2.2\, a - b \geq -1$

In solving the problem for least absolute value estimation, feasible value of a , b and r_1 to r_4 will be found first. Then these values will be changed in order to iterative by decrease the value of the cost function . When the cost function reaches its minimum value, subject to satisfying the constraints, the estimation procedure is complete and the values of a and b represent the least absolute value estimate . The value of r_1 and r_4 represent the absolute value of residuals.

There are many other formulations of linear the programming problem based on least absolute value. However linear programming is the most popular method of calculating the least absolute values . But it also has some drawbacks as :

i- it requires excessive memory storage

ii- it is an iterative technique and thus may use considerable time and be computationally inefficient

iii- The solution obtained may not be unique

Some of the more recent algorithms [13,23 and 33] have attempted to overcome these difficulties and researches continue in this area.

1.7.2.2. Schlossmacher Iterative Algorithm

In 1973 Schlossmacher [6] presented an iterative technique which uses successive weighted least squares estimation to find least absolute value estimation.

The algorithm has the following steps :

1- obtain a weighted least squares estimate with all of the weighting factors set to one

2 - use the generated weighted least squares solution to calculate the residuals [r_i , i= 1 , 2 , 3 ,

```
.... , m ] ;
```

$$3 - \text{set} \quad W_i = \frac{1}{|r_i|} \quad , \quad i = 1 , 2 , 3 , .. , m$$

if any $r_i \simeq 0$ set $W_i = 0$;

4 - repeat step 2 and 3 until the change in the $r_i's$ between successive iteration approach zero.

Although   Schlossmacher's   technique   gives approximate estimates,   it is an iterative   technique and   has   been   criticized   as   been   computationally inefficient.

1.7.2.3. Sposito Algorithm

In   1976-1977   Sposito   et al. [8]   [14] suggested that least squares estimates can be used as starting points for linear programming based on least absolute value estimators.   Their   research   indicates that starting a linear programming problem based on a least absolute value estimator at the   least   squares estimate saves many iterations. They found   that   the total computing time equals to the time to   calculate the least squares estimate plus the   time   needed   to calculate the least   absolute   estimation   by   linear programming,   this computing time is   less   than   the time needed to calculate the linear programming based on least absolute value estimation from a flat   start.

The main drawback of their   technique   is   that   it still requires   a   linear   programming   algorithm   to calculate the least absolute value estimation.

In 1987 Soliman and Christensen [31],   [32], [35], developed a new technique for solving the least absolute value state estimation problem. Their algorithm will be discussed in detail in the next section.

1.7.3. A New Technique for Least Absolute Value Estimation

The new technique developed by Soliman and Christensen (1987) is characterized by its ease and simplicity to apply, therefore the computing time and storage required is small compared to other techniques.

As mentioned before, the least absolute value estimation for a given m measurements and n unknown parameters is based on minimizing the absolute value of the residuals as given in (1.36).

$$J_1(\theta) = \sum_{i=1}^{m} |r_i| = \sum_{i=1}^{m} |z_i - H_i \theta| \qquad (1.36)$$

The nature of interpolation of the new technique is given by the following theorem.

Theorem 1 [8,32]

If the column rank of the m×n matrix H is n, then there will exist a vector $\hat{\theta}$ corresponding to a best approximation that interpolates at least n *points of the measurements.*

This theorem simply states that , if we have m measurements z_i , i = 1 , 2 , ... , m and n unknowns then the optimal hyperplane based on LAV will pass through at least n points of the measurements set. This is in contrast to the least squares approximation which does not necessarily pass through any of the measurement points of the set z.

It can be noticed from this theorem that the least absolute value estimate interpolates n points. Thus the problem is reduced to selecting these points to minimize the cost function. The new technique assumes that H is a full column rank matrix

Given the measurements equation (1.37) the first step of the new algorithm is to calculate state parameters θ by using least squares technique which is given by equation (1.38)

$$z = H \, \theta + v \qquad\qquad\qquad (1.37)$$

$$\theta^* = [\ H^T\ H\ ]^{-1}\ H^T\ z \qquad\qquad (1.38)$$

The least squares residuals is calculated by equation (1.39)

$$r_i = z_i - H_i \, \theta^* , \quad i = 1 , 2 , ... , m \quad (1.39)$$

where r_i is the ith residual of the ith measurement.

z_i is the ith measurement

H_i is the ith row of the matrix H

The obtained estimate from least squares solution is more affected by erroneous measurements. To obtain estimates parameters less affected by the erroneous measurements, we need to reject the most erroneous measurements. This is carried out by rejecting the measurements which have residuals greater than the standard deviation given by equation (1.40).

$$\sigma^2 = \frac{1}{m - n + 1} \sum_{i=1}^{m} (r_i - r_{av})^2 \qquad (1.40)$$

where σ^2 is the variance of measurements

r_i is the ith residual

r_{av} is the average value of the least squares residuals

$$r_{av} = \frac{1}{m} \sum_{i=1}^{m} r_i$$

$$\sigma = \sqrt{\text{variance}} = \text{Standard deviation} \qquad (1.41)$$

After we reject the measurements which have residuals greater than the standard deviation, then the least squares solution is recalculated for the rest of measurements to have a new parametr estimate parameters which is less affected by the erroneous measurements.

Recalculate the least squares residuals using equation (1.39) The new residuals, for the rest of measurements generated from this solution are then ranked in ascending order by their absolute values and stored in an m×1 vector with the smallest residual coming first as r_1 and the largest residual comes at the end as r_m.

$$
r = \begin{bmatrix} r_1 \\ r_2 \\ r_n \\ \\ r_m \end{bmatrix} = \begin{bmatrix} \hat{r} \\ \overline{} \\ r^* \end{bmatrix} \begin{array}{l} n \times 1 \\ \\ \quad\quad m \times 1 \\ (m-n) \times 1 \end{array} \qquad (1.42)
$$

Also we rank the rows of measurements vector z and H matrix in the same order as the residual matrix.

$$
z = \begin{bmatrix} z_1 \\ z_2 \\ z_n \\ \\ z_m \end{bmatrix} = \begin{bmatrix} \hat{z} \\ \rule{2cm}{0.4pt} \\ z^* \end{bmatrix} \quad \begin{array}{c} n \times 1 \\ \\ (m-n) \times 1 \end{array} \right\} \; m \times 1 \qquad (1.43)
$$

$$
H = \begin{bmatrix} H_1 \\ H_2 \\ H_n \\ \\ H_m \end{bmatrix} = \begin{bmatrix} \hat{H} \\ \rule{2cm}{0.4pt} \\ H^* \end{bmatrix} \quad \begin{array}{c} n \times n \\ \\ (m-n) \times n \end{array} \right\} \; m \times n \qquad (1.44)
$$

From equations (1.43), (1.44), we select a subset vector of z and H corresponding to the n smallest residuals which are $\hat{z}$ and $\hat{H}$.

The final step of the algorithm is to calculate the least absolute value estimation by using equations (1.45) and (1.46)

$$
\hat{z} = \hat{H} \, \hat{\theta} \qquad\qquad (1.45)
$$

or

$$
\hat{\theta} = \hat{H}^{-1} \, \hat{z} \qquad\qquad (1.46)
$$

Once the final estimates are calculated the least absolute value measurements residuals can be calculated by using (1.47)

$$\hat{r} = z - H \hat{\theta}$$ (1.47)

From equations (1.37) to (1.46)   it can easily be seen that the first n residuals of (1.46) will be equal zero . The other m-n residuals will be either zero or non zero.

It can be also noticed that if we have m measurements with no errors , the least squares and the least absolute value solutions will be the same. But   if the measurement set contains bad data,   then in calculating the residuals of least squares in (1.39)   the bad measurements will have the   largest residuals   then in the arranging step these residuals will not be included in the sub vector $\hat{r}$. Thus   the least absolute value estimates   will not contain bad data.   This is a nice   property of the new   technique   that it rejects   the   outlier   measurements   in   the solution process without any prior   knowledge   of   the bad measurements and their locations.

The steps in the new technique can be   summarized as :

1- Calculate the least squares estimates θ using :

$$\theta^* = [H^T H]^{-1} H^T z$$

2- Calculate the least squares estimation residuals using :

$$r_i = z_i - H_i \theta^* ,   i = 1 , 2 , \ldots , m$$

3- Reject the outlier which having residuals greater than the standard deviation σ and recalculate the least squares solution and the new corresponding residuals for the rest of measurements

4- Select the n measurements that corresponding to the smallest least squares residuals and form the corresponding $\hat{z}$ and $\hat{H}$

5- Solve for the least absolute value estimates $\hat{\theta}$ using :

$$\hat{\theta} = \hat{H}^{-1} \hat{z}$$

There is another useful property of least absolute value estimation in theorem 2.

Theorem 2 [8,32]
_____

For a least absolute value estimate of n parameters, if n_1 is the number of positive residuals and n_2 is the number of negative residuals then an optimal least absolute value estimate $\hat{\theta}$ obeys the following equation

$$| n_1 - n_2 | \leq n \qquad\qquad (1.48)$$

Theorem 3.2 gives necessary but not sufficient conditions for optimality. Thus least absolute value estimates that are produced by the new technique or by any other technique can be checked for optimality using theorem 3.2. If $\mid n_1 - n_2 \mid \leq n$, then the estimates may be optimal However , if $\mid n_1 - n_2 \mid >$ n, then the estimates are definitely not optimal.

1.7.3.1. Non-Linear Estimation Using the New Technique

The non linear parameter estimation is introduced in section (1.5). Recall the LAV cost function of the non linear parameters as:

$$J_1(\theta) = \sum_{i=1}^{m} \mid \Delta z_i - H_i \, \Delta\theta_i \mid \qquad (1.49)$$

where $\qquad \Delta z_i = z_i^o - H_i \, \hat{\theta}^o \qquad\qquad (1.50)$

To solve the non linear parameter estimation problem the following steps are used :

Step 1   Assume an initial guess of θ^o

Step 2   Linearize the state equations around the initial estimate θ^o

Step 3   Calculate the change in estimation parameter $\Delta\theta$ using the LS algorithm as :

$$\Delta\theta^* = [\ H^T \ H \ ]^{-1} \ H^T \ \Delta z$$

Step 4   Update the state estimation parameters

$$\theta_{nev} = \theta_{old} + \Delta\theta^*$$

Step 5   If $\Delta\theta^*$ is less than a predetermined terminating criteron go to step 6 otherwise go to step 2

Step 6   Calculate the least squares estimation residuals using

$$r_i = \Delta z_i - H \Delta\theta_i^* \ , \quad i = 1 , 2 , \dots , m$$

Step 7   Calculate the standard deviation of these residuals

Step 8   Reject the outliers having residuals greater than the standard deviation .

Step 9   Repeat step 2 to step 6 . for the rest of measurements

Step 10   Rank the new residuals in ascending order

Step 11   Select the n measurements that correspond to the smallest n residuals and form $\hat{\Delta z}$ and $\hat{H}$

Step 12   Calculate the change in least absolute value estimates using

$$\hat{\Delta\theta} = \hat{H}^{-1} \Delta z$$

Step 13   Update the state estimation parameters

$$\theta_{new} = \theta_{old} + \hat{\Delta\theta}$$

Step 14   If $\Delta\hat{\theta}$ is less than a prescribed terminating
_____ criteria then terminate the iteration,   go
           to step 15, otherwise calculate the change
           in Δz vector   and   H   matrix   then
           recalculate the new residuals and go to step
           10

Step 15   Print the least absolute value estimates
_____

The above steps of solution is summarized in the
flow chart which is   given in Fig. 1.1 .

1.7.3.2. Constrained   Estimation   Using   the   New Technique

The constrained state estimation problem can be
handled by the new proposed technique [31].   If we
have an m measurements and ℓ constraints , n > ℓ , the
new technique will interpolate at least n points of
the given measurements.   The constraints represent a
good measurements so that the residuals of the least
squares solution for these constraints will be zero.
The least absolute value technique   must interpolate
the ℓ constraints before interpolation of (n $-\ell$)   of
the other measurements.   The   total   number   of   the
interpolated points will equal (n $-$ ℓ + n = n).
Thus the   new   method   will   select directly the ℓ
constraints and the (n $-$ ℓ)   measurements
corresponding to the smallest LS residuals.

Note that the number of constraints should be
less than the number of unknowns otherwise the least

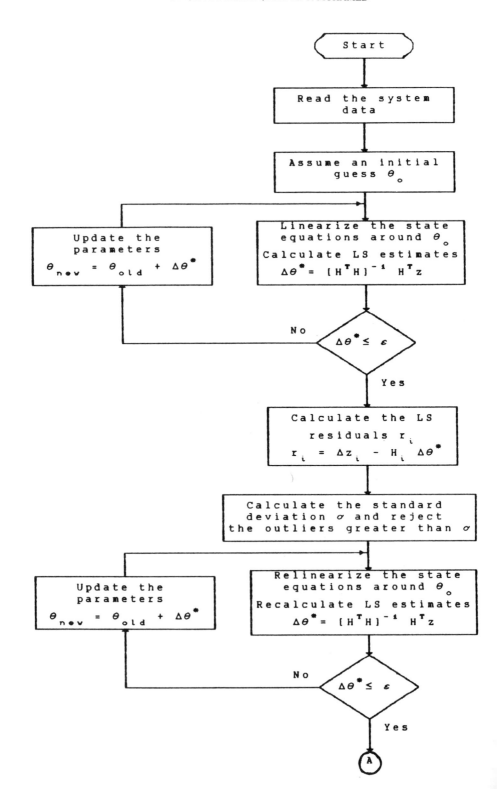

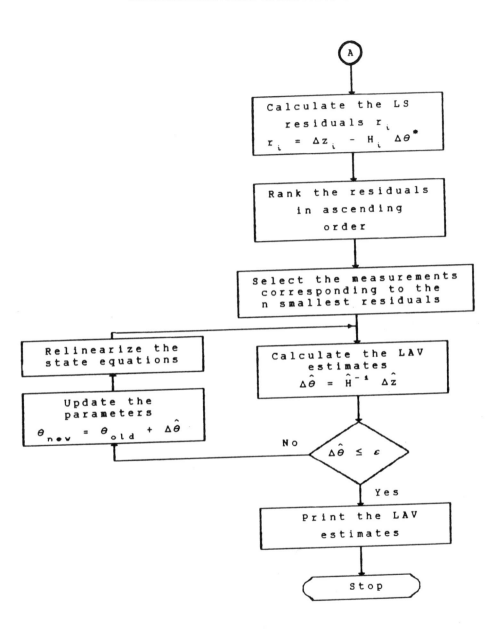

Fig. 1·1   Flow Chart of Non-Linear Estimation
Using the New Technique

absolute value will interpolate the n points from the constraints equations only.

The solution technique may use the method proposed in the previous section for LS parametersestimation with constraints and then proceed in the same manner as the LAV technique to obtain the least absolute value optimal solution.

1.7.4. Comparison Between Least Squares Estimation and The New Technique of Least Absolute Value Estimation

After studying the main properties of least squares parameter estimation and the least absolute value parameter estimation technique we can conclude the following :

The main difference in solution of the least squares and least absolute value parameter estimation technique is that the least absolute value interpolates a subset of the available measurements, where the least squares interpolates the mean of the available measurements [i.e., the least squares may not fit any of the available measurements if the measurements set is contaminated with bad data].

If there is an erroneous measurement, the least squares will give an optimal estimation when the error distribution obeys Gaussian distribution or when there is a great redundancy in the available measurements. But if the measurements error obeys another distribution and the sample of measurements is small, the solution of least squares will not be

optimal. On the other hand, the least absolute value
gives the optimal solution for any measurements error
distribution and   small sample of measurements.
Thus the least absolute value will be   the best when
the error distribution is unknown and   the sample of
the measurements is small [20].

The least squares technique is always biased by
the presence of bad data, the least squares technique
does not detect or identify the bad data during the
solution process but it can be done separately after
the least squares solution. In the new least absolute
value technique the bad data are identified and
rejected automatically during the solution process.

Now   it is clear that the new technique of least
absolute value is more suitable for power systems
static state estimation problem.   To support this
conclusion we will introduce some numerical examples
from curve fitting to compare the least error squares
and least absolute value parameter estimation
algorithms.

Example 2.
_____

Fit the data point $\left\{ (1 , 3) , (3 , 6) , (4, 6) , (6 , 12) , (9 , 11) \right\}$ with a straight line
of form $y = a_1 x + a_2$ (the data represent exactly the
line $y = x + 2$. Except points 2 and 4)

$$
z = \begin{bmatrix} Y_1 \\ Y_2 \\ Y_3 \\ Y_4 \\ Y_5 \end{bmatrix} = \begin{bmatrix} 3 \\ 6 \\ 6 \\ 12 \\ 11 \end{bmatrix} , \quad H = \begin{bmatrix} 1 & 1 \\ 3 & 1 \\ 4 & 1 \\ 6 & 1 \\ 9 & 1 \end{bmatrix} , \quad \theta = \begin{bmatrix} a_1 \\ a_2 \end{bmatrix}
$$

Step 1   Calculate the least error squares estimates
θ^* by using

$$
\theta^* = [\ H^T\ H\ ]^{-1}\ H^T z = \begin{bmatrix} a_1 \\ a_2 \end{bmatrix} = \begin{bmatrix} 1.10753 \\ 2.50538 \end{bmatrix}
$$

Step 2   Calculate the least error squares residuals
generated from this solution

$$
r = z - H\ \theta^*
$$

$$
r = \begin{bmatrix} -0.6129 \\ 0.1720 \\ -0.9355 \\ 2.8495 \\ -1.4731 \end{bmatrix}
$$

Step 3   Calculate the standard deviation of
residuals by using

$$
\sigma = \sqrt{ \frac{1}{m - n + 1} \sum_{i=1}^{n} (\ r_i - r_{av}\ )^2 }
$$

where $r_{av} = \sum\limits_{i=1}^{m} \dfrac{r_i}{m}$ to obtain :

$\sigma = 1.70073$

Step 4   Reject the measurements which have residuals greater than standard deviation σ. From the values of r the residual greater than the standard deviation belong to point 4.   If this measurement is rejected then,   the new measurement vector z and H matrix will be as follows :

$$z = \begin{bmatrix} 3 \\ 6 \\ 6 \\ 11 \end{bmatrix} \quad , \quad H = \begin{bmatrix} 1 & 1 \\ 3 & 1 \\ 4 & 1 \\ 9 & 1 \end{bmatrix}$$

Step 5   Recalculate the least squares estimates θ^*.

$$\theta^*_{new} = [\ H^T\ H\ ]^{-1}\ H^T\ z\ = \begin{bmatrix} 0.96403 \\ 2.40288 \end{bmatrix}$$

Step 6   Recalculate the least error squares residuals generated from this solution.

$$r_{new} = z - H\ \theta^*_{new}, \quad \text{which gives ;}$$

$$r_{new} = \begin{bmatrix} -0.36691 \\ 0.70504 \\ -0.25899 \\ -0.07913 \end{bmatrix}$$

Step 7   *Since the rank of the matrix H is 2,   the estimator fits at least two points   and these two points having the smallest LS residuals. These smallest residuals are r_4 and r_3 . Form $\hat{z}$ and $\hat{H}$ as :*

$$\hat{z} = \begin{bmatrix} 11 \\ 6 \end{bmatrix} \quad , \quad \hat{H} = \begin{bmatrix} 9 & 1 \\ 4 & 1 \end{bmatrix}$$

Step 8   *Solve for the least absolute value estimates by using :*

$$\hat{\theta} = \hat{H}^{-1} \hat{z}$$

which gives :

$$\hat{\theta} = \begin{bmatrix} 9 & 1 \\ 4 & 1 \end{bmatrix}^{-1} \begin{bmatrix} 11 \\ 6 \end{bmatrix}$$

$$\hat{\theta} = \begin{bmatrix} 1 \\ 2 \end{bmatrix}$$

Step 9   *Calculate the least absolute value residuals by using :*

$$\hat{r} = z - H \hat{\theta}$$

$$\hat{r} = \begin{bmatrix} 0.0 \\ 1.0 \\ 0.0 \\ 4.0 \\ 0.0 \end{bmatrix} \quad , \text{ with } \| r \| = 5.0$$

From this example we notice that the first least squares solution is more affected by the presence of bad data (point 2 and 4)   but the second solution is less affected by bad data and gives more acceptable solution than the first solution.   The rejection of bad data by using the standard deviation gives an effective method for rejecting the bad data and determinating the number of bad data which must be rejected to obtain a suitable solution. All residuals of least absolute value are zero except the measurement number 2 and if we calculate the residual for point 4   which was rejected before the second solution of least squares,   it will be equal to 4 [i.e., the least absolute value technique fits all but the 2nd and 4th measurements]. Thus the two bad data points which do not fall on the line $y = x + 2$ are rejected by the new algorithm. In contrast the two solutions of the least squares estimates do not interpolate any of the data points and are affected by bad data. Fig. 1.2   demonstrate the data points, the least absolute value estimates, and least squares estimates for the first and the second solution.

Example 3

Fit the set of measurements $\{$ (0 , 1) , (1 , 3) , (2 , - 5) , (3 , - 5) , (4 , - 3) , (5 , 1) , (6 , 1) , (7 , 15) $\}$ with a polynomial of the form $y = a_1 x^2 + a_2 x + a_3$ (the data represent the polynomial $y = x^2 - 5 x + 1$ except for points 2 and 7)

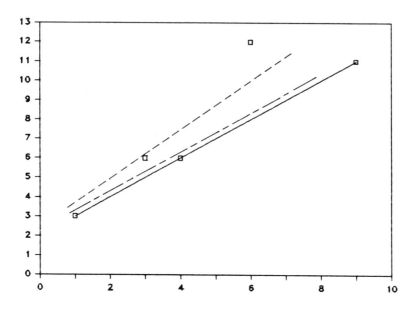

Legend		
1ˢᵗ least squares estimates $y = 1.107\ x + 2.51$		– – – – –
2ⁿᵈ least squares estimates $y = 0.964\ x + 2.40$		—— – ——
Least absolute value estimates $\quad y = x + 2$		——————
Data points		▫

Fig. 1·2   Least Absolute Value and Least Squares

Estimates For Example   2

$$
z = \begin{bmatrix} 1 \\ 3 \\ -5 \\ -5 \\ -3 \\ 1 \\ 1 \\ 15 \end{bmatrix} , \quad H = \begin{bmatrix} 0 & 0 & 1 \\ 1 & 1 & 1 \\ 4 & 2 & 1 \\ 9 & 3 & 1 \\ 16 & 4 & 1 \\ 25 & 5 & 1 \\ 36 & 6 & 1 \\ 49 & 7 & 1 \end{bmatrix} , \quad \theta = \begin{bmatrix} a_1 \\ a_2 \\ a_3 \end{bmatrix}
$$

Step 1   Calculate the least error squares estimates θ^* by using

$$
\theta^* = [\ H^T H\ ]^{-1} H^T z = \begin{bmatrix} 1.00595 \\ -5.69643 \\ 3.45830 \end{bmatrix}
$$

Step 2   Calculate the least error ssquares residuals using

$$
r = z - H \theta^*
$$

$$
r = \begin{bmatrix} -2.4583 \\ 4.2321 \\ -1.0893 \\ -0.4226 \\ 0.2321 \\ 0.875 \\ -3.494 \\ 2.125 \end{bmatrix}
$$

Step 3   Calculate the standard deviation σ using :

$$\sigma = \sqrt{\frac{1}{m - n + 1} \sum_{i=1}^{8} (r_i - r_{av})^2}$$

$\sigma = 2.6727$

Step 4   Reject the measurements having residuals greater than the standard deviation σ. Measurements number 2 and 7 have residuals greater than the standard deviation and they will be rejected to obtain :

$$z = \begin{bmatrix} 1 \\ -5 \\ -5 \\ -3 \\ 1 \\ 15 \end{bmatrix} \qquad H = \begin{bmatrix} 0 & 0 & 1 \\ 4 & 2 & 1 \\ 9 & 3 & 1 \\ 16 & 4 & 1 \\ 25 & 5 & 1 \\ 49 & 7 & 1 \end{bmatrix}$$

Step 5   Recalculate the least error squares estimates θ^*

$$\theta^*_{new} = [H^T H]^{-1} H^T z = \begin{bmatrix} 1 \\ -5 \\ 3 \end{bmatrix}$$

Step 6   Recalculate the least error squares residuals

$$r^T = [0.0 \quad 0.0 \quad 0.0 \quad 0.0 \quad 0.0 \quad 0.0]$$

Step 7   *Select any three measurement to solve for*

least   absolute   value.   Since   all

measurements have zero residuals.

$$\hat{\theta} = \begin{bmatrix} 0 & 0 & 1 \\ 4 & 2 & 1 \\ 9 & 3 & 1 \end{bmatrix}^{-1} \begin{bmatrix} 1 \\ -5 \\ -5 \end{bmatrix} = \begin{bmatrix} 1 \\ -5 \\ 1 \end{bmatrix}$$

with $\hat{r}^{T} = [0.0 , 6.0 , 0.0 , 0.0 , -6.0 , 0.0]$,

$\| \hat{r} \| = 12$

This example demonstrates the powerful use of the rejection of bad data by using the standard deviation method . Where all bad data (points 2 and 7) are rejected and the measurements used to solve the least squares for the 2nd time do not contain any bad data, so we expected that the least squares estimates will fit all the data points as well as the least absolute value estimates.

When we rank the residuals there are some cases in which the absolute value of two residuals are equal, but there is only one position for the corresponding measurement in the interpolated measurements set. In this case we implement the tie-breaking procedure [35]. The two measurements which are equal constitute a tie. Then we have two least absolute value estimates which are corresponding to two different interpolated sets of measurements. Each set contains the same measurement (corresponding to the n smallest least squares residuals) with the only difference being that each set contains one of the two measurements that was involved in the tie. The cost function of both estimates are then calculated and compared. The estimate with smaller value of cost function is the unique least absolute value estimate, that is produced by the new technique and the tie-breaking procedure.

If more than two residuals are equal, a tie-breaking procedure can be implemented but there will be more than two cost functions.

In the following example we will demonstrate the case when the tie-breaking procedure must be implemented.

Example 4

Fit the data $\left\{ \; (\; 0 \; , \; 2 \;) \; , \; (\; 1 \; , \; 5 \;) \; , \; (\; 2 \; , \; 7 \;) \right.$ $\left. , \; (\; 3 \; , \; 7 \;) \; , \; (\; 4 \; , \; 11 \;) \; , \; (\; 5 \; , \; 15 \;) \; \right\}$, with the line of form $y = a \; x + b$

$$
z = \begin{bmatrix} 2 \\ 5 \\ 7 \\ 7 \\ 11 \\ 15 \end{bmatrix} \quad , \quad H = \begin{bmatrix} 0 & 1 \\ 1 & 1 \\ 2 & 1 \\ 3 & 1 \\ 4 & 1 \\ 5 & 1 \end{bmatrix} \quad , \quad \theta = \begin{bmatrix} a \\ b \end{bmatrix}
$$

Step 1   Calculate the least error squares estimates θ^*

$$\theta^* = [\ H^T\ H\ ]^{-1}\ H^T\ z = \begin{bmatrix} a \\ b \end{bmatrix} = \begin{bmatrix} 2.3714 \\ 1.9047 \end{bmatrix}$$

Step 2   Calculate the least error squares residuals

$$r = z - H\ \theta^*$$

$$r = \begin{bmatrix} 0.0952 \\ 0.7238 \\ 0.3524 \\ -2.0190 \\ -0.3905 \\ 1.2381 \end{bmatrix}$$

Step 3   Calculate the standard deviation

$$\sigma = \sqrt{\frac{1}{m-n+1} \sum_{i=1}^{6} (\ r_i - r_{av}\ )^2}$$

$$\sigma = 1.133$$

Step 4   Reject the measurement having residuals greater than the standard deviation value and form the new z vector and H matrix. The points to be rejected are points #4 and #6 .

$$z = \begin{bmatrix} 2 \\ 5 \\ 7 \\ 11 \end{bmatrix} \qquad H = \begin{bmatrix} 0 & 1 \\ 1 & 1 \\ 2 & 1 \\ 4 & 1 \end{bmatrix}$$

Step 5   Recalculate   the   least   error   squares

estimates θ^*

$$\theta^*_{new} = \begin{bmatrix} a \\ b \end{bmatrix} = \begin{bmatrix} 2.2 \\ 2.4 \end{bmatrix}$$

Step 6   Recalculate   the   least   error   squares

residuals

$$r_{new} = \begin{bmatrix} -0.4 \\ 0.4 \\ 0.2 \\ -0.2 \end{bmatrix}$$

Step 7   Rank the residuals in ascending order and
select the two measurements corresponding
to the smallest residuals.   Here   the 3$\underline{rd}$
and 4$\underline{th}$ measurements havethesame absolute
value of residual. Thus   the   tie-breaking
procedure must be implemented. Also   the
1$\underline{st}$ and 2$\underline{nd}$ measurements have   equal
absolute value of residual. In this case
four estimates must be calculated as :

$$\hat{\theta}_1 = \begin{bmatrix} 2 & 1 \\ 0 & 1 \end{bmatrix}^{-1} \begin{bmatrix} 7 \\ 2 \end{bmatrix} = \begin{bmatrix} 2.5 \\ 2.0 \end{bmatrix}$$

$$\hat{\theta}_2 = \begin{bmatrix} 2 & 1 \\ 1 & 1 \end{bmatrix}^{-1} \begin{bmatrix} 7 \\ 5 \end{bmatrix} = \begin{bmatrix} 2 \\ 3 \end{bmatrix}$$

$$\hat{\theta}_3 = \begin{bmatrix} 4 & 1 \\ 0 & 1 \end{bmatrix}^{-1} \begin{bmatrix} 11 \\ 2 \end{bmatrix} = \begin{bmatrix} 2.25 \\ 2.00 \end{bmatrix}$$

$$\hat{\theta}_4 = \begin{bmatrix} 4 & 1 \\ 1 & 1 \end{bmatrix}^{-1} \begin{bmatrix} 11 \\ 5 \end{bmatrix} = \begin{bmatrix} 2 \\ 3 \end{bmatrix}$$

Step 8   Calculate the least absolute value residuals and calculate the cost function in each case

J_1 = 1.50
J_2 = 1.00
J_3 = 1.25
J_4 = 1.00

The cost function J_2 and J_4 are equal since $\hat{\theta}_2$ and $\hat{\theta}_4$ are identical. Therefore the value of $\hat{\theta}_2$ and $\hat{\theta}_4$ arethe unique estimatesproduced by the new technique and the tie-breaking procedure.

If a case occur in which the two cost functions are equal and the corresponding estimates are different then both estimates are equally valid. Under such circumstances the estimates produced by the new technique will not be unique. Extensive testing of the new technique has demonstrated that this case rarely occur and that the estimates are always unique.

In the next example we will apply the new technique for a problem that contains equality constraints.

Example 5

Fit the point $\Big\{$ (-1 , 2) , (0 , 5) , (1 ,

6) , (2 , 13) , (3 , 20) , (4 , 29) $\Big\}$, with

the polynomial of form $y = a_1 x^2 + a_2 x + a_3$,
subjected to the constraint $y (- 2) = 5$

$$
z = \begin{bmatrix} 2 \\ 5 \\ 6 \\ 13 \\ 20 \\ 29 \end{bmatrix} \quad , \quad H = \begin{bmatrix} 1 & -1 & 1 \\ 0 & 0 & 1 \\ 1 & 1 & 1 \\ 4 & 2 & 1 \\ 9 & 3 & 1 \\ 16 & 4 & 1 \end{bmatrix} \quad , \quad \theta = \begin{bmatrix} a_1 \\ a_2 \\ a_3 \end{bmatrix}
$$

The equality constraint can be written as :

$$C \theta = d$$
$$C = [4 \qquad -2 \qquad 1] \qquad , \qquad d = [5]$$

The constrained problem can be solved by least squares
as mentioned earlier.

Step 1
Calculate the least error squares estimates
θ^* by using

$$
\theta^* = [H^T H]^{-1} \left[H^T z - C^T \left[C [H^T H]^{-1} C^T \right]^{-1} \left[C [H^T H]^{-1} H^T z - d \right] \right]
$$

$$\theta^* = \begin{bmatrix} a_1 \\ a_2 \\ a_3 \end{bmatrix} = \begin{bmatrix} 1.136 \\ 1.796 \\ 4.049 \end{bmatrix}$$

Step 2   Calculate the least error squares residuals

$$r = z - H\,\theta^*$$

$$r = \begin{bmatrix} -1.388 \\ 0.951 \\ -0.982 \\ 0.812 \\ 0.355 \\ -0.415 \end{bmatrix}$$

Step 3   Calculate the standard deviation of the residuals

$$\sigma = \sqrt{ \frac{1}{m - n + 1} \sum_{i=1}^{6} (r_i - r_{av})^2 }$$

Where $r_{av} = \dfrac{1}{m} \displaystyle\sum_{i=1}^{6} r_i$

$$\sigma = 1.0796$$

Step 4   Reject measurements having residuals greater than the standard deviation   and form the new z vector and H matrix for the rest of measurements. Measurement number 1 will be rejected.

$$z = \begin{bmatrix} 5 \\ 6 \\ 13 \\ 20 \\ 29 \end{bmatrix} \quad , \quad H = \begin{bmatrix} 0 & 0 & 1 \\ 1 & 1 & 1 \\ 4 & 2 & 1 \\ 9 & 3 & 1 \\ 16 & 4 & 1 \end{bmatrix}$$

Step 5   *Recalculate the least error squares*
estimates by using the same expression of
step 1 as:

$$\theta^{*}_{new} = \begin{bmatrix} a_1 \\ a_2 \\ a_3 \end{bmatrix} = \begin{bmatrix} 1.092 \\ 1.851 \\ 4.333 \end{bmatrix}$$

Step 6   *Recalculate the least error squares*
residuals

$$r_{new} = \begin{bmatrix} 0.666 \\ -1.277 \\ 0.594 \\ 0.282 \\ -0.215 \end{bmatrix}$$

Step 7   *Rank the residuals and select the two*
measurements corresponding to the smallest
residuals together withthe constrainet
since the rank of H is 3.   Form $\hat{z}$ and $\hat{H}$ The
smallest residuals are the 5th and the 4th
residuals.

$$\hat{z} = \begin{bmatrix} 29 \\ 20 \\ 5 \end{bmatrix} \quad , \quad \hat{H} = \begin{bmatrix} 16 & 4 & 1 \\ 9 & 3 & 1 \\ 4 & -2 & 1 \end{bmatrix}$$

Step 8   *Solve for the least absolute value estimates $\hat{\theta}$ by using*

$$\hat{\theta} = \hat{H}^{-1} \hat{z}$$

$$\hat{\theta} = \begin{bmatrix} a_1 \\ a_2 \\ a_3 \end{bmatrix} = \begin{bmatrix} 1 \\ 2 \\ 5 \end{bmatrix}$$

Step 9   Calculate the LAV residuals vector generated from this solution as :

$$\hat{r}^T = [\ 2.0 \quad 0.0 \quad 2.0 \quad 0.0 \quad 0.0 \quad 0.0 \ ] \ , \ \| \hat{r} \| = 4$$

II. POWER SYSTEM MODELING AND OBSERVABILITY

2.1. Introduction

The concept of power system state estimation means the ability to characterize completely and reliably the power system at any given moment. The power system state estimation is based on collecting measurements from the network. These measurements are filtered and used in determination of the system states. Thus, before applying any estimation technique, a complete power system model must be developed. Specifically the relationship between the measurements, the state variable, and the system parameters must be described by equations.

Consider the π-section network elements of the transmission line model between buses i and j and the corresponding simplified model in fig. 2.1.

The injection active and reactive power at bus i is given by the following equations :

$$P_i = \sum_{j=1}^{N} V_i V_j Y_{ij} \cos (\delta_i - \delta_j - \beta_{ij}) , \quad i = 1 ,.., N \quad (2.1)$$

$$Q_i = \sum_{j=1}^{N} V_i V_j Y_{ij} \sin (\delta_i - \delta_j - \beta_{ij}) , \quad i = 1 ,.., N \quad (2.2)$$

The flow of active and reactive power from bus i to bus j is given by the following equations :

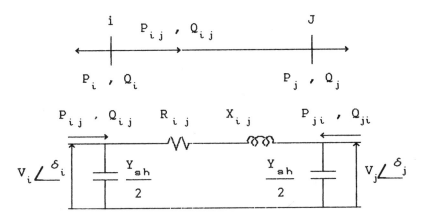

Fig. 2.1 Transmission Line Model

$$P_{ij} = - V_i^2 \ Y_{ij} \ \cos \ (\ \beta_{ij} \ ) \ +$$

$$+ \ V_i \ V_j \ Y_{ij} \ \cos \ (\ \delta_i \ - \ \delta_j \ - \ \beta_{ij} \ ) \qquad (\ 2.3 \ )$$

$$Q_{ij} = V_i^2 \ Y_{ij} \ \sin \ (\ \beta_{ij} \ ) \ - \ V_i^2 \ Y_{sh} \ +$$

$$+ \ V_i \ V_j \ Y_{ij} \ \sin \ (\ \delta_i \ - \ \delta_j \ - \ \beta_{ij} \ ) \qquad (\ 2.4 \ )$$

Where   N   is the number of buses in power system

P$_i$ is the active power injected into bus i

Q$_i$ is the reactive power   injected   into
            bus i

P$_{ij}$ is the active power flow from bus   i   to
            bus j

Q$_{ij}$ is the reactive power flow from bus i   to
            bus j

V$_i$ is the magnitude of the voltage of bus i

δ_i is the phase angle of   the   voltage   of
            bus i

Y$_{ij}$ is the magnitude of element (i,j) of the
            bus admittance matrix

β_{ij} is the phase angle of element (i,j)   of
            the bus admittance matrix

Y$_{sh}$ is the total of the magnitude of the
            shunt capacitance of line i , j

At this point we have complete models of the
system injection and flow powers which are functions
of system parameters. These parameters are always
available and need not to be determined such as Y_{ij},

β_{ij} , and Y_{sh} . The states which are to be determined from the estimation process are the voltage magnitude and phase angle at each bus. The parameters which can be obtained by the measuring devices on the system network are the active injected power, reactive injected power, active power flow and reactive power flow.

In power system the measured quantities P and Q are related to the state variables V and θ that are to be estimated by the set of non linear equations of (2.1) to (2.4). These non linear form are not suitable for the state estimation process that requires linear relationships. Therefore these equations must be linearized. Let z_i , i = 1 , ..,m define the measured variables, while the state variables are defined by x_j , j = 1 , .. , n (m > n). Each measurements z_i is related to the state variables x_j by the non linear equation

$$z_i = F_i (x) , i = 1 , .. , m \qquad (2.5)$$

Where   x   is the nx1   states vector to be estimated. This vector includs the N - 1 voltage magnitude vector | V | and N - 1 voltage phase angle δ of each bus

n   is the number of the states to be estimated which equals to 2(N - 1)

Assume that x_i^o , i = 1 , .. , n is the initial estimates of the system states, then by using the first order Taylor series expansion, equation (2.5) can be written as :

$$z_j = z_j^o + \sum_{i=1}^{n} \frac{\partial F(x)}{\partial x_i}\bigg|_{x_i = x_i^o} \Delta x + v_j \quad ;$$

$$j = 1, \ldots, m \quad (2.6)$$

The above equation can be written as :

$$\Delta z_j = z_j - z_j^o = \sum_{i=1}^{n} H_{ji} \Delta x_i + v_j \quad ;$$

$$j = 1, \ldots, m \quad (2.7)$$

Where v_j is the error to be minimized which contains the noise in the available measurements and the negligible higher terms of Taylor expansion

H_{ji} is the elements of the Jacobian matrix which are given by :

$$H_{ji} = \frac{\partial F(x)}{\partial x_i}, \quad i = 1, \ldots, n, \quad j = 1, \ldots, m \quad (2.8)$$

In a vector form equation (2.7) can be written as :

$$\Delta z = H \Delta x + v \quad (2.9)$$

Where Δz is an m×1 vector of elements $\Delta z_i = z_i - z_i^o$

Δx is an n×1 vector of states $\Delta x_j = x_j - x_j^o$

v is an m×1 vector of elements v_j

H   is an m×n matrix represents the   Jacobian
matrix of the power system with   elements
H_{ji} of equation (2.8)

Before applying any estimation   techniques   the
question now is   how do we know that the taken
measurements   can be used in solving the power system
state estimation problem?   The   answer   of   this
question   is   known   by   studying   the   system
observability.

2.2. Observability

The observability concept is an important factor
in studying the power system state estimation problem
since it decides   if   the   taken   measurements   are
suitable for solving   the   power   system   state
estimation problem or not [9 , 11 , 24 and 27].

The power system with specified measurements   is
said to be observable if the entire state   vector   of
the bus voltage magnitude and phase angle through out
the system can be estimated from the set of available
measurements   else the system is unobservable.

For   least   squares   estimation   the   entire
measurement set is used to determine the state
variables,   therefore the entire measurement set   is
tested for observability.   In   least   absolute   value
estimation only n of m measurements   are   used   to
calculate the final state vector,   and   consequently
only the subset of the n measurements must be   tested
for observability.

Thus the observability problem is equivalent to testing whether the set of equations solved for the state variables have a non trivial solution or not.

The power system is observable if the matrix H is of full rank equal to n. However computing the rank of large matrix is at least a time consuming activity and no information is obtained about the adequate system measurements to be introduced in order to achieve full system observability if H is not a matrix of full rank.

In 1980, Krumpholz, Clements, and Davies [15] discovered and proved an extremely useful relationship that exists between the measurement set and the jacobian matrix. The relationship that they discovered made it unnecessary to calculate the rank of the Jacobian matrix. Instead they showed that the power system measurement set or subset can be tested for observability by examining the measurement set and the structure of the power system.

Before examining the observability conditions that were established by Krumpholz et al. we will discuss some observability terminology.

2.2.1. Observability Terminology

There are several key words that must be defined and explained before we can examine the observability The power system contains a set of buses (nodes) and a set of lines (branches). A bus at which the real and reactive power injection are

measured is called a measured bus,   and a bus at
which these quantities are not measured is an
unmeasured bus. A measured line is a line whose
active and reactive flow are measured.

A tree consists of any connected loops.   These
loops are free collection of measured lines and all
the buses that the lines are connected to.   A
critical tree is a tree which contains all the
unmeasured buses in the power system.   If a power
system measurement set contains more than one tree,
the collection of trees is called a forest.   A
boundary injection of measurement is an active or
reactive power injection measurement at a bus that is
part of a tree and is also connected to one or more
unmeasured lines. Fig. 2.2.a shows a measurement
that is not a boundary injection measurement,   and
Fig. 2.2.b shows a measurement that is a boundary
injection measurement.

A line flow measurement is redundant if its
addition to the measurement set does not increase the
number of unknowns that can be solved for.   Consider
the four buses and four line flow measurements shown
in fig. 23 , (note that all voltages in this chapter
are in per unit and all phase angles are in radians).

If P_{2-1} , P_{3-2} , and P_{4-3} are measured δ_2 , δ_3
and δ_4 can be calculated. The addition of P_{1-4} to
the measurement set does not add any information
since that no new unknowns can be calculated.
Therefore measurement P_{1-4} is redundant with   respect

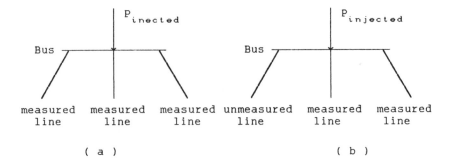

(a)   (b)

Fig. 2.2. $P_{injected}$ (a) is not a boundary
injection measurement because the bus
is not connected to an unmeasured
line.

$P_{injected}$ in (b) is a boundary
injection measurement because the bus
is connected to a tree by both
measured lines and unmeasured line.

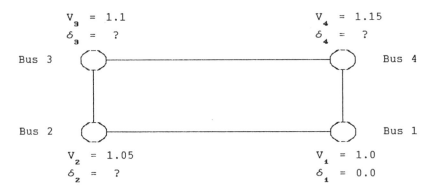

Fig. 2.3   A Four Bus Power System

to the measurement set formed by the other three measurements. Note that the three measurements formed a tree which contains four buses and three lines. The addition of the fourth measurement creates a loop. In general a measurement that forms a loop from a tree or part of a tree is redundant measurement.

2.2.2. Necessary Conditions for Observability

Krumpholz et al. proved that a measurement set is observable if all unmeasured buses are connected by a tree [i.e. the system is observable if and only if there exists a critical tree of full rank]. This is an important simplification of the problem because now we can check the observability of the system by looking for the critical tree only which is much smaller than looking for the spanning tree of full rank. They also demonstrated that an unobservable measurement set can be made observable by addition of boundary injection to the measurement set. The following conditions must be satisfied before a boundary injection measurement can be added to the measurement set.

a - The bus at which the measurement is made must be connected to at least one unmeasured line.

b - A path of unmeasured line(s) that lead from the injected bus to an unmeasured bus must be available. The unmeasured bus may be part of another tree or may not belong to any tree.

c - The path of unmeasured lines must not pass through any other measured buses

If all three conditions are satisfied, the boundary injection is added to the measurement set. The lines in the path between injected bus and the unmeasured bus are added to the connection diagram.

If enough boundary injections are added to the measurement set, the forest will eventually become a critical tree, and the measurement set will thus be observable.

The following four examples illustrate the application of the observability conditions. The power system used for all examples is shown in fig. 2.4.

Example 1

Given the power system shown in fig. 2.4 and the measurement set (P_{1-2} , P_{2-5} , P_3 , P_4 , P_6) determine if the measurement set is observable.

The three buses (3 , 4 , 6) that have injection measurements are measured buses. The three unmeasured buses (1 , 2 , 5) are connected by the line flows P_{1-2} and P_{2-5} . Therefore a critical tree that connects the three unmeasured buses exists and consequently the measurement set is observable.

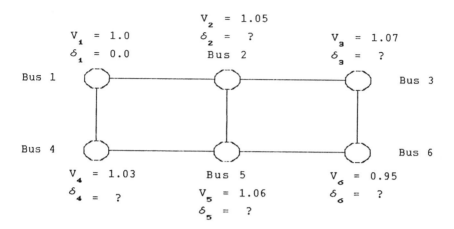

$V_2 = 1.05$
$\delta_2 = ?$
Bus 2

$V_1 = 1.0$
$\delta_1 = 0.0$

$V_3 = 1.07$
$\delta_3 = ?$

Bus 1

Bus 3

Bus 4

Bus 6

$V_4 = 1.03$
$\delta_4 = ?$

Bus 5
$V_5 = 1.06$
$\delta_5 = ?$

$V_6 = 0.95$
$\delta_6 = ?$

Fig. 2.4   Power System For Examples 2.1 to 2.4

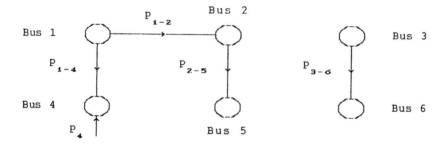

Bus 2

Bus 1 P_{1-2}

Bus 3

P_{1-4} P_{2-5} P_{3-6}

Bus 4   Bus 6

P_4 Bus 5

Fig. 2.5   Connection Diagram For Example 2.2

Example   2

Given the power system shown in fig. 2.4   and
the measurement set is $(P_{1-2}$, P_{1-4} , P_{2-5} , P_{3-6} ,
P_4)   determine if the measurement set is observable.

The only measured bus is bus 4.   In   order   for
the   measurement   set   to   be   observable   the   five
unmeasured   buses   must   be   connected.   From   the
connection diagram shown in fig. 2.5 , it can be seen
that buses 1 , 2 , 4 , and 5 are   connected   together
and buses 3 and 6 are connected together.

However   the two trees   that   connect   the   two
sets of buses   are not connected   to   each   other.
Thus a forest composed of two trees exists.   Since   a
critical tree does not exist,   the measurement set is
unobservable.

Note that bus   4   is   a   measured   bus   and   also
belongs to a tree.   Since   a   measured   bus   does   not
need to be part of a   tree   to   satisfy   the
observability conditions,   the line flow measurements
connect bus 1 and 4 is redundant and should be
deleted from the measurement set.   Alternatively,
the power injection measurement at bus 4 can be
deleted from the measurement set,   thus making bus   4
an unmeasured bus.   The non-redundant measurement set
will not be observable   but contains one less
measurement than the redundant measurement set.

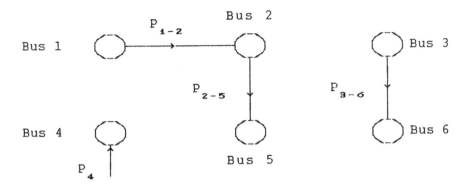

Fig. 2.6 Connection Diagram For Example 2.3
with the Reduced Measurement Set

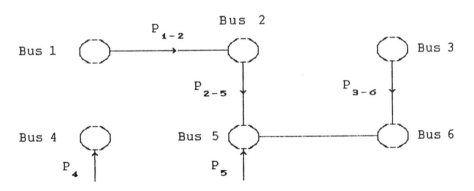

Fig. 2.7 Connection Diagram For Example 2.3
with New Measurement Set

Example   3

Add a boundary injection measurement to the measurement set that is given in example 2.2   so that it becomes observable.

It was mentioned in the previous example that one of the two measurements P_4 and P_{1-4} should be deleted from the interpolated measurement set. Arbitrarily delete P_{1-4} from the interpolated measurement set. The connection diagram for the reduced measurement set is shown in fig. 2.6. A boundary injection measurement which makes the measurement set observable must now be added to the measurement set. First consider adding a power injection at bus 1. Bus 1 is part of a tree and is connected to an unmeasured line (line 1-4) therefore, the measurement of power injection at bus 1 is a boundary injection measurement. However a path of unmeasured lines that run from bus 1 to an unmeasured bus that belongs to the other tree in the forest (containing buses 3 and 6)   does not exist. Therefore, adding the power injection measurement at bus 1 will not make the measurement set observable. Now consider adding a power injection measurement at bus 5. Bus 5 is part of a tree and is connected to two unmeasured lines, therefore a power injection measurement at bus 5 is a boundary injection measurement. One of the two unmeasured lines connects bus 5 and bus 6, which belong to

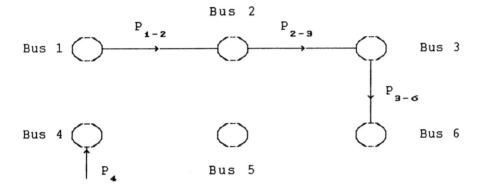

Fig. 2.8   Connection Diagram For Example 2.4

another component of the forest.   Therefore   adding
the injection measurement at bus 5 will add line 5-6
to the connection diagram.   The connection diagram of
the new measurement set is shown in fig. 2.7 . All of
the unmeasured buses   are   connected   by   a   critical
tree, consequently the measurement set (P_{1-2} , P_{2-5}
, P_{3-6} , P_4 , P_5) is observable.   Note   that   adding
P_5 to the measurement set does not add line 4-5 to
the connection diagram   because   line   4-5   does   not
connect two trees.

Example   4
___

   List all the measurements which can be added   to
the measurement set (P_{1-2} , P_{2-3} , P_{3-6} , P_4)   to
make   it   observable.   Fig.   2.8   contains   the
connection diagram for the four measurements.

   If a power injection measurement   at   bus   5   is
added to the measurement set,   then bus 5 will become
a measured bus.   All the unmeasured buses   will   then
be connected   and consequently,   the measurement   set
will be observable.   If   a   power   flow   measurement
along either line 2-5 or line 5-6   is   added   to   the
measurement set, all the   unmeasured   buses   will   be
connected and the measurement set will be   observable.
   If a power flow measurement along line 4-5 is added
to the measurement set,   line 4-5 will   be   added   to
the measurement set,   line 4-5 will be added   to   the
connection diagram.   However,   the   measurement   set

will remain unobservable since the unmeasured buses (1 , 2 , 3 , 5 , 6) will remain unconnected. If a boundary injection measurement at bus 2 is added to the measurement set, line 2-5 will be added to the connection diagram. The unmeasured buses will then be connected by a critical tree and the measurement set will be observable. Similarly a boundary injection measurement at bus 6 will add line 5-6 to the connection diagram and a critical tree will be formed.

Thus adding an injection measurement at bus 6 makes the measurement set observable. So that individual measurements that can be added to the measurement set to make it observable are (P_5 , P_{5-6} or P_{6-5} , P_{2-5} or P_{5-2} , P_2 and P_6) .

2.2.3 A New Observability Algorithm

Krumpholz et al. developed an algorithm that can be used to test measurements set for observability [15]. If the measurement set is observable a state estimation can be calculated. If the measurement set is not observable their algorithm returns the observable sub-networks of the measurement set.

Although their algorithm is well suited to least squares estimation it is not feasible for LAV estimation. In least squares estimation , every measurement is used to calculate the estimation. Thus when a measurement set is tested for observability all m of measurements are available to fulfill the observability conditions. For least absolute value estimation only n of m measurements are used to calculate the estimate. Consequently for a least absolute value estimation, the n measurement subset must satisfy the same observability condition that the entire measurements had to fulfill for the least squares estimation. Also, for least squares estimation, if the set m measurement is not observable, the least squares estimation can not be obtained until additional measurements are added to the measurement set. Where as for least absolute value estimation, if a particular subset of n measurements is not observable, another subset may be observable and addition of more measurements to the measurement set may be not required.

There are $\binom{m}{n}$ (i.e. $\dfrac{m!}{n!(m-n)!}$) possible subsets of n measurements many of which are not observable. If the algorithm that was developed by Krumpholz et al. is applied to observability test for least absolute estimation, many subsets of measurements may have to be tested before an observable subset is found. If the first subset tested is not observable another subset would have to be formed and tested for observability. If the second subset is not observable the process would continue until an observable subset is found or all possible subset have been tested.

Rather than using the observability algorithm that was developed by Krumpholz et al. an observability algorithm better suited for least absolute value was devised. The new observability algorithm constructs an observable subset of n measurements.

As seen before, the least absolute value method interpolate a set of n measurements corresponding to the n measurements which have the smallest least squares residuals. To ensure the set n measurement of smallest residuals can solve the state estimation problem, they must be checked for observability, so the procedure of choosing the interpolated measurements must be modified.

In a power system, a change in the voltage does not have a great effect on the active power but it has a large effect on the reactive power, also a change in phase angle δ has a great effect on the active power and a small effect on the reactive power. Thus, there is a weak coupling between real power and the bus voltage, also between reactive power and the bus phase angle. So the observability conditions that mentioned previously are applied to the two observability sub-problems, namely P - δ and Q - V observability. This means that for complete observability of the measurement set both the subset of the real power measurements and the subset of reactive power measurements must be observable.

The observability algorithm process one measurement at a time. If a measurement is redundant it is rejected, but if a measurement contains non redundant information it is added to the interpolated measurement set. Once the interpolated measurement set contains n measurements the least absolute value estimation can be calculated.

The order of processing the measurements is as follows :

1 - The measurement are ranked according to their least error squares residuals.
2 - The measurements are then processed according to this ranking starting from the measurement that corresponds to the smallest least error squares residuals.
3 - The process is continued until n measurements have been accepted into the interpolated set, the process is stopped and the least absolute estimation is calculated.

The flow charts that are given in figs. 2.9 and 2.10. demonstrate how real power measurements are processed. Note that for the new observability algorithm a line is considered to be unmeasured unless a measurement of power flow along the line belongs to the interpolated measurement set. Similarly a bus is an unmeasured bus unless a power injection measurements at the bus belongs to the interpolated measurement set. Consequently before any measurements are processed all lines and all buses are considered to be unmeasured.

If the entire set of real power measurements is processed as in fig. 29, and an observable set of measurement is not found the real power boundary injection measurements are processed in the manner described in fig. 2.10. The order of the processing of real power boundary injection measurements is determined according to the least error squares residuals of each measurements (i.e., the boundary injection measurement with the smallest residual is processed first followed by the boundary injection measurement with the next smallest residual if necessary etc.).

Consider the flow chart in fig. 2.9. An injection measurement is accepted only if it is taken at a bus that is not part of a tree. Thus an unmeasured bus that belongs to a tree will remain an unmeasured bus and can not become measured bus. Line flow measurements are accepted only when they connect two unmeasured buses that do not belong to the same tree. A line flow measurement that connects two unmeasured buses which are already connected by a tree is redundant and forms a loop.

Reactive power measurements are handled in the same manner as real power measurements.

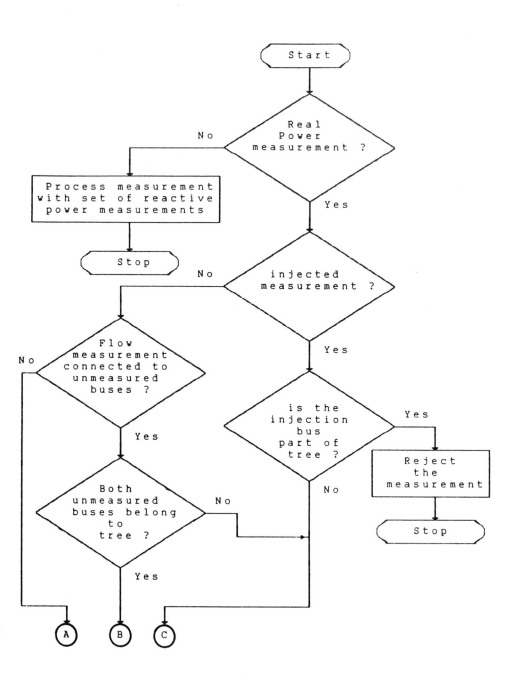

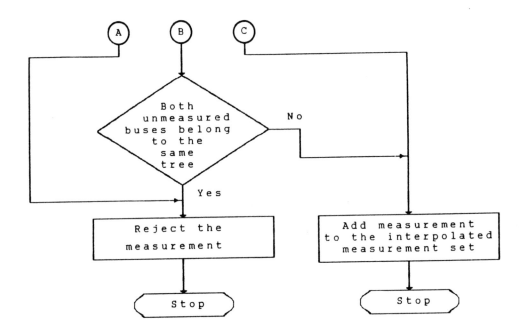

Fig. 2·9 Flow Chart of Observability for Active

Power Measurements

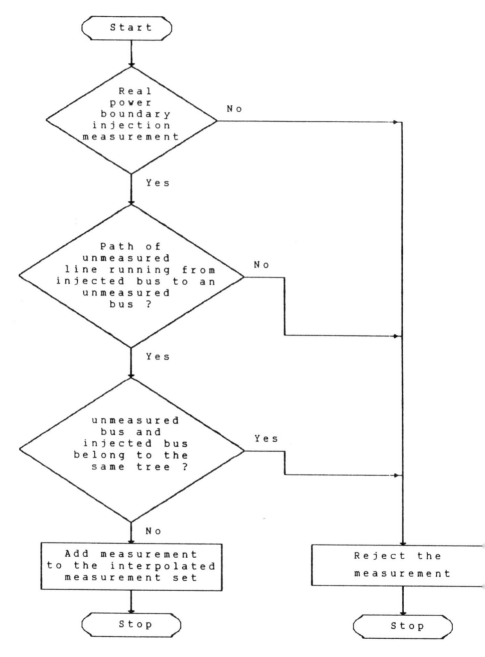

Fig. 2·10   Flow Chart for Real Power Boundary
Injection Measurements

Example   5

Given the power system shown in fig. 2.11, apply the new observability algorithm to the measurement set $\{ P_{1-2} , P_2 , P_6 , P_5 , P_{4-1} , P_{2-5} , P_{3-6} , P_3 , P_1 , P_{5-6} , Q_{1-2} , Q_{1-4} , Q_{4-5} , Q_1 , Q_{2-5} , Q_5 , Q_6 , Q_{2-3} , Q_2 , Q_3 , Q_{5-6} \}$ and determine an observable subset of measurements if one exists. Process the measurements in the order given.

First consider the real power measurement subset $\{ P_{1-2} , P_2 , P_6 , P_5 , P_{4-1} , P_{2-5} , P_{3-6} , P_3 , P_1 , P_{5-6} \}$. The first measurement P_{1-2} is accepted into the interpolated set. The next measurement P_2 is not accepted because bus 2 is already part of tree. The next two measurements P_5 and P_6 are accepted because neither bus 5 nor bus 6 belongs to a tree. Thus, bus 5 and 6 becomes measured buses and do not have to belong to a critical tree. The next measurement P_{4-1} is accepted because bus 4 is unmeasured bus that does not belong to a tree. The set of unmeasured buses now contains buses 1 to 4. Three of the unmeasured buses (1 , 2 , and 4) are connected by a tree. The only unmeasured bus that is not connected by a tree is bus 3. Consequently , bus 3 must either join the set of measured buses connected to the tree by a line flow measurement. The next measurement P_{2-5} does not involve bus 3

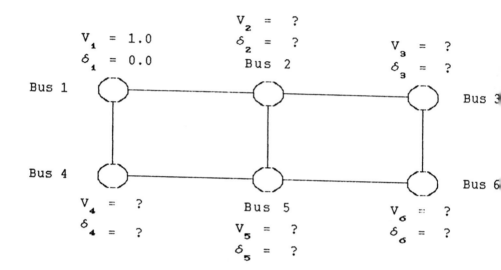

Fig. 2.11   Power System For Example 2.5 and 6

and is rejected   The next measurement to be processed is P_{3-6} since it connects bus 3 to a measured bus instead of an unmeasured bus   it is also rejected. The following measurement P_3 is accepted into the measurement set because it transform bus 3 into a measured bus. The acceptance of P_3 into a measurement set leaves the remaining three unmeasured buses connected.   The accepted subset of real power measurements $\left\{ P_{1-2} , P_6 , P_5 , P_{4-1} , P_3 \right\}$ is thus observable and processing of real power measurement need not continue.

Now consider the reactive power measurements $\left\{ Q_{1-2} , Q_{1-4} , Q_{4-5} , Q_1 , Q_{2-5} , Q_5 , Q_6 , Q_{2-3} , Q_2 , Q_3 , Q_{5-6} \right\}$. The first three measurements that connected four unmeasured buses are accepted into the interpolated measurement set. The next measurement Q_1 is an injection measurement at an unmeasured bus that already belongs to a tree   and is rejected. The following measurement Q_{2-5} connects buses 2 and 5 which are already connected by the tree that is formed from the first three measurements.   So the measurement is rejected. The measurement of Q_5 is rejected for the same reason that Q_1 was rejected for.

The measurement of Q_6 is accepted since bus 6 is an unmeasured bus and is not part of a tree.   The next measurement Q_{2-3} is also accepted because it

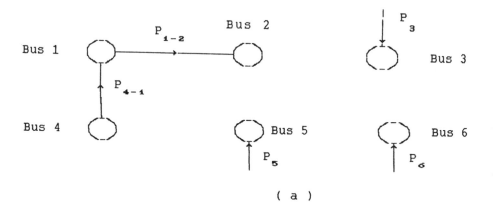

(a)

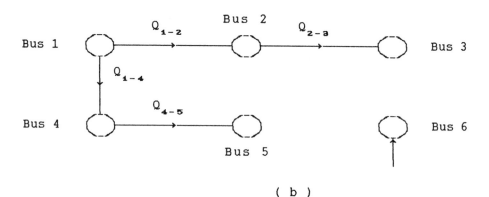

(b)

Fig. 2.12   (a) Connection Diagram For Interpolated
Real Power Measurements
(b) Connection Diagram For Interpolated
Reactive Power Measurements

Example   6
_____

Given the power system shown in fig. 4.11, apply the new observability algorithm to measurement set $\left\{ P_{1-2} , P_2 , P_6 , P_5 , P_{4-1} , P_{2-5} , P_{3-6} , Q_{1-2}, Q_{4-5} , Q_4 , Q_{2-5} , Q_{1-4} , Q_6 , Q_1 \right\}$ and determine an observable subset of measurements if one exists Process the measurements in the order given.

First consider the real power measurement subset $\left\{ P_{1-2} , P_2 , P_6 , P_5 , P_{4-1} , P_{2-5} , P_{3-6} \right\}$. This measurement subset is the same as the first seven measurements of the real power measurement set in example 4.5   The measurements are thus processed in the same manner as the first seven real power measurements were processed in example 4.5.   So the measurements $\left\{ P_{1-2} , P_6 , P_5 , P_{4-1} \right\}$ constitute the interpolated measurement set after all seven real power measurements have been processed.   The interpolated measurement set is not observable because bus 3 is an unmeasured bus that does not belong to a tree.   The only boundary injection measurement P_2 is then processed in the manner described by the flow chart given in fig. 4.10. Since there is an unmeasured line from bus 2 to bus 3 P_2 is accepted into the measurement set and line

2-3 is added to the tree   All unmeasured buses are now connected by a tree and therefore the interpolated measurement set $\left\{ P_{1-2} , P_6 , P_5 , P_{4-1}, P_2 \right\}$, is observable.

Now consider the reactive power measurements $\left\{ Q_{1-2} , Q_{4-5} , Q_4 , Q_{2-5} , Q_{1-4} , Q_6 , Q_1 \right\}$ After all the measurements have been processed once, the interpolated measurement set of reactive power measurements is $\left\{ Q_{1-2} , Q_{4-5} , Q_{2-5} , Q_6 \right\}$. The connection diagram is shown in fig. 4.13.

The boundary injection measurements (Q_4 and Q_1) are then processed in the manner illustrated in fig. 4.10. Since a path of unmeasured lines between bus 3 and bus 1 or bus 4 does not exist neither boundary injection will make the set of reactive power measurements observable. Since the set of reactive power measurements is unobservable, the entire measurement set is unobservable.

Now after a detailed discussion of the system observability problem to insure that the interpolated set of measurement can be used for solving power system state estimation, we will go on to design our state estimator by applying the new technique of least absolute value state estimates in the next section.

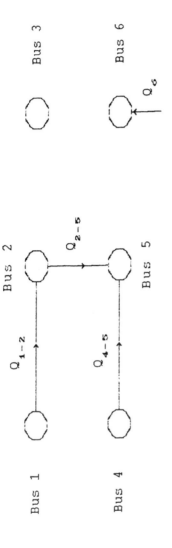

Fig. 2.13 Connection Diagram For the Interpolated
Reactive Power Measurements of Example2.6

III. DESIGN OF THE NEW STATE ESTIMATOR BASED ON LEAST ABSOLUTE VALUE APPROXIMATION (LAV)

3.1. Introduction

The power system state estimator is one of the vital links in the day to day control of power system. The primary function of the state estimator is to obtain a reliable estimate of power system states consisting of the voltage magnitude and voltage phase angle at each system node from acquired measurements and known physical characteristics.

The state estimator is used for a number of purposes such as security assessment and determining future operating strategies and hence it is essential that the information it provides must be reliable.

At the heart of state estimator is an algorithm that finds an estimate of the state using available measurements and the linearized model of power system.

Thus we can describe the static estimator as a data processing algorithm that transform the metered measurements and information about system structure into an accurate estimate of the system state after making a filtering out process to the erroneous measurements before it enters into calculating processes. Fig. 3.1 demonstrate the state estimator function.

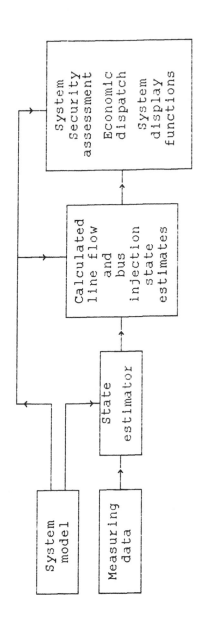

Fig. 3.1   Power System State Estimator

The measurement sets processed by power system state estimator usually contain measurements of real power and reactive power flow along system transmission lines and measurements of active and reactive power injection at the system buses. There are many other algorithms which use the previous measurements and add the voltage magnitude measurements. In other algorithms they process the active and reactive power flow measurements and the voltage measurements only. However in our case the state estimator processes active and reactive power flow measurements also active and reactive injection power measurements.

3.2. The bad data Problem

There are two types of bad data, one results from erroneous measurements and the others are caused by having wrong information about the system modeling. The main difference between measurement errors and modeling errors is not necessarily in the identification and detection schemes but rather in the manner in which the error is corrected. Measurement errors can be easily corrected by either removing the measurements or replacing it with its expected value (pseudomeasurement) Topological errors are more difficult to correct because a time consuming complete relinearization of the system is usually required. There are some algorithms used to detect and identify the topological error [16 and 34]

As mentioned before   the measurement errors take place due to inaccurate reading device or transducers may be wired incorrectly or the transducer itself may be malfunctioning so that it simply no longer gives accurate readings   or due to reading of transmission systems.   The presence of bad data among the measurements results in poor state estimates accuracy.

Therefore, there is a need to develop a technique to detect whether abnormally erroneous measurement errors are present among the measurement set or not. Moreover   it is also necessary to identify the faulty measurements.

Some bad data treatment techniques were developed to be used before carrying out the estimation process [28].   These methods mainly depend upon the consistency checks of the measured quantities   Some other methods were developed to be used after the estimation process [12 , 21 , 25 and 26].   They mainly depend upon the residual investigation techniques.

In our state estimator there is no need to develop another algorithm to detect and identify the presence of bad data   since this is done during the solution process.

3.3. Practical Consideration

The major factors in design and development of any state estimator are those pertaining to practical considerations [17 and 18].   The effectiveness of any proposed method must be evaluated in light of the

basic requirements of the state estimator.   As
mentioned before   the primary role of the state
estimator is to expand the system data base for
operation and security related calculations. In this
regard   there are three basic aspects of the state
estimator that will determine its practicality :
i - the reliability of the estimator
ii - the accuracy of the estimator
iii - the computational efficiency of the estimator

The reliability of the estimator is essential if
there is to be any confidence in the data that creates
the data base.   The reliability is a measure of
confidence one has in a state estimate. It gives the
possibility that the system model may not true and
that some measurements may have significant errors.
On the other hand, the accuracy assumes the correctness
of the given model also the accuracy is a measure of
the degree to which the the state estimate reflects
the true state of the given model.   For example, if
the state estimator can not detect topological errors
or bad data,   then no matter to what degree of
accuracy the state estimate is obtained,   it is an
unreliable estimator. Thus   the state estimator must
exhibit a high degree of robustness in its ability to
handle the modeling and a measurement errors.

The accuracy of the state estimate obtained is a
function of a number of factors such as the quality of
the measuring and data transmission devices and the
number of approximations made in the estimator itself.

Although one of the purposes of the state estimator is to filter out the noise in measurements, the amount of noise in each measurement does limit the minimum accuracy to which the estimate can be obtained. Another limitation to the degree of accuracy that can be obtained is the fact that error is introduced by linearization of the system equations. Other approximations to increase the computational efficiency such as decoupling the system equations further reduce the accuracy of the state estimates.

The third aspect of the computational efficiency merits considerable attention. It is possible to design a state estimator that gives extremely accurate and highly reliable estimates. However there is a fourth aspect which depends on the computational efficiency that is the estimator execution time. So the use of approximations can reduce the computation time without adversely effecting the quality of the estimates. However care must be taken not to compromise the reliability in making approximations.
The primary reason why state estimators need to be fast is that state estimation is only a small part of the power system computer control functions that must be done on-line. Furthermore the state estimator must be able to adapt to changing system conditions as they occur if a reliable estimator is to be available.

An additional aspect is the storage requirements so there must be a concern on the manner in which the information required to calculate the state estimate is stored. The sparse nature of the power system

structure permits the use of sparsity oriented
programming techniques to reduce the storage
requirements as well as computational time.

3.4. New Least Absolute Value State Estimator Algorithm

The steps in the new least absolute value power
system state estimation algorithm are given as the
following :

Step 1   Read the power flow and injection
                measurements and all system data

Step 2   Assume an initial estimate θ_o of the state
                vector . Usually a flat start is assumed
                (all voltages magnitudes = 1 , all phase
                angles = 0.0) . However , any measured bus
                voltages and phase angles may be used as
                part of initial estimate

Step 3   Linearize the system equations about θ_o

Step 4   Calculate the least error squares solution
                ($\Delta\theta^{*}$) of the linearized equations

Step 5   Update the state estimate vector

$$\theta_{new} = \theta_{old} + \Delta\theta^{*}$$

Step 6   Check the convergence of the solution   if
                $\Delta\theta$ less than a predetermined tolerance then
                continue to step 7   else repeat from
                step 3

Step 7   Calculate the least error squares residuals using

$$r = \Delta z - H \Delta\theta^*$$

Step 8   Calculate the standard deviation of the residuals

$$\sigma = \sqrt{\frac{1}{m - n + 1} \sum_{i=1}^{m} (r_i - r_{av})^2}$$

$$r_{av} = \frac{1}{m} \sum_{i=1}^{m} r_i$$

Step 9   Reject the measurements which have absolute residual greater than the standard deviation

Step 10   Recalculate the least error squares estimates $\Delta\theta^*$

Step 11   Update the state estimate vector

$$\theta_{new} = \theta_{old} + \Delta\theta^*$$

Step 12   Relinearize the system equation about the new value of the state vector. Check if $\Delta\theta^*$ is smaller than the convergence criterion go to step 13 else go to step 10

Step 13   Recalculate the least error squares residuals

Step 14   Rank the residuals in ascending order from smallest to largest residual by their absolute values

Step 15   Process the measurements to the observability subprogram to determine which measurement should be interpolated. Begin with the measurement that corresponds to the smallest residual and continue once n measurements have been interpolated stop processing of the measurements and move to step 16

Step 16   Using only the n interpolated measurements solve for the LAV estimates

Step 17   Update the state vector

Step 18   Check if $\Delta\hat{\theta}$ is smaller than the convergence criterion go to step 19 else relinearize the state equation about the new value of θ and go to step 13

Step 19   The least absolute value solution is $\hat{\theta}$

A flow chart of the previous procedure is given in fig.

Note that the new algorithm does not weight the measurements. All measurements have a weight of 1. This is in contrast to weighted least squares

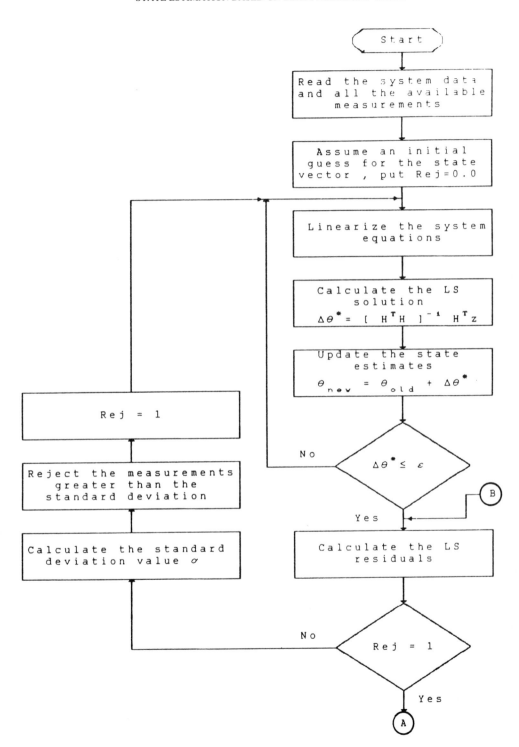

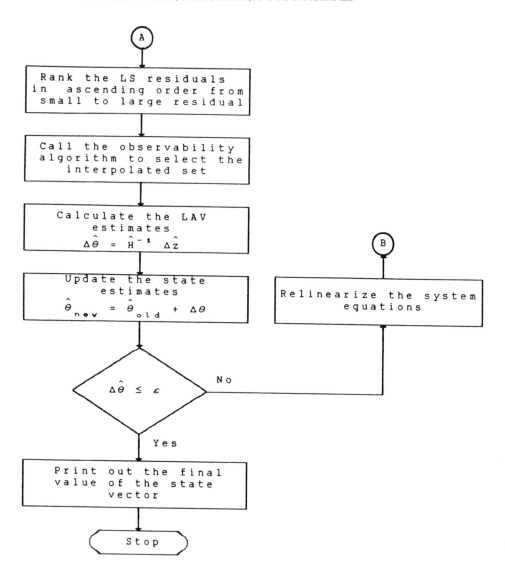

Fig. 3·2 Flow Chart of the New Least
Absolute Value Algorithm

estimation which usually uses a weighting matrix that is equal to the inverse of the covariance matrix. Since the entire measurement set is used to calculate the least squares estimates, a weighting matrix which emphasizes the more accurate measurements is necessary. However in least absolute value estimation only n of good measurements are used to calculate the estimates. Since all n measurements are interpolated a weighting matrix which pulls the estimates closer to more accurate measurements is unnecessary. In this way we avoid an important reason which cause ill-conditioning in the gain matrix especially when there are some constraints, we must give the constraints a large weight where the bad measurements take a small weight which make the weighting matrix contain very distinctive numbers, this will createill-conditioning of the gain matrix consequently the convergence will be slow or there is no convergence at all.

3.5. Constrained Power System State Estimation

In most power system network there are many buses which have no generation and also no load. These buses are just used in connection of transmission lines. Since there is no injected active or reactive power, these buses are called zero injection buses.

The zero injection power can be treated as a good measurement, this increases the number of available measurements , therefore it increases the redundancy factor ($\dfrac{\text{number of measurements}}{\text{number of states}}$) without any additional metering devices.

In the power system state estimation these zero injection power are treated as constraints on the solution of the state estimator algorithm. We can solve the constrains least squares as mentioned in section. Then the zero injection measurements will be automatically made a member of the interpolated measurement set. Thus the least absolute value estimator will be forced to interpolate all of the zero injection measurements.

The last flow chart can be used with small modifications to solve for zero injection measurements. However we can summarize the handling of the zero injection measurements as follows :

a - The zero values of real and reactive power injected at the zero injection buses are treated as measurements and are added to the measurement set

b -- The least squares solution is obtained by using the procedure mentioned in chapter II

c - After the least squares estimates have been calculated, n measurements must be selected for interpolation. All of the zero injection measurements are placed in the interpolated measurement set at first. The other measurements are then processed in the same manner as in the unconstrained state estimation [i.e. assuming there are ℓ zero injection measurements] the interpolated measurement set will contain ℓ of the zero injection measurements and ($n-\ell$) of the actual measurements.

d - The least absolute value estimates can be then calculated. It is now forced to interpolate the zero injection measurement.

The new Proposed technique is applied to the IEEE 5 , 10 and 14 bus systems. The results and discussion of these systems are given in section IV.

IV RESULTS AND DISCUSSIONS

In this document the results of testing the new power system state estimation algorithm are presented and discussed.

The algorithm is applied to three standard power systems: the 5 bus system [13], the 10 bus system [13], and the IEEE 14 bus system [13], lines parameters, generation and load data for all three systems are given in Appendix A.

The exact value of the bus voltages and phase angles are determined by performing the Newton-Raphson A.C. load flow method. In each test the estimates of the LS technique and the new technique estimates are present besidesthe exact value of the estimates. The convergence criteria used is 1×10^{-3} per unit (P.U.) bus voltages state variables and 1×10^{-3} radians for the phase angles state variables. The zero injection measurements are treated as fixed measurements so they do not affect the redundancy ratio where redundancy ratio is defined as the number of actual measurements divided by the number of state variables to be estimated (redundancy ratio = $\frac{m}{n}$) . The following sets of measurements are used :

4.1. Tested Measurements Sets

i) For the 5 bus system

(1) Measurement set (5 A)
Redundancy ratio = 2.875
Active power injection   : P_2 , P_4 , P_5

Reactive power injection : Q_2 , Q_3 , Q_4

Active Power flow   : P_{1-2} , P_{1-3} , P_{2-3} , P_{3-1} , P_{3-2} ,

P_{3-4} , P_{4-3} , P_{5-4}

Reactive power flow   : Q_{1-2} , Q_{1-3} , Q_{2-5} , Q_{3-2} , Q_{4-2} ,

Q_{4-3} Q_{4-5} , Q_{5-2} , Q_{5-4}

(2) Measurement set (5 B)

Redundancy ratio = 2.625

Active power injection   : P_2 , P_3 , P_4 , P_5

Reactive power injection : Q_2 , Q_3 , Q_4 , Q_5

Active Power flow   : P_{1-2} , P_{2-3} , P_{2-4} , P_{3-4} , P_{5-2} ,

P_{5-4}

Reactive power flow   : Q_{1-2} , Q_{1-3} , Q_{2-5} , Q_{3-2} ,

Q_{3-4} , Q_{4-2}

ii) For the 10 bus system

(1) Measurement set (10 A)

Redundancy ratio = 2.944444

Active power injection   : P_2 , P_3 , P_4 , P_5 , P_6 , P_7 , P_8

P_9 , P_{10}

<u>Reactive power injection</u> : Q_2 , Q_3 , Q_4 , Q_5 , Q_6 , Q_7 , Q_8

Q_9 , Q_{10}

<u>Active Power flow</u>   : P_{1-3} , P_{2-3} , P_{2-4} , P_{3-4} , P_{4-3} ,

P_{4-5} , P_{5-1} , P_{5-6} , P_{5-7} , P_{6-7} , P_{7-8} , P_{8-9} , P_{9-4} ,

P_{9-10} , P_{10-4}

<u>Reactive power flow</u>   : Q_{1-3} , Q_{2-3} , Q_{2-4} , Q_{3-2} , Q_{3-4} ,

Q_{4-2} , Q_{4-9} , Q_{4-10} , Q_{5-1} , Q_{5-4} , Q_{5-6} , Q_{6-7} , Q_{7-5} ,

Q_{7-6} , Q_{8-7} , Q_{8-9} , Q_{9-4} , Q_{9-8}

Q_{10-4} , Q_{10-9}

(2) Measurement set (10 B)

Redundancy ratio = 3.000000

<u>Active power injection</u>   : P_2 , P_3 , P_5 , P_6 , P_8 , P_9 , P_{10}

<u>Reactive power injection</u> : Q_2 , Q_3 , Q_5 , Q_6 , Q_8 , Q_9 , Q_{10}

<u>Active Power flow</u>   : P_{1-3} , P_{1-5} , P_{2-3} , P_{2-4} , P_{3-4} ,

P_{4-2} P_{4-3} , P_{4-5} , P_{4-9} , P_{4-10} , P_{5-4} , P_{5-7} P_{6-5} ,

P_{6-7} , P_{7-5} , P_{7-6} , P_{7-8} , P_{8-7} , P_{8-9} , P_{9-4} , P_{9-10}

P_{9-10} , P_{10-4}

<u>Reactive power flow</u>   : Q_{1-3} , Q_{1-5} , Q_{2-4} , Q_{3-2} , Q_{3-4} ,

Q_{4-5} Q_{4-9} , Q_{4-10} , Q_{5-6} , Q_{5-7} , Q_{7-5} , Q_{7-6} , Q_{7-8} , Q_{8-9}

, Q_{9-4} , Q_{9-10} , Q_{10-}

iii) For 14 bus system

(1) Measurement set (14 A)

Redundancy ratio = 2.076930

Active power injection   : P_3 , P_{10}, P_{11}, P_{12}, P_{13}, P_{14}

Reactive power injection : Q_3 , Q_8 , Q_{10}, Q_{11}, Q_{12}, Q_{13}, Q_{14}

Active Power flow   : P_{1-2} , P_{1-5} , P_{2-3} , P_{2-4} , P_{2-5} ,

P_{3-4}, P_{4-2} , P_{4-5} , P_{4-7} , P_{4-9} , P_{5-6} , P_{6-11} P_{6-12}, P_{6-13}, P_{7-8} , P_{7-9} , P_{9-10}, P_{9-14}, P_{10-11}, P_{12-13}, P_{12-14}

Reactive power flow   : Q_{1-2} , Q_{1-5} , Q_{2-3} , Q_{2-4} , Q_{2-5} ,

Q_{3-4} Q_{4-5} , Q_{4-7} , Q_{4-9} , Q_{5-6} , Q_{6-11}, Q_{6-12}, Q_{6-13}, Q_{7-8} , Q_{7-9} , Q_{9-10}, Q_{9-14}, Q_{10-11}, Q_{12-13} Q_{13-14}

Zero power injections   : P_7 , Q_7 , P_8

(2) Measurement set (14 B)

Redundancy ratio = 2.615385

Active power injection   : P_2 , P_3 , P_4 , P_6 , P_9 , P_{10} , P_{11}

P_{12} , P_{13} , P_{14}

Reactive power injection : Q_2 , Q_3 , Q_4 , Q_6 , Q_8, Q_9, Q_{10}

Q_{11} , Q_{12} , Q_{13} , Q_{14}

Active Power flow   : P_{1-2} , P_{1-5} , P_{2-3} , P_{2-4} , P_{2-5} ,

P_{3-2}, P_{3-4} , P_{4-5} , P_{4-7} , P_{4-9} , P_{5-2} , P_{5-4} , P_{5-6} ,

P_{6-11}, P_{6-12}, P_{6-13}, P_{7-8} , P_{9-7} , P_{9-10} , P_{9-14} , P_{10-11},

P_{12-6} , P_{13-6} , P_{3-12}, P_{13-14}

Reactive power flow
$\quad : Q_{1-2} \;,\; Q_{1-5} \;,\; Q_{2-1} \;,\; Q_{2-3} \;,\; Q_{2-4}$

$$Q_{2-5}, \; Q_{3-2} \;,\; Q_{3-4} \;,\; Q_{4-5} \;,\; Q_{4-7} \;,\; Q_{4-9} \;,\; Q_{5-6}$$

$$Q_{6-11}, \; Q_{6-12}, \; Q_{6-13}, \; Q_{7-8} \;,\; Q_{7-9} \;,\; Q_{9-10}$$

$$Q_{9-14} \;,\; Q_{10-11}, \; Q_{12-13}, \; Q_{13-14}$$

Zero power injections
$\quad : P_7 \;,\; Q_7 \;,\; P_8$

GROUP (5 A)						
The Redundancy Ratio = 2.875						
Bad Data Points P_2 is reversed						
BUS	EXACT SOLUTION		LS SOLUTION		LAV SOLUTION	
	E	THETA	E	THETA	E	THETA
1	1.06000	.00000	1.06000	.00000	1.06000	.00000
2	1.04740	-.04897	1.04582	-.05445	1.04740	-.04896
3	1.02440	-.08728	1.02674	-.08215	1.02440	-.08728
4	1.02374	-.09305	1.02590	-.08839	1.02374	-.09305
5	1.01797	-.10735	1.01962	-.10331	1.01797	-.10735

Table (1)

BUS	ERROR IN LEAST SQUARES		ERROR IN LAV	
	E	THETA	E	THETA
1	.00000	.00000	.00000	.00000
2	.00158	.00549	.00000	.00000
3	.00234	.00512	.00000	.00000
4	.00216	.00466	.00000	.00000
5	.00165	.00404	.00000	.00000

Table (2)

GROUP (5 A)					
The Redundancy Ratio = 2.875					
Bad Data Points P_{3-1} is reversed $\quad Q_{5-4}$ is halved					
BUS	EXACT SOLUTION		LS SOLUTION		LAV SOLUTION
	E	THETA	E	THETA	E	THETA
1	1.06000	.00000	1.06000	.00000	1.06000	.00000
2	1.04740	-.04897	1.05092	-.03943	1.04740	-.04897
3	1.02440	-.08728	1.03275	-.06239	1.02440	-.08728
4	1.02374	-.09305	1.03153	-.06947	1.02374	-.09305
5	1.01797	-.10735	1.02307	-.09237	1.01797	-.10735

Table (3)

BUS	ERROR IN LEAST SQUARES		ERROR IN LAV	
	E	THETA	E	THETA
1	.00000	.00000	.00000	.00000
2	.00352	.00953	.00000	.00000
3	.00834	.02488	.00000	.00000
4	.00779	.02359	.00000	.00000
5	.00510	.01498	.00000	.00000

Table (4)

GROUP (5 A)					
The Redundancy Ratio = 2.875					
Bad Data Points P_{3-1} is reversed , Q_{5-4} is halved P_{5-4} = zero					
BUS	EXACT SOLUTION		LS SOLUTION		LAV SOLUTION	
	E	THETA	E	THETA	E	THETA
1	1.06000	.00000	1.06000	.00000	1.06000	.00000
2	1.04740	-.04897	1.05094	-.03938	1.04740	-.04897
3	1.02440	-.08728	1.03247	-.06309	1.02440	-.08728
4	1.02374	-.09305	1.03130	-.07008	1.02374	-.09305
5	1.01797	-.10735	1.02346	-.09131	1.01797	-.10734

Table (5)

BUS	ERROR IN LEAST SQUARES		ERROR IN LAV	
	E	THETA	E	THETA
1	.00000	.00000	.00000	.00000
2	.00354	.00958	.00000	.00000
3	.00807	.02419	.00000	.00000
4	.00756	.02297	.00000	.00000
5	.00549	.01604	.00000	.00000

Table (6)

GROUP (5 B)					
The Redundancy Ratio = 2.625					
Bad Data Points P_2 is reversed , P_{3-4} is reversed Q_3 is halved					
BUS	EXACT SOLUTION		LS SOLUTION		LAV SOLUTION
	E	THETA	E	THETA	E	THETA
1	1.06000	.00000	1.06000	.00000	1.06000	.00000
2	1.04740	-.04897	1.04658	-.05459	1.04740	-.04897
3	1.02440	-.08728	1.02823	-.08670	1.02440	-.08728
4	1.02374	-.09305	1.02802	-.08908	1.02374	-.09306
5	1.01797	-.10735	1.02063	-.10329	1.01797	-.10735

Table (7)

BUS	ERROR IN LEAST SQUARES		ERROR IN LAV	
	E	THETA	E	THETA
1	.00000	.00000	.00000	.00000
2	.00082	.00562	.00000	.00000
3	.00383	.00058	.00000	.00000
4	.00428	.00397	.00000	.00000
5	.00266	.00405	.00000	.00000

Table (8)

GROUP (5 B)						
The Redundancy Ratio = 2.625						
Bad Data Points Q_4 = zero , P_{3-4} is doubled P_5 is halved , Q_{3-4} = zero						
BUS	EXACT SOLUTION		LS SOLUTION		LAV SOLUTION	
	E	THETA	E	THETA	E	THETA
1	1.06000	.00000	1.06000	.00000	1.06000	.00000
2	1.04740	-.04897	1.04821	-.04719	1.04740	-.04897
3	1.02440	-.08728	1.02567	-.08703	1.02440	-.08728
4	1.02374	-.09305	1.02468	-.09385	1.02374	-.09305
5	1.01797	-.10735	1.02199	-.09438	1.01797	-.10735

Table (9)

BUS	ERROR IN LEAST SQUARES		ERROR IN LAV	
	E	THETA	E	THETA
1	.00000	.00000	.00000	.00000
2	.00081	.00177	.00000	.00000
3	.00127	.00024	.00000	.00000
4	.00094	.00079	.00000	.00000
5	.00402	.01297	.00000	.00000

Table (10)

GROUP (5 B)					
The Redundancy Ratio = 2.625					
Bad Data Points Q_4 = zero , P_{3-4} is doubled P_5 is halved , Q_{3-4} = zero Q_{4-2} is reversed , P_{2-4} is doubled					
BUS	EXACT SOLUTION		LS SOLUTION		LAV SOLUTION	
	E	THETA	E	THETA	E	THETA
1	1.06000	.00000	1.06000	.00000	1.06000	.00000
2	1.04740	-.04897	1.04812	-.04866	1.04740	-.04897
3	1.02440	-.08728	1.02569	-.09305	1.02440	-.08728
4	1.02374	-.09305	1.02475	-.10018	1.02374	-.09305
5	1.01797	-.10735	1.02195	-.09688	1.01797	-.10735

Table (11)

BUS	ERROR IN LEAST SQUARES		ERROR IN LAV	
	E	THETA	E	THETA
1	.00000	.00000	.00000	.00000
2	.00072	.00030	.00000	.00000
3	.00129	.00577	.00000	.00000
4	.00101	.00712	.00000	.00000
5	.00398	.01047	.00000	.00000

Table (12)

GROUP (10 A)					
The Redundancy Ratio = 2.944444					
Bad Data Points P_{10-4} is reversed , Q_{5-6} is reversed					
BUS	EXACT SOLUTION		LS SOLUTION		LAV SOLUTION	
	E	THETA	E	THETA	E	THETA
1	1.03000	.00000	1.03000	.00000	1.03000	.00000
2	1.04500	.08890	1.04526	.09129	1.04500	.08890
3	1.04397	.06145	1.04405	.06310	1.04397	.06145
4	1.03637	.06834	1.03764	.07671	1.03637	.06834
5	1.03967	-.03899	1.03905	-.03584	1.03967	-.03899
6	1.04649	-.07894	1.03694	-.07005	1.04649	-.07894
7	1.03046	-.12617	1.03391	-.10562	1.03046	-.12617
8	1.03059	-.11769	1.03674	-.09061	1.03059	-.11769
9	1.02807	-.06047	1.03658	-.03014	1.02807	-.06047
10	1.02028	-.00472	1.03327	.03532	1.02028	-.00472

Table (13)

BUS	ERROR IN LEAST SQUARES		ERROR IN LAV	
	E	THETA	E	THETA
1	.00000	.00000	.00000	.00000
2	.00026	.00239	.00000	.00000
3	.00008	.00165	.00000	.00000
4	.00128	.00836	.00000	.00000
5	.00061	.00315	.00000	.00000
6	.00955	.00890	.00000	.00000
7	.00345	.02055	.00000	.00000
8	.00615	.02708	.00000	.00000
9	.00851	.03033	.00000	.00000
10	.01299	.04004	.00000	.00000

Table (14)

GROUP (10 A)					
The Redundancy Ratio = 2.944444					
Bad Data Points P_3 is doubled and reversed P_9 = zero , P_{2-4} is halved					
BUS	EXACT SOLUTION		LS SOLUTION		LAV SOLUTION	
	E	THETA	E	THETA	E	THETA
1	1.03000	.00000	1.03000	.00000	1.03000	.00000
2	1.04500	.08890	1.04957	.11438	1.04500	.08890
3	1.04397	.06145	1.04885	.09233	1.04397	.06145
4	1.03637	.06834	1.04081	.09263	1.03637	.06834
5	1.03967	-.03899	1.03944	-.03689	1.03967	-.03899
6	1.04649	-.07894	1.04591	-.07775	1.04649	-.07894
7	1.03046	-.12617	1.03114	-.11559	1.03046	-.12617
8	1.03059	-.11769	1.03314	-.09816	1.03060	-.11769
9	1.02807	-.06047	1.03907	-.01939	1.02807	-.06047
10	1.02028	-.00472	1.02498	.02035	1.02028	-.00472

Table (15)

BUS	ERROR IN LEAST SQUARES		ERROR IN LAV	
	E	THETA	E	THETA
1	.00000	.00000	.00000	.00000
2	.00457	.02548	.00000	.00000
3	.00489	.03087	.00000	.00000
4	.00444	.02429	.00000	.00000
5	.00022	.00210	.00000	.00000
6	.00058	.00120	.00000	.00000
7	.00068	.01057	.00000	.00000
8	.00254	.01953	.00000	.00000
9	.01100	.04108	.00000	.00000
10	.00470	.02507	.00000	.00000

Table (16)

GROUP (10 A)					
The Redundancy Ratio = 2.944444					
Bad Data Points Q_{5-6} is reversed , Q_{8-9} = zero Q_{1-3} is doubled , Q_7 is halved P_{10-4} is reversed , P_9 is halved					
BUS	EXACT SOLUTION		LS SOLUTION		LAV SOLUTION	
	E	THETA	E	THETA	E	THETA
1	1.03000	.00000	1.03000	.00000	1.03000	.00000
2	1.04500	.08890	1.04850	.09103	1.04500	.08890
3	1.04397	.06145	1.04690	.06290	1.04397	.06145
4	1.03637	.06834	1.04152	.07764	1.03637	.06834
5	1.03967	-.03899	1.03896	-.03556	1.03967	-.03899
6	1.04649	-.07894	1.03499	-.06944	1.04649	-.07894
7	1.03046	-.12617	1.03737	-.10438	1.03046	-.12617
8	1.03059	-.11769	1.03734	-.08487	1.03059	-.11769
9	1.02807	-.06047	1.04486	-.01775	1.02807	-.06047
10	1.02028	-.00472	1.03870	.03814	1.02028	-.00472

Table (17)

BUS	ERROR IN LEAST SQUARES		ERROR IN LAV	
	E	THETA	E	THETA
1	.00000	.00000	.00000	.00000
2	.00350	.00213	.00000	.00000
3	.00293	.00144	.00000	.00000
4	.00516	.00929	.00000	.00000
5	.00070	.00343	.00000	.00000
6	.01150	.00950	.00000	.00000
7	.00691	.02178	.00000	.00000
8	.00675	.03282	.00000	.00000
9	.01679	.04272	.00000	.00000
10	.01842	.04286	.00000	.00000

Table (18)

GROUP (10 B)					
The Redundancy Ratio = 3.0000					
Bad Data Points P_6 is doubled and reversed P_{10} is reversed , $Q_{4-5} = 0$					
BUS	EXACT SOLUTION		LS SOLUTION		LAV SOLUTION	
	E	THETA	E	THETA	E	THETA
1	1.03000	.00000	1.03000	.00000	1.03000	.00000
2	1.04500	.08890	1.04517	.08883	1.04500	.08890
3	1.04397	.06145	1.04411	.06140	1.04397	.06145
4	1.03637	.06834	1.03663	.06826	1.03637	.06834
5	1.03967	-.03899	1.03967	-.03751	1.03967	-.03899
6	1.04649	-.07894	1.05171	-.06371	1.04649	-.07894
7	1.03046	-.12617	1.02830	-.12980	1.03046	-.12616
8	1.03059	-.11769	1.02898	-.12036	1.03060	-.11769
9	1.02807	-.06047	1.02822	-.05934	1.02807	-.06047
10	1.02028	-.00472	1.02706	.01111	1.02028	-.00472

Table 19

BUS	ERROR IN LEAST SQUARES		ERROR IN LAV	
	E	THETA	E	THETA
1	.00000	.00000	.00000	.00000
2	.00017	.00007	.00000	.00000
3	.00014	.00005	.00000	.00000
4	.00027	.00009	.00000	.00000
5	.00001	.00148	.00000	.00000
6	.00522	.01524	.00000	.00000
7	.00216	.00363	.00000	.00000
8	.00161	.00267	.00000	.00000
9	.00015	.00113	.00000	.00000
10	.00678	.01583	.00000	.00000

Table (20)

GROUP (10 B)					
The Redundancy Ratio = 3.000000					
Bad Data Points P$_2$ is reversed , P$_{5-7}$ is reversed P$_6$ is halved , Q$_{4-9}$ = zero					
BUS	EXACT SOLUTION		LS SOLUTION		LAV SOLUTION
	E	THETA	E	THETA	E	THETA
1	1.03000	.00000	1.03000	.00000	1.03000	.00000
2	1.04500	.08890	1.04628	.09038	1.04500	.08890
3	1.04397	.06145	1.04501	.06275	1.04397	.06145
4	1.03637	.06834	1.03845	.07050	1.03637	.06834
5	1.03967	-.03899	1.03892	-.03708	1.03967	-.03899
6	1.04649	-.07894	1.04637	-.07234	1.04649	-.07894
7	1.03046	-.12617	1.03481	-.09992	1.03046	-.12617
8	1.03059	-.11769	1.03323	-.09796	1.03059	-.11769
9	1.02807	-.06047	1.02921	-.04954	1.02807	-.06047
10	1.02028	-.00472	1.02176	.00118	1.02028	-.00472

Table (21)

BUS	ERROR IN LEAST SQUARES		ERROR IN LAV	
	E	THETA	E	THETA
1	.00000	.00000	.00000	.00000
2	.00128	.00148	.00000	.00000
3	.00104	.00130	.00000	.00000
4	.00208	.00215	.00000	.00000
5	.00075	.00191	.00000	.00000
6	.00012	.00660	.00000	.00000
7	.00435	.02624	.00000	.00000
8	.00264	.01972	.00000	.00000
9	.00114	.01093	.00000	.00000
10	.00148	.00591	.00000	:00000

Table (22)

GROUP (10 B)					
The Redundancy Ratio = 3.000000					
Bad Data Points P_9 is reversed , P_{6-7} = 0 Q_5 is reversed , P_{9-10} = 0 Q_{9-10} = 0 , Q_{7-5} is doubled					
BUS	EXACT SOLUTION		LS SOLUTION		LAV SOLUTION	
	E	THETA	E	THETA	E	THETA
1	1.03000	.00000	1.03000	.00000	1.03000	.00000
2	1.04500	.08890	1.04546	.08761	1.04500	.08890
3	1.04397	.06145	1.04437	.06039	1.04397	.06145
4	1.03637	.06834	1.03708	.06655	1.03637	.06834
5	1.03967	-.03899	1.03249	-.03513	1.03967	-.03899
6	1.04649	-.07894	1.04883	-.07712	1.07172	-.08777
7	1.03046	-.12617	1.04170	-.11046	1.05613	-.13275
8	1.03059	-.11769	1.04160	-.09216	1.04423	-.12008
9	1.02807	-.06047	1.04828	-.01188	1.04982	-.01612
10	1.02028	-.00472	1.02616	-.00053	1.02028	-.00472

Table (23)

BUS	ERROR IN LEAST SQUARES		ERROR IN LAV	
	E	THETA	E	THETA
1	.00000	.00000	.00000	.00000
2	.00046	.00129	.00000	.00000
3	.00040	.00107	.00000	.00000
4	.00072	.00180	.00000	.00000
5	.00718	.00386	.00000	.00000
6	.00234	.00182	.02523	.00882
7	.01124	.01571	.02567	.00659
8	.01100	.02553	.01363	.00239
9	.02021	.04859	.02175	.04435
10	.00589	.00419	.00000	.00000

Table (24)

GROUP (14 A)						
The Redundancy Ratio = 2.076923						
Bad Data Points P_{9-14} = 0 $\quad Q_{6-11}$ is reversed $\quad P_{2-3}$ is halved						
BUS	EXACT SOLUTION		LS SOLUTION		LAV SOLUTION	
	E	THETA	E	THETA	E	THETA
1	1.06000	.00000	1.06000	.00000	1.06000	.00000
2	1.04500	-.08701	1.04485	-.08748	1.04500	-.08701
3	1.01000	-.22235	1.01349	-.20749	1.01000	-.22235
4	1.01547	-.17948	1.01740	-.17216	1.01547	-.17948
5	1.01832	-.15289	1.02001	-.14622	1.01832	-.15289
6	1.07000	-.24816	1.06960	-.23729	1.07000	-.24816
7	1.06053	-.23279	1.06319	-.22631	1.06053	-.23279
8	1.09000	-.23279	1.09258	-.22631	1.09000	-.23279
9	1.05498	-.26038	1.05794	-.25531	1.05498	-.26038
10	1.05020	-.26320	1.05361	-.25776	1.05020	-.26320
11	1.05650	-.25796	1.06025	-.25103	1.05651	-.25796
12	1.05512	-.26307	1.05556	-.25160	1.05512	-.26307
13	1.05024	-.26444	1.05113	-.25279	1.05024	-.26444
14	1.03492	-.27962	1.03957	-.26661	1.03492	-.27962

Table (25)

BUS	ERROR IN LEAST SQUARES		ERROR IN LAV	
	E	THETA	E	THETA
1	.00000	.00000	.00000	.00000
2	.00015	.00047	.00000	.00000
3	.00349	.01486	.00000	.00000
4	.00192	.00732	.00000	.00000
5	.00169	.00667	.00000	.00000
6	.00040	.01087	.00000	.00000
7	.00266	.00647	.00000	.00000
8	.00258	.00647	.00000	.00000
9	.00296	.00507	.00000	.00000
10	.00341	.00544	.00000	.00000
11	.00375	.00694	.00000	.00000
12	.00044	.01147	.00000	.00000
13	.00089	.01165	.00000	.00000
14	.00465	.01301	.00000	.00000

Table (26)

GROUP (14 A)					
The Redundancy Ratio = 2.076923					
Bad Data Points $P_{4-9} = 0$, P_{3-4} is halved Q_{13-14} is halved , Q_{12-13} is reversed Q_{9-14} is doubled , P_{4-2} is doubled					
BUS	EXACT SOLUTION		LS SOLUTION		LAV SOLUTION
	E	THETA	E	THETA	E	THETA
1	1.06000	.00000	1.06000	.00000	1.06000	.00000
2	1.04500	-.08701	1.04573	-.08557	1.04500	-.08701
3	1.01000	-.22235	1.00932	-.23059	1.01000	-.22235
4	1.01547	-.17948	1.00972	-.20291	1.01088	-.20019
5	1.01832	-.15289	1.01312	-.17361	1.01375	-.17335
6	1.07000	-.24816	1.06349	-.26587	1.06570	-.26944
7	1.06053	-.23279	1.05530	-.24878	1.05607	-.25394
8	1.09000	-.23279	1.08494	-.24878	1.08566	-.25394
9	1.05498	-.26038	1.04978	-.27515	1.05060	-.28175
10	1.05020	-.26320	1.04477	-.27841	1.04580	-.28459
11	1.05650	-.25796	1.05050	-.27438	1.05214	-.27932
12	1.05512	-.26307	1.04789	-.28028	1.05076	-.28447
13	1.05024	-.26444	1.04324	-.28179	1.04586	-.28585
14	1.03492	-.27962	1.02728	-.29511	1.03046	-.30116

Table (27)

BUS	ERROR IN LEAST SQUARES		ERROR IN LAV	
	E	THETA	E	THETA
1	.00000	.00000	.00000	.00000
2	.00073	.00144	.00000	.00000
3	.00068	.00825	.00000	.00000
4	.00575	.02343	.00459	.02070
5	.00520	.02072	.00457	.02046
6	.00651	.01771	.00430	.02128
7	.00522	.01599	.00445	.02115
8	.00506	.01599	.00434	.02115
9	.00520	.01477	.00439	.02137
10	.00543	.01521	.00440	.02139
11	.00600	.01641	.00437	.02136
12	.00723	.01721	.00436	.02141
13	.00700	.01735	.00438	.02141
14	.00764	.01549	.00446	.02153

Table (28)

GROUP (14 B)					
The Redundancy Ratio = 2.615385					
Bad Data Points P_{9-14} = 0 Q_{6-11} is reversed P_{2-3} is halved					
BUS	EXACT SOLUTION		LS SOLUTION		LAV SOLUTION	
	E	THETA	E	THETA	E	THETA
1	1.06000	.00000	1.06000	.00000	1.06000	.00000
2	1.04500	-.08701	1.04575	-.08514	1.04500	-.08701
3	1.01000	-.22235	1.01289	-.21079	1.01000	-.22235
4	1.01547	-.17948	1.01672	-.17486	1.01547	-.17948
5	1.01832	-.15289	1.01924	-.14886	1.01832	-.15289
6	1.07000	-.24816	1.07085	-.24179	1.07000	-.24816
7	1.06053	-.23279	1.06248	-.22788	1.06053	-.23279
8	1.09000	-.23279	1.09186	-.22788	1.09000	-.23279
9	1.05498	-.26038	1.05714	-.25598	1.05498	-.26038
10	1.05020	-.26320	1.05266	-.25948	1.05020	-.26320
11	1.05650	-.25796	1.05989	-.25398	1.05651	-.25796
12	1.05512	-.26307	1.05568	-.25563	1.05512	-.26307
13	1.05024	-.26444	1.05124	-.25692	1.05024	-.26444
14	1.03492	-.27962	1.03841	-.26983	1.03492	-.27962

Table 29

BUS	ERROR IN LEAST SQUARES		ERROR IN LAV	
	E	THETA	E	THETA
1	.00000	.00000	.00000	.00000
2	.00075	.00187	.00000	.00000
3	.00289	.01156	.00000	.00000
4	.00125	.00462	.00000	.00000
5	.00091	.00403	.00000	.00000
6	.00085	.00637	.00000	.00000
7	.00195	.00491	.00000	.00000
8	.00186	.00491	.00000	.00000
9	.00216	.00440	.00000	.00000
10	.00246	.00372	.00000	.00000
11	.00338	.00398	.00000	.00000
12	.00056	.00743	.00000	.00000
13	.00100	.00752	.00000	.00000
14	.00349	.00979	.00000	.00000

Table (30)

GROUP (14 B)					
The Redundancy Ratio = 2.615385					
Bad Data Points $P_{4-9} = 0$, P_{3-4} is halved Q_{13-14} is halved , Q_{12-13} is reversed Q_{9-14} is doubled , P_{2-4} is doubled					
BUS	EXACT SOLUTION		LS SOLUTION		LAV SOLUTION
	E	THETA	E	THETA	E	THETA
1	1.06000	.00000	1.06000	.00000	1.06000	.00000
2	1.04500	-.08701	1.04392	-.08995	1.04500	-.08701
3	1.01000	-.22235	1.00864	-.22789	1.01000	-.22235
4	1.01547	-.17948	1.01150	-.19233	1.01547	-.17948
5	1.01832	-.15289	1.01420	-.16648	1.01832	-.15289
6	1.07000	-.24816	1.06471	-.26312	1.07000	-.24816
7	1.06053	-.23279	1.05608	-.24037	1.06053	-.23279
8	1.09000	-.23279	1.08559	-.24037	1.09000	-.23279
9	1.05498	-.26038	1.04967	-.27374	1.05498	-.26038
10	1.05020	-.26320	1.04393	-.27781	1.05020	-.26320
11	1.05650	-.25796	1.04958	-.27314	1.05651	-.25796
12	1.05512	-.26307	1.05100	-.27799	1.05512	-.26307
13	1.05024	-.26444	1.04481	-.27963	1.05024	-.26444
14	1.03492	-.27962	1.02733	-.29504	1.03492	-.27962

Table (31)

BUS	ERROR IN LEAST SQUARES		ERROR IN LAV	
	E	THETA	E	THETA
1	.00000	.00000	.00000	.00000
2	.00108	.00294	.00000	.00000
3	.00136	.00555	.00000	.00000
4	.00398	.01285	.00000	.00000
5	.00412	.01359	.00000	.00000
6	.00529	.01496	.00000	.00000
7	.00445	.00758	.00000	.00000
8	.00441	.00758	.00000	.00000
9	.00532	.01336	.00000	.00000
10	.00626	.01461	.00000	.00000
11	.00693	.01518	.00000	.00000
12	.00411	.01492	.00000	.00000
13	.00544	.01519	.00000	.00000
14	.00759	.01542	.00000	.00000

Table (32)

4.2. Discussion of the Results

In the last section the new LAV technique algorithm was tested to obtain the estimates for the bus voltage magnitudes and phase angles with initially being set 1.0 p.u. and 0.0 radian for the voltage magnitude and phase angle at each bus respectively. The given tables are the results obtained for testing the new LAV and LS techniques on the IEEE 5 , 10 and 14 bus systems using the set of measurements mentioned earlier at the beginning of this chapter.

Tables 1 , 2 , 3 , 4 , 5 and 6 give the results obtained for the first group of the 5 bus system. These results indicate that the LS estimates are biased by the presence of the bad data . In Tables 1 and 2 there is only one bad data measurement (P_2 is reversed) the LS estimate of the voltage magnitude and phase angle of buses (2 , 3 , 4 and 5) are affected by this bad data point. The greatest effect is at the magnitude of voltage and phase angle of bus 2. As the number of the bad data measurements increases the error in the LS estimate is also increase (Tables 4 and 6) where the largest error at Table 6 , due to the presence of three bad measurements. On contrast to the LS the proposed technique is able to reject the bad measurements. The new LAV estimates are exactly the same as the value of the voltage magnitudes and phase angles at each bus obtained form the load flow (Tables 1 , 3 and 5).

In Tables 7 , 8 , 9 , 10 , 11 and 12 the proposed LAV technique is applied to the 5 bus system with a new set of measurements having small redundancy ratio (2.625). In Tables 7 and 8 there are three bad data measurements, the results obtained by LS estimate are much better than the results of Table 5 and 6 for group 5-A for the same number of bad measurements and lower redundancy ratio. This is due to the dependance of the estimate on the number of the measurements taken from the region of the bad data measurements (Local redundancy ratio). As the number of the measurements taken from the region of the bad data increases, the effect of the bad data measurements on the LS estimate decreases.

In Tables 11 and 12, 6 bad measurements are applied the LS estimator produces poor estimates on contrast to the LAV estimator which rejects all the bad measurements and produce more accurate estimates.

Tables 13 , 14 , 15 , 16 , 17 and 18 give the results of the IEEE 10 bus system . In Table 13 there are two bad measurements (P_{10-4} and Q_{5-6} are reversed). It can be noticed that the voltage magnitudes and phase angles at buses 4 , 5 , 6 and 10 are more affected by bad data where the two bad data points connect these buses. The other buses are less affected by these two bad measurements i.e. the bad data points have the greatest effect on the buses connected these points and less effect on the other buses. In most cases the proposed LAV gives the exact estimates.

In some cases the proposed technique interpolates some bad data measurements to satisfy the system observability requirements. This case may occur if the local redundancy ratio for the area of the bad measurements is small. Tables 23 and 24 give the results obtained for the IEEE 10 bus system group 10-B. It can be noticed from these tables that the proposed new LAv technique does not produce the exact values for the voltage magnitudes and phase angles and the error obtained in this case are slightly greater than that of the LS estimates. This simply can be explained as the least error squares estimators uses the entire measurements to calculate the estimates while the LAV uses only n of m measurements. Even through both estimator use the same bad measurements the ratio between the bad data and the given measurements in case of the LAV technique more larger than of the LS technique. Consequently the bad data has a greater impact on the least absolute value estimate. However the overall estimates of the new LAV technique is slightly better than the overall estimates of the LS where the interpolated bad data measurements in LAV usually affect directly on the buses connected only the bad data measurements but in LS estimate the bad data measurements affect on all buses of the system (Tables 23 and 24). For this case the proposed technique interpolates two bad data measurements P_9 and P_{6-7} and the buses 6 , 7 , 8 and 9 are mostly affected by these two bad measurements.

Tables 25 , 26 , 27 and 28 give the results obtained for IEEE 14 bus system with constrained measurements. There are three bad measurements in Tables 25 and 26 the LS estimates biased by the presence of these bad measurements in contrast to the LAV which produces the exact estimates.

In Tables 27 and 28 the new LAV interpolates three bad measurements. As the number of measurements increases the redundancy ratio is increased the LS estimates are tend to be more accurate as well as the new proposed LAV technique which produces the exact estimates (increasing the redundancy ratio improves the rejection properties of the new LAV technique).

It can be noticed from Tables 29 , 30 , 31 and 32 that as the redundancy ratio increases to 2.615385 the LS estimates obtained in Tables 29 and 30 are more accurate than that obtained in Tables 25 and 26 for the same bad measurements. Also in Tables 31 and 32 the new LAV technique rejects all the bad measurements that are interpolated in Tables 27 and 28 for the same case and gives the exact estimates.

From the previous discussions and results obtained for the tested systems we can conclude the following :
1 - The proposed new LAV technique produces estimates better than the LS estimates when the measurement set is contaminated with bad measurements. This is due to the interpolation properties of the LAV technique. In contrast to the LS which is biased by the presence of bad measurements.
2 - In some case when the local redundancy is small the proposed algorithm interpolates some of the bad measurements to overcome this drawback the redundancy ratio should be increased as discussed earlier. However if the redundancy ratio is not increased and the algorithm is interpolates some bad measurements this interpolated bad measurements will affect the estimate of the buses connect these bad measurements only.

REFERENCES

1 - Fred C. Schweppe and J. Wildes " POWER SYSTEM STATIC STATE ESTIMATION , PART I : EXACT MODEL " IEEE Transactions on Power Appartus and Systems , Vol. PAS-89 , No. 1 , January 1970 , PP. 120 - 125 .

2 - Fred C. Schweppe and Douglas B. Rom " POWER SYSTEM STATIC STATE ESTIMATION , PART II : APPROXIMATE MODEL " IEEE , Transactions on Power Appartus and Systems , Vol. PAS-89 , No. 1 , January 1970 , PP. 125 - 130 .

3 - Fred C. Schweppe " POWER SYSTEM STATIC STATE ESTIMATION , PART III : IMPLEMENTATION " IEEE , Transactions on Power Appartus and Systems , Vol. PAS-89 , No. 1 , January 1970 , PP. 130 - 135 .

4 - Robert E. Larson and William F. Tinney " STATE ESTIMATION IN POWER SYSTEMS , PART I : THEORY AND FEASIBILITY " IEEE , Transactions on power appartus and systems , Vol. PAS - 89 , No. 3 , March 1970 , PP. 345 - 352 .

5 - Robert E. Larson and William F. Tinney " STATE ESTIMATION IN POWER SYSTEMS , PART II : IMPLEMENTATION AND APPLICATIONS " IEEE , Transactions on power appartus and systems , Vol. PAS - 89 , No. 3 , March 1970 , PP. 353 - 363 .

6 - E.J.Schlossmacher " AN ITERATIVE TECHNIQUE FOR ABSOLUTE DEVIATION CURVE FITTING " Journal of the American Statistical Association , Vol. 68 , No. 344 , December 1973 , PP. 857 - 859 .

7 - Fred C. Schweppe , and Edemund J. Handschin "
STATIC STATE ESTIMATION IN ELECTRIC POWER SYSTEMS "
Processing of the IEEE, Vol. 02 , No. 7 , July 1974 ,
PP. 972 - 980 .

8 - V.A.Sposito , M.L.Hand and McCromick " USING   AN
APPROXIMATE L_1 ESTIMATOR   " Communication   in
Statistical - Simulations and Computations , B6(3) ,
1977 , PP. 263 - 268 .

9 - J. S. Horton , and R. D. Masiello " ON-LINE
DECOUPLED OBSERVABILITY PROCESSING " Power Industry
Computer Application Conference , 1977 , PP. 420 - 426

10 - F. C. Aschmoneit , N. M. Peterson , and E. C.
Adrian " STATE ESTIMATION WITH EQUALITY CONSTRAINTS "
1977 Power Industry Computer Application Conference ,
PP. 427 - 430 .

11 - K. Phua and T. S. Dillon " OPTIMAL CHOICE OF
MEASUREMENTS FOR STATE ESTIMATION " 1977 Power
Industry Computer Application Conference .

12 - F. Broussolle " STATE ESTIMATION IN POWER SYSTEMS:
DETECTING BAD DATA THROUGH THE SPARSE INVERSE MATRIX
METHOD " IEEE , Transaction on Power Appartus and
Systems , Vol. PAS - 97 , No. 3 , May / June 1978 ,
PP. 678 - 682 .

13 - M. R. Lrving , R. C. Owen , and M. J. H. Sterling
" POWER SYSTEM STATE ESTIMATION USING LINEAR
PROGRAMMING " Processing of IEE , Vol. 125 , No. 9
September 1978 , PP. 879 - 885 .

14 - G.F.McCormick and V.A.Sposito " USING THE L_2 -
ESTIMATOR IN L_1 - ESTIMATION " SIAM , Journal on
Scientific and Statistical Computing , Vol. 1 , No. 2
June 1980 , PP. 290 - 301 .

15 - G.R.Krumpholz , K.A.Clements and P.W.Daves " POWER SYSTEM OBSERVABILITY : A PRACTICAL ALGORITHM USING NETWORK TOPOLOGY " IEEE , Transactions on Power Appartus and Systems Vol. PAS - 99 , No. 4 , July - August 1980 , PP. 1534 - 1542

16 - R. L. Lugtu , D. F. Hakett , K. C. Liu , and D. D. Might " POWER SYSTEM STATE ESTIMATION : DETECTION OF TOPOLOGICAL ERRORS " IEEE , Transactions on Power Appartus and Systems , Vol. PAS - 99 , No. 6 , Nov./Dec. 1980 , PP. 2406 - 2412 .

17 - R. F. Bischke " POWER SYSTEM STATE ESTIMATION : PRACTICAL CONSIDERATION " IEEE , Transactions on Power Appartus and Systems , Vol. PAS - 100 , No. 12 , Dec. 1981 , PP. 361 - 369

18 - E. Handschine and Bongers " THEORETICAL AND PRACTICAL CONSIDERATIONS IN THE DESIGN OF STATIC STATE ESTIMATORS FOR ELECTRIC POWER SYSTEMS " Proceedings of the International Symposium "Computerized Operation of Power Systems" (COPOS ' 75) , August 18 - 20 , SAO Carlos , S.P.Brazil , PP. 104 - 136 .

19 - J.J.Allemong , L.Radu and A.M.Sassom " A FAST AND RELIABLE STATE ESTIMATION ALGORITHM FOR AEP'S NEW CONTROL CENTER " IEEE , Transactions on Power Appartus and Systems , Vol. APS - 101 , No. 4 , April 1982 , PP. 933 - 944 .

20 - Willy W. Kotuiga and M.Vidyasagar " BAD DATA REJECTION PROPERTIES OF WEIGHTED LEAST ABSOLUTE VALUE TECHNIQUES APPLIED TO STATIC STATE ESTIMATION " IEEE , Transactions on Power Appartus and Systems , Vol. PAS - 101 , No. 4 , April 1982 , PP. 844 - 853 .

21 - V. H. Quintana , A. Simoes - Costa , and M. Mier
" BAD DATA DETECTION AND IDENTIFICATION TECHNIQUE
USING ESTIMATION ORTHOGONAL METHODS " IEEE ,
Transactions on Power Appartus ans Systems , Vol. PAS
- 101 , No. 9 , September 1982 , PP. 3356 - 3364 .

22 - W.W.Kotiuga " POTENTIAL OF LEAST ABSOLUTE VALUE
APPROXIMATION IN SOLVING POWER SYSTEM PLANING AND
OPERATION PROBLEMS " Transactions of Candian Electric
Association , Vol. 24 , Part 4 , 1985 , Paper 85 - SP
- 162 .

23 - K. L. Lo and Y. M. Mahmoud " A DECOUPLED LINEAR
PROGRAMMING TECHNIQUE FOR POWER SYSTEM STATE
ESTIMATION " IEEE , Transactions on Power Systems ,
Vol. PWRS - 1 , No. 1 , February 1986 , PP. 154 - 160

24 - A. Bargiela , M. R. Lrving , and M. J. H.
Sterling " OBSERVABILITY DETERMINATION IN POWER SYSTEM
STATE ESTIMATION USING A NETWORK FLOW TECHNIQUE " IEEE
    Transactions on Power Systems , Vol. PWRS - 1 , No.
2 , May 1986 , PP. 108 - 114 .

25 - A. Monticelli , Felix F. Wu , and Maosong Yen "
MULTIPLE BAD DATA IDENTIFICATION FOR STATE ESTIMATION
BY COMBINATORIAL OPTIMIZATION " IEEE , Transactions on
Power Delivery , Vol. PWRD - 1 , No. 3 , July 1986 ,
PP. 361 - 369 .

26 - Flin Zhuang and R. Balasubramanian " BAD DATA
PROCESSING IN POWER SYSTEM STATE ESTIMATION BY DIRECT
DATA DETECTION AND HYPOTHESIS TESTS " IEEE ,
Transactions on Power Systems , Vol. PWRS - 2 , No. 2
    May 1987 , PP. 321 - 330 .

27 - Ilya W. Slutskar and Jon M. Scudder " NETWORK OBSERVABILITY ANALYSIS THROUGH MEASUREMENT JACOBEAN MATRIX REDUCTION " IEEE , Transactions on Power Systems , Vol. PWRS - 2 , No. 2 May 1987 , PP. 331 - 338 .

28 - A. Abur , A. Keyhani , and H. Bakhtiari " AUTOREGRESSIVE FILTERS FOR THE IDENTIFICATION AND REPLACEMENT OF BAD DATA IN POWER SYSTEM STATE ESTIMATION " IEEE , Transaction on Power Systems , Vol. PWRS - 2 , No. 3 , August 1987 , PP. 552 - 560

29 - R.Gonin and A.H.Money " A REVIEW OF COMPUTATIONAL METHODS FOR SOLVING THE NONLINEAR L_1-NORM ESTIMATION PROBLEM " Elsevier Science Publishers B.V.(North Holand) , PP. 117-129 , 1987

30 - Felix F. Wu , Wen Hsiung E. Liu , and Shau Ming Lun " OBSERVABILITY ANALYSIS AND BAD DATA PROCESSING FOR STATE ESTIMATION WITH EQUALITY CONSTRAINTS " IEEE Transactions on Power Systems , Vol. 3 , No. 2 , May 1988 , PP. 541 - 548 .

31 - S.A.Soliman and G.S.Christensen " A NEW TECHNIQUE FOR DISCRETE LINEAR PARAMETER ESTIMATION WITH LINEAR CONSTRAINTS BASED ON WLAV APPROXIMATION " Submitted For Review International Journal Of Control .

32 - S.A.Soliman , G.S.Christensen and A.Rouhi " A NEW TECHNIQUE FOR CURVE FITTING BASED ON MINIMUM ABSOLUTE DEVIATION " Compution al Statistics and Data Analysis Vol. 6 , No. 4 , May 1988 , PP. 341 - 351

33 - Felix F. Wu and Wen Hsiung E. Liu " DETECTION OF TOPOLOGY ERRORS BY STATE ESTIMATION " IEEE , Transaction on Power Systems , Vol. 4 , No. 1 , PP. 176-183 , Feb. 1989 .

34 - S.A.Soliman , and G.S.Christensen " A NEW TECHNIQUE FOR CURVE FITTING BASED ON WEIGHTED LEAST ABSOLUTE VALUE ESTIMATION " Journal of Optimization Theory and Applic. Vol. 61 , No. 2 , PP 281-299 , April 1989 .

35 - G.S.Christensen , S.A.Soliman , and Rouhi " AN OBSERVABILITY ALGORITHM FOR SEQUENTIAL POWER SYSTEM STATE ESTIMATION " Accepted for Publication to Electric Machines and Power System Journal , 1989 .

36 - Allen J. Wood and Bruce F.Wooltenberg " POWER GENERATION OPERATION AND CONTROL " John Wiley and Sons New York , 1984

APPENDIX [A]

THE 5 BUS TEST SYSTEM

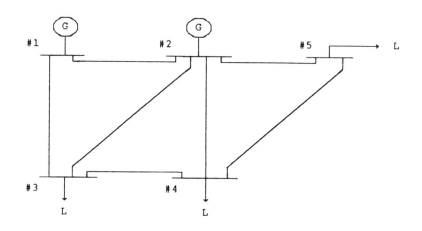

Fig. A.1   The 5 Bus Test System Connection Diagram

Line	Resistance (P.U.)	Reactance (P.U.)	Line Charging (P.U.)
1 - 2	0.02	0.06	0.030
1 - 3	0.08	0.24	0.025
2 - 3	0.06	0.18	0.020
2 - 4	0.06	0.18	0.020
2 - 5	0.04	0.12	0.015
3 - 4	0.01	0.03	0.010
4 - 5	0.08	0.24	0.025

Table A.1   Impedances and Line Charging Data for the

5 Bus Test System (Base = 100 MVA)

Bus No	Net Generation	
	MW	MVAR
1*	–	–
2**	20.0	20.0
3	– 45.0	– 15.0
4	– 40.0	– 05.0
5	– 60.0	– 10.0

* Slack Bus

** Generation Bus

Table A.2. Operating Conditions for the 5 Bus Test System

THE 10 BUS TEST SYSTEM

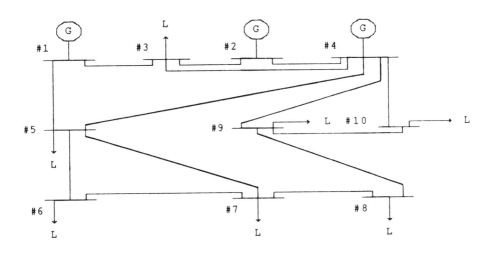

Fig. A.2. The 10 Bus Test System Connection Diagram

Line	Resistance (P.U.)	Reactance (P.U.)	Line Charging (P.U.)
1 - 3	0.004	0.032	0.000
1 - 5	0.005	0.042	0.000
2 - 3	0.001	0.010	0.000
2 - 4	0.003	0.028	0.000
3 - 4	0.054	0.151	0.000
4 - 5	0.143	0.364	0.000
4 - 9	0.044	0.112	0.000
4 - 10	0.029	0.073	0.000
5 - 6	0.055	0.140	0.000
5 - 7	0.073	0.185	0.000
6 - 7	0.132	0.336	0.000
7 - 8	0.029	0.073	0.000
8 - 9	0.033	0.084	0.000
9 - 10	0.033	0.084	0.000

Table A.3. Impedances and Line Charging Data for the 10 Bus
Test system (Base = 100 MVA)

Bus No	Net Generation	
	MW	MVAR
1*	–	–
2**	380.0	– 70.0
3	– 90.0	55.0
4**	160.0	– 80.0
5	– 50.0	25.0
6	– 10.0	– 15.0
7	– 70.0	20.0
8	– 50.0	25.0
9	– 100.0	50.0
10	– 40.0	– 100.0

* Slack Bus

** Generation Bus

Table A.**4**. Operating Conditions for the 10 Bus Test System

THE 14 BUS TEST SYSTEM

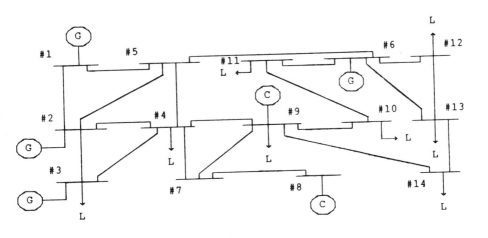

Fig. A.3. The 14 Bus Test System Connection Diagram

G. S. CHRISTENSEN, S. A. SOLIMAN, AND M. Y. MOHAMED

Line	Resistance (P.U.)	Reactance (P.U.)	Line Charging (P.U.)
1 - 2	0.01938	0.05917	0.02640
1 - 5	0.05403	0.23304	0.02460
2 - 3	0.04699	0.19797	0.02190
2 - 4	0.05811	0.17632	0.01870
2 - 5	0.05695	0.17388	0.01700
3 - 4	0.06701	0.17103	0.01730
4 - 5	0.01335	0.04211	0.00640
4 - 7	0.00000	0.20912	0.00000
4 - 9	0.00000	0.55618	0.00000
5 - 6	0.00000	0.25202	0.00000
6 - 11	0.09498	0.19890	0.00000
6 - 12	0.12291	0.25581	0.00000
6 - 13	0.06615	0.13027	0.00000
7 - 8	0.00000	0.17615	0.00000
7 - 9	0.00000	0.11001	0.00000
9 - 10	0.03181	0.08450	0.00000
9 - 14	0.12711	0.27038	0.00000
10 - 11	0.08205	0.19207	0.00000
12 - 13	0.22092	0.19988	0.00000
13 - 14	0.17093	0.34802	0.00000

Table A.5. Impedances and Line Charging Data for the 14 Bus Test system (Base = 100 MVA)

Bus No	Net Generation	
	MW	MVAR
1*	-	-
2**	18.3	-
3**	- 94.2	-
4	- 47.8	03.9
5	- 07.6	- 01.6
6**	- 11.2	-
7	- 00.0	00.0
8**	- 00.0	-
9	- 29.0	- 16.6
10	- 09.0	- 05.8
11	- 03.5	- 01.8
12	- 06.1	- 01.6
13	- 13.5	- 05.8
14	- 14.9	- 05.0

* Slack Bus

** Generation Bus

Table A.6. Operating Conditions for the 14 Bus Test System

Transformer designation	Tap setting
5 - 6	0.932
4 - 7	0.978
4 - 9	0.969

Table A.7. Transformer Data for the 14 Bus Test System

Bus No	Suscptance (P.U.)
9	0.19

Table A.8 Static Capacitor Data for the 14 Bus Test System

INDEX

A

Accuracy, least absolute value state estimator, 436–437
A/D converters, *see* Analog-to-digital converters
Aging
  polymeric insulators, dry band arcing, 146
  polymers, 134–135
  polymer watershed material
    crystallization effects, 173, 179–180
    dry band arcing, 171–173, 180
    effects of low-molecular-weight polymer chains, 171–173, 177–178
    energy dispersive X-ray analysis, 174–175, 177–178
    Fourier transform infrared spectroscopy, 175–176, 178–179
    mechanisms, 170–174
    surface roughness effects, 172–173, 176, 179–182
    X-ray diffraction, 176, 179–182
Algorithms
  fast power flow, 296–297
    block decoupling and rotations, 316–319
    BX and XB, 301–305
    critically coupled XB and BX, 321–325
    Newton, 297–301
    novel rules, 319–321
      test systems, 326–327
    performance
      on 14, 30, 57, and 118 bus IEEE networks, 330–339
      on 1655 bus networks, 339–340
      of Carpentier's method, 328–330
      CCBX and CCXB, 330–336
    quasi-Newton, 297–301
    super decoupled, 313–316

least absolute value state estimation
  least absolute value based on linear programming, 364–368
  Schlossmacher iterative, 368–369
  Sposito, 369–370
novel
  for least absolute value state estimator, 438–443
  for power system observability, 417–430
power system observability, 417–430
power transformer
  current-based restraints, 47–49
  voltage-based restraints, 49–52
    digital tripping suppressor, 49
    flux restraint, 50–51
production cost model
  hydro subproblem, 215–219
  pumped-storage, 228–231
transmission line
  differential equations, 27–31
  Fourier, 23–25
  notations for, 13–18
  phasors, 26–27
  recursive, 25–26
  sources of error, 18–23
  tree-modification, 270–273
Aliasing, anti-aliasing filters, 8
Alumina trihydrate
  filler, 135, 154–158
  heat transfer, 156
Analog-to-digital converters, 10
  quantization errors, 10
  voltage input, 7
Analytical production cost model, *see* Production cost model
Arcing, dry band, 146, 160, 171–173, 180
ASTM tests, aging of polymeric insulators, 146–147

Attributes
  in ID3, 266
  power system contingencies, 285–286
  in VCES, 277
Autoregressive Moving Average Exogeneous
  (ARMAX) model, 81–85, 110–112

B

Backward chaining, 241–242
Bad data, in LAV state estimator, 434–435
Boundary injection measurements, 407
Branches, of a power system, 406
Brushless motors
  components, 64
  model of, 63–66
  variable structure tracking applications,
    92–100
Buses
  measured, 407
  power system, 406
  protection with digital relays, 52–53
  unmeasured, 407
  voltage, LAV technique algorithm
    5 bus system, 451–456, 472–473
    10 bus system, 457–463, 473–474
    14 bus system, 464–471, 474–475
Butterworth filters, 8
BX algorithm, 301–305
  critically coupled, 321–325
  generalized, convergence of, 305–309
  for nonradial networks, 309–313
  for radial networks, 304
  update rules, 302–303

C

Cable terminations, polymer
  construction, 137–138
  fog chamber tests, 152–153
  service performance, 142–143
Classes, in ID3, 266
Composite insulators, *see* Polymeric insulators
Computer hardware, *see* Hardware
Computer relays, *see also* Transmission line
  relaying
  analog-to-digital converters, 10
  architecture, 5–10
  benefits of
    cost, 3
    self-checking and reliability, 3–4

system integration, 4–5
  bus protection, 52–53
  protective, 11–12
  substation computer hierarchy, 11
  transmission line algorithms, *see*
    Transmission line algorithms
Computer software, *see* Software
Constrained least-squares estimation,
  352–354
Constrained power system state estimation,
  443–445
Contact angle, silicone rubber and EPR
  polymers, 166–168
Contaminated areas, polymeric insulators used
  in, 142
Contamination
  polymers, 134
  porcelain and glass, 133–134
Contingency selection
  attributes selected, 285
  decision tree for, 286–287
  false alarms, 284
  hits, 284
  performance measures with/without tree
    modification, 288–290
  test scenarios, 286–288
  training scenarios, 286–288
Continuous attributes, in ID3, 266
Controllers
  adaptive, 73
  mathematical formulation, 81–85
  recursive parameter estimation, 85–87
  robust, 73
  robustness, 75
  selection of reference tracks, 70–72
  self-tuning, 79–81
  variable structure system control, 74–78
Cost, computer relays, 3
CRAFT, *see* Customer Restoration and Fault
  Testing
Critically coupled XB and BX algorithms,
  321–325
  performance, 330–336
Critical trees, 407
Cross over voltage, silicone rubber and EPR
  polymers, 162, 165–166
Crystallization, polymer insulators, 173,
  179–180
Cumulative charge, polymeric materials,
  149–154
Current
  restraints based on, 47–49
  transformer, for digital relays, 7

Customer Restoration and Fault Testing
    (CRAFT), 254–257
    explanation facility, 260–262
    inference engine, 259–260
    knowledge base, 257–259
    maintenance facility, 262–264

                    D

Data base management, UWRIM system, 259
Data structures, in inference engines, 246–247
Data windows, in relay samples, 16–17
DC offset, removal, 37–38
Deadends, in inductive learning, 267
Decision trees
    building, 267–268
    leaf nodes, 266
    for power system contingencies
        attribute selection, 285–286
        generation of test sets, 286–288
        generation of training sets, 286–288
        performance measures, 284
        tests, 288–290
    tree-modification algorithm, 270–273
    in VCES, 278–280
Degree of uncertainty, emitted by leaf nodes,
    266
Dependability, computer relays, 12
Differential-equation algorithms, for
    transmission lines, 27–31
Diffusion, low-molecular-weight polymer
    chains, 164–165, 168–169
Digital relaying, see Computer relays
Diphontine equations, 84
Direct current motors
    brushless, see Brushless motors
    model of, 61–63
    self-tuning control applications, 110–120
    transfer function, 112
    variable structure tracking applications,
        100–103
Discrete attributes, in ID3, 266
Dry band arcing, 146, 160, 171–173, 180

                    E

Electrical testing, silicone rubber and EPR
    polymers, 160–161
Electron spectroscopy for chemical analysis,
    silicone rubber and EPR polymers,
    161–165

Energy demand, expected annual, 199–200
Energy dispersive X-ray analysis
    filler dispersion, 158
    watershed polymers, 174–175, 177–178
Energy management system alarm model,
    251–252
Entropy, information gain and, 266
EPDM, 135
    accelerated aging tests, 148–154
    filler effects, 154–159
    surface hydrophobicity
        contact angle measurement, 166–168
        cross over voltage, 162, 165–166
        electrical testing, 160–161
        electron spectroscopy for chemical
            analysis, 161–165
EPM, 135
Epoxy, 135
    filler effects on degradation, 155–158
EPR
    filler effects on degradation, 155–158
    heat transfer, 156–157
Erasable PROM (EPROM), for digital relays, 6
Error analysis
    differential equation algorithms, 29–31
    transmission line algorithms, 18–23
Ethylene propylene rubber, 135
Ethyl vinyl acetate, 135
Expected annual cost, in production cost
    model, 205–206
Expected annual energy demand, in production
    cost model, 199–200
Expected marginal values, of annual cost,
    206–207
Expert systems
    application areas, 244–245
    benefits of, 241
    Customer Restoration and Fault Testing
        (CRAFT), 254–257
        data base, 259
        explanation facility, 260–262
        inference engine, 259–260
        knowledge base, 257–259
        maintenance facility, 262–264
    definition, 240
    development of, 244–247
    domain expert — knowledge engineer
        interactions, 245
    elements of, 240
    energy management system alarm model,
        251–252
    extended power system model, 249–251
    human tasks modeled by, 252–254

Expert systems (*continued*)
  implementation, 245–246
  inference engine, 241–244
  knowledge representation, 240
  operational constraints, 252
  performance evaluation, 247
  rule-based, 240
  voltage control (VCES), 273–276
    attribute selection and classification, 277
    capabilities and limitations, 274–275
    decision tree, 278–280
    learning module (VCES/LM system), 276–277
    simulations without tree-modification algorithm, 282
    simulations with tree-modification algorithm, 282–283
    test cases, 280–282
    training set, 277–278
Extended power system model, 249–251

F

False alarms, of decision-tree outcomes, 284
Fast power flow algorithms, 296–297
  block decoupling and rotations, 316–319
  BX and XB, 301–305
  critically coupled XB and BX, 321–325
  Newton, 297–301
  novel rules, 319–321
    test systems, 326–327
  performance
    on 14, 30, 57, and 118 bus IEEE networks, 330–339
    on 1655 bus networks, 339–340
    of Carpenter's method, 328–330
    CCBX and CCXB, 330–336
  quasi-Newton, 297–301
  super decoupled, 313–316
Faults, classification (transmission line), 38–41
Field oriented control, for induction motors, 67
Figure of merit data, Brighton tests of polymer insulators, 144
Fillers, for polymeric insulators
  effect on tracking and erosion, 154–158
  types of, 135
Filtering, before A/D conversion, 7–8
Flashover, contamination-related, 133–134
Fog chamber, for polymer aging tests, 146–147
Forests, of trees, 407
Forward chaining, 241–242
Fourier algorithms, for transmission lines, 23–25

Fourier transform infrared spectroscopy, watershed polymers, 175–176, 178–179

G

Gaussian random distributions, of load curves, 198–200
Generalized minimum variance adaptive control, 79–82
Generation expansion planning, optimal long-term (OLGEP)
  analytical production cost model, *see* Production cost model
  assumptions, 196
  hydro-thermal system
    problem statement, 212–214
    subproblem statement, 214–215
  problem statement, 195–197
  pumped storage system
    problem statement, 225–227
    supply-shortage cost, 227–228
  simulation results
    accuracy of computation algorithm, 215–219
    with hydro plants, 221–225
    for pumped-storage system, 228–235
    without hydro plants, 219–220
Glass outdoor insulation, 132–133

H

Hardware
  for computer relays, 4–5
  premature fatality, 70
Heat transfer, fillers used for polymeric insulators, 156–157
High performance drives
  components, 60
  controllers for, 60
  control strategies, 73, 87–90
    mathematical formulation, 81–85
    recursive parameter estimation, 85–87
    self-tuning, 79–81
    variable structure system control, 74–78
  definition, 59–60
  laboratory implementation, 87–90
    setup, 91–92
    software, 90–91
  motors, *see* specific motor
High-temperature vulcanized rubber, *see* HTV silicone rubber
Hits, of decision-tree outcomes, 284

HTV silicone rubber, 135
  accelerated aging tests, 148–154
  aging
    energy dispersive X-ray analysis,
      174–175, 177–178
    Fourier transform infrared spectroscopy,
      175–176, 178–179
    mechanisms, 170–174
    X-ray diffraction, 176, 179–182
  filler effects on degradation, 155–158
Hydro-thermal systems, long-term generation
    planning
  accuracy of computation algorithm, 215–219
  problem statement, 212–214
  simulation results, 221–225
  subproblem statement, 214–215
Hyperplanes, in VSTC technique, 77–78

I

ID3 method
  classes and attributes, 266
  continuous attributes, 266
  deadends, 267
  decision trees, building, 267–268
  discrete attributes, 266
  information gain, 266
  tests as queries, 267
  training sets, 265–266, 270
  tree-modification algorithm, 270–273
IEEE networks
  fast power flow algorithm performance,
    326–339
    Carpentier's method, 328–330
    CCBX and CCXB algorithms, 330–336
  14 bus system, LAV algorithm applications,
    464–471, 474–475, 485–487
Induction motors
  equations for, 68–69
  model of, 66–70
  self-tuning control applications, 120–124
  variable structure tracking applications,
    103–109
Inductive learning
  classes and attributes, 266
  continuous attributes, 266
  deadends, 267
  decision trees, building, 267–268
  discrete attributes, 266
  information gain, 266
  tests as queries, 267
  training sets, 265–266, 270
  tree-modification algorithm, 270–273

Inference engines, expert systems
  backward chaining, 242–244
  CRAFT OPS83, 259–260
  forward chaining, 241–242
Information gain, in ID3, 266
Insulators, polymeric, *see* Polymeric insulators
Integration, computer relays, 4–5

K

Kalman filtering, 86
Kalman filters, 31–36
  algorithms for transmission lines, 31–36
Knowledge base, CRAFT, 257–25940
Knowledge representation, in expert systems,
  240

L

LAV, *see* Least absolute value state estimator
Leaf nodes, in decision trees, 266
Leakage current, polymeric materials, 149–154
Learning module, of VCES, 276–277
Least absolute value state estimation, 360–361
  algorithms
    least absolute value based on linear
      programming, 364–368
    Schlossmacher iterative, 368–369
    Sposito, 369–370
  history of, 361–363
  least-squares estimation and, 382–399
  novel technique for, 370–382
Least absolute value state estimator
  accuracy of, 436–437
  applications
    5 bus system, 451–456, 472–473,
      482–483
    10 bus system, 457–463, 473–474,
      483–485
    14 bus system, 464–471, 474–475,
      485–487
  bad data problem, 434–435
  computational efficiency of, 437
  constrained state estimation, 443–445
  design of, 432–424
  novel algorithm for, 438–443
  practicality of, 435–438
  reliability of, 436
Least error squares estimation, 349–350,
  358–359
  nonlinear, 354–358

Least-squares estimation, 346–347
  constrained, 352–354
  least absolute value estimation and, 382–399
  parameter estimation problem, 348–349
  weighted linear, 350–352
Line flow measurements, redundant, 407
Load modeling, in production cost model,
  197–200
  direct (Gaussian) use of load curves, 198
  random load fluctuations, 199–200
Loops, connected, 407
Loss-of-load probability, 203–204

M

Measured buses, 407
Metal oxide varistor (MOV) assemblies, 139
Modeling, power systems, 400–405
Models, *see* specific model

N

Networks, fast power flow algorithm
  performance
  1665 bus, 326–339
  IEEE 14, 30, 57, and 118 bus, 326–339
Newton algorithm, 297–301
Nodes, of a power system, 406
Nonlinear least error squares estimation,
  354–358

O

Observability, power systems, 405–406
  necessary conditions for, 409–417
  novel algorithm for, 417–430
  terminology for, 406–409
OLGEP, *see* Generation expansion planning,
  optical long-term
One-step-ahead predictors, 84
OPS83 system language, 259–260
Outdoor insulation
  glass, problems with, 132–133
  polymeric, *see* Polymeric insulators
  porcelain, problems with, 132–133

P

Percentage differential protection, 46–49
Phase angles, LAV technique algorithm

5 bus system, 451–456, 472–473
14 bus system, 464–471, 474–475
10 bus system, 457–463
Phasors, quantities, sign convention and
  normalization, 26–27
Polyethylene, 135
Polymer chains, low-molecular-weight
  and aging of watershed polymers, 171–173
  diffusion, 164–165, 168–169
Polymeric insulators
  aging tests, laboratory, 146–154
    fog chamber for, 147–149
    leakage current, 149–154
  basic and highest pulse currents (Anneberg),
    145
  choice of, 140
  construction, 136–137
  contact angle measurement, 166–168
  cross over voltage, 162, 165–166
  crystallization effects, 173, 179–180
  effects of low-molecular-weight polymer
    chains, 171–173, 177–178
  electrical testing, 160–161
  electron spectroscopy for chemical analysis,
    161–165
  failure, 141–142
  figure of merit data (Brighton), 144
  inorganic fillers, 135
  material degradation
    role of filler, 154–159
    surface hydrophobicity, 159–169
  materials for, 135–136
  number used, 140
  outdoor test sites, 143–145
  special outdoor locations, 140
  surface roughness effects, 176, 179–182
  use of
    in contaminated areas, 142
    at different voltages, 141
    USA, 141
    worldwide, 140
Polymers
  aging, 134–135
  surface hydrophobicity, 134
Porcelain
  accelerated aging tests, 148–154
  surge arrestors, 138
Porcelain outdoor insulation, 132–133
Power systems
  contingencies, decision trees for
    attribute selection, 285–286
    generation of test sets, 286–288
    generation of training sets, 286–288
    performance measures, 284

tests, 288–290
expert systems
  CRAFT, *see* Customer Restoration and
    Fault Testing
  energy management system alarm model,
    251–252
  extended power system model, 249–251
  human tasks modeled by, 252–254
  operational constraints, 252
  voltage control, 273–276
    attribute selection and classification,
      277
    capabilities and limitations, 274–275
    decision tree, 278–280
    learning module (VCES/LM system),
      276–277
    simulations without tree-modification
      algorithm, 282
    simulations with tree-modification
      algorithm, 282–283
    test cases, 280–282
    training set, 277–278
generation expansion planning, *see*
    Generation expansion planning
modeling, 400–405
observability, 405–406
  necessary conditions for, 409–417
  novel algorithm for, 417–430
  terminology for, 406–409
state estimation
  constrained least-squares, 352–354
  least absolute value, 360–361
  algorithms
    least absolute value based on linear
      programming, 364–368
    Schlossmacher iterative, 368–369
    Sposito, 369–370
  history of, 361–363
  least-squares estimation and, 382–399
  novel technique for, 370–382
  least error squares, 349–350, 358–359
    nonlinear, 354–358
  least squares, 346–347
    constrained, 352–354
    least absolute value estimation and,
      382–399
    parameter estimation problem, 348–349
    weighted linear, 350–352
  nonlinear least-squares, 354–358
  parameter estimation problem, 348–349
  properties of least error squares estimates,
    358–359
  weighted linear least-squares, 350–352
Power transformer algorithms

current-based restraints, 47–49
voltage-based restraints, 49–52
  digital tripping suppressor, 49
  flux restraint, 50–51
Premature fatality, of hardware, 70
Production cost model
  algorithms
    hydro subproblem, 215–219
    pumped-storage, 228–231
  generation modeling, 200–203
    annual energy generation, 203
    available capacities, 200–201
    expected plant output, 201–203
  load modeling, 197–200
    direct use of load curves, 198
    random load fluctuations representation,
      199–200
  production cost
    expected annual, 205–206
    expected marginal values, 206–207
  reliability measures, 203–204
    annual loss-of-load probability, 204
    expected annual unserved energy, 204
  simulation results, 207–211
Programmable Read Only Memory (PROM),
  for digital relays, 6
Pulse-width modulation, 88–89, 93
Pumped-storage systems, long-term generation
  planning
  problem statement, 225–226
  simulation results
    accuracy of computational algorithm,
      228–231
    case studies, 231–235
    supply-shortage costs, 227–228

Q

Quantization errors, in A/D conversion, 10
Quasi-Newton algorithm, 297–301

R

Random Access Memory (RAM), for digital
  relays, 6
Read Only Memory (ROM), for digital relays, 6
Recursive algorithms, transmission line, 25–26
Recursive estimation, 85–87
Redundant line flow measurements, 407
Reference tracks
  selection, 70–72
  sigmoid function, 71–72

Reliability
    computer relays, 3–4, 12
    least absolute value state estimator, 436
    production cost model, 203–204
Robustness, of controllers, 73
Room temperature vulcanized rubber, *see* RTV
    silicone rubber
RTV silicone rubber, 135
    aging
        energy dispersive X-ray analysis,
            174–175, 177–178
        Fourier transform infrared spectroscopy,
            175–176, 178–179
        mechanisms, 170–174
        X-ray diffraction, 176, 179–182
Rule structures, in inference engines, 246–247

S

Samarium-cobalt brushless motors, 63
Sampling
    in digital relays, 8–9
    simultaneity, 9
Security, computer relays, 12
Self-tuning control, 79–81
Service performance
    polymer cable terminations, 142–143
    polymeric insulators, 140–141
    polymer surge arrestors, 143
Sigmoid function (SF), 71–72
Silica fillers, 135
    heat transfer, 156–157
Silicone rubber, 135
    aging of watershed material
        energy dispersive X-ray analysis,
            174–175, 177–178
        Fourier transform infrared spectroscopy,
            175–176, 178–179
        mechanisms, 170–174
        X-ray diffraction, 176, 179–182
    forms, 135–136
    heat transfer, 156–157
Sliding hyperplanes, 77–78
Soft transition, property, 70
Software, for computer relays, 4–5
Speed–reach relationship, for digital relays,
    43–46
State estimation, power systems
    constrained least-squares, 352–354
    least absolute value, 360–361
        algorithms
            least absolute value based on linear
                programming, 364–368

        Schlossmacher iterative, 368–369
        Sposito, 369–370
    history of, 361–363
    least-squares estimation and, 382–399
    novel technique for, 370–382
least error squares, 349–350, 358–359
    nonlinear, 354–358
    properties of estimates, 358–359
least squares, 346–347
    constrained, 352–354
    least absolute value estimation and,
        382–399
    parameter estimation problem, 348–349
    weighted linear, 350–352
parameter estimation problem, 348–349
Super decoupled algorithm, 313–316
Surface hydrophobicity
    low-molecular-weight polymer chains and,
        171–173
    of polymers, 134
        contact angle measurement, 162
        cross over voltage, 162
        electrical testing, 160–161
        electron spectroscopy for chemical
            analysis, 161–165
Surface roughness, and aging of watershed
    polymers, 179–182
Surge arresters, polymer
    construction, 138–139
    porcelain, construction, 138
    service performance, 144
Switching hyperplanes, 77–78
Symmetrical component discrete Fourier
    transform, 41–42
Symposium on Expert Systems Application to
    Power Systems, 248–249

T

Teething factor, 201
Teflon, 135
Test sets, for contingency selection, 286–288
Training sets
    for contingency selection, 286–288
    in ID3, 265–266
    representativeness of, 270
    in VCES, 278–280
Transformers, for digital relays, 7
Transient monitor function, 42–43
Transmission line algorithms
    differential equations, 27–31
    Fourier, 23–25
    notations for, 13–18

phasors, 26–27
recursive, 25–26
sources of error, 18–23
Transmission line relaying, 38–41, *see also*
    Computer relays
    fault classification, 38–41
    speed reach limitations, 43–46
    symmetrical component calculation, 41–42
    transient monitor, 42–43
Tree-modification algorithm, 273–274
Trees
    critical, 407
    and forests, 407
    in a power system, 407
Tripping suppressor, 49

                    U

University of Washington Relational
    Information Management System
    (UWRIM), 259
Unmeasured buses, 407
UWRIM data base management system, 259

                    V

Variable Structure System Control (VSC),
    74–78
Variable structure tracking
    application
        to brushless motors, 92–100
        to DC motors, 100–103
        to induction motors, 103–110
    Variable Structure System Control, 74–78
    Variable Structure Tracking Control, 77
Variable Structure Tracking Control (VSTC),
    77
Vector control, for induction motors, 67
Vector transformation, induction motors with,
    67–70
Voltage
    bus, LAV technique algorithm
        5 bus system, 451–456, 472–473,
            482–483

10 bus system, 457–463, 473–474,
    483–485
14 bus system, 464–471, 474–475,
    485–487
restraints based on, 49
transformers, for digital relays, 7
Voltage Control Expert System (VCES),
    273–276
    attribute selection and classification, 277
    capabilities and limitations, 274–275
    decision tree, 278–280
    learning module (VCES/LM system),
        276–277
    simulations
        with tree-modification algorithm,
            282–283
        without tree-modification algorithm, 282
    test cases, 280–282
    training set, 277–278
VSC, *see* Variable Structure System Control
VSTC, *see* Variable Structure Tracking Control

                    W

Weighted linear least-squares estimation,
    350–352

                    X

XB algorithm, 301–305
    critically coupled, 321–325
    generalized, convergence of, 305–309
    for nonradial networks, 309–313
    for radial networks, 304
    update rules, 302–303
X-ray diffraction, watershed polymers, 176,
    179–182

                    Z

Zones of protection, 12